Palgrave Studies in the History of Science and Technology

Series Editors
James Rodger Fleming
Colby College
Waterville, ME, USA

Roger D. Launius
Auburn, AL, USA

Designed to bridge the gap between the history of science and the history of technology, this series publishes the best new work by promising and accomplished authors in both areas. In particular, it offers historical perspectives on issues of current and ongoing concern, provides international and global perspectives on scientific issues, and encourages productive communication between historians and practicing scientists.

More information about this series at
http://www.palgrave.com/gp/series/14581

Kathleen L. Housley

The Scientific World of Karl-Friedrich Bonhoeffer

The Entanglement of Science, Religion, and Politics in Nazi Germany

palgrave
macmillan

Kathleen L. Housley
Glastonbury, CT, USA

Palgrave Studies in the History of Science and Technology
ISBN 978-3-030-07096-0 ISBN 978-3-319-95801-9 (eBook)
https://doi.org/10.1007/978-3-319-95801-9

Cover credit: United States Holocaust Memorial Museum, courtesy of Leorah Kroyanker

This Palgrave Macmillan imprint is published by the registered company
Springer Nature Switzerland AG
The registered company address is: Gewerbestrasse 11, 6330 Cham, Switzerland

Karl-Friedrich Bonhoeffer in the physical chemistry laboratory at the University of Göttingen during the early 1950s. Photo Credit: Paul Harteck Papers, MC17, Institute Archives and Special Collections, Rensselaer Polytechnic Institute, NY

Preface

My research did not begin with Karl-Friedrich Bonhoeffer but with his brother Dietrich, specifically his extensive reading list while in a Nazi prison. On that list was *The World View of Physics* by the German physicist Carl Friedrich von Weizsäcker that Karl-Friedrich had sent to Dietrich in 1944. That book was of interest to me because it had an impact on Dietrich's ideas about science and religion. I decided to cast a broader research net because of a comment made by Roald Hoffmann, winner of the Nobel Prize in Chemistry and professor emeritus at Cornell University. Hoffmann spent a year and a half of his childhood hiding in an attic in German-occupied Poland with his mother, barely surviving World War II. In his reply to an email I had sent him about my research, Hoffmann asked if I knew about Weizsäcker's involvement in Nazi atomic bomb research. That question was like a warning flare. I realized that I had to expand my research exponentially into nuclear science during the Third Reich, a very difficult subject.

In doing so, I came across numerous references to Karl-Friedrich's research on heavy water. Some were troubling. It was not until I found several exceptional letters written after the war by Paul Rosbaud, the spy known by the code name *The Griffin*, as well as letters from prominent Jewish scientists to Karl-Friedrich, expressing their gratitude and offering their support, that I began to comprehend what had occurred. I am grateful to Hoffmann for his comment on Weizsäcker that changed the course of my research. So also, I am grateful to him for reading an early draft and challenging me to look carefully at changes in the field of physical chemistry in the early 1930s.

Glastonbury, USA Kathleen L. Housley

Acknowledgements

My thanks go to Victoria Barnett, Director of the Program of Ethics, Religion, and the Holocaust, the United States Holocaust Memorial Museum in Washington, D.C., and Clifford Green, past president of the International Dietrich Bonhoeffer Society, English Language Section, for their encouragement and guidance. As editors for *Dietrich Bonhoeffer Works*, their knowledge of the Bonhoeffers is of the highest order. Victoria read an early draft of *The Scientific World of Karl-Friedrich Bonhoeffer: The Entanglement of Science, Religion, and Politics in Nazi Germany* and made suggestions. Clifford helped me find an excellent researcher and translator, Saskia Triesscheijn, who obtained copies of many of Karl-Friedrich Bonhoeffer's letters in the Max-Planck-Gesellschaft Archives in Berlin. Synergistically, Saskia has a degree in biology and is studying theology at Humboldt University in Berlin.

In regard to translation, I would not have gotten very far without the help of Margaret Ludwig who grew up in Germany, coming to the USA as a young woman in the early 1950s. Sitting with Margaret at her dining room table, I would hand her letters in faded handwriting between Paul Harteck, Michael Polanyi, and Karl-Friedrich Bonhoeffer and she would set to work, puzzling over which words most accurately conveyed the meaning and explaining to me colloquialisms that resisted translation.

I am also indebted to the libraries and librarians of Wesleyan University in Middletown, Connecticut. Wesleyan's libraries are open-stack, enabling the important research skill of browsing. I could peruse older, seemingly tangential books on science, which turned out to be significant, for example, *The Born-Einstein Letters* by Max Born (1971). Browsing also led me to newer books that I would have overlooked such as *Stalin's Captive: Nikolaus Riehl and the Soviet Race for the Bomb* by Nikolaus Riehl and Frederick Seitz (1996). In the world of Internet research and ebooks, actual hard-copy books and helpful librarians remain invaluable. Special thanks to Melissa Behney, science librarian at Wesleyan.

Thanks also to Jenifer Monger, archivist at the Institute Archives and Special Collections, Folsom Library, Rensselaer Polytechnic Institute, for helping me locate correspondence and photographs in the Paul Harteck Papers.

My husband, Timothy J. Housley, encouraged me to keep going, patiently reading numerous drafts. I also found encouragement in the Bonhoeffers themselves. Their letters to each other were warm and supportive even when they disagreed. More than once when I questioned whether I should continue, it would be their words about standing firm in the face of difficulties that would restore my belief that a book could—and should—be written.

Contents

1 Introduction 1

2 The Father's Scientific World 7

3 The First World War 29

4 The Promise of Pure Science 51

5 Traveling with Polanyi 75

6 Turbulence and Conformity 87

7 Seizing the Wheel 113

8 The Beginning of Resistance 133

9 Heavy Water and the Atomic Bomb 147

10 The Summer of Decision 165

11 The Uranium Club 191

12 Steadfast to the End 213

13 Rebuilding the World 239

Appendixes 267

Selected Bibliography 309

Index 319

About the Cover Image

In this photograph taken around 1930, Karl-Friedrich Bonhoeffer (middle) and his doctoral assistants, Paul Harteck (top), Ladislaus Farkas (bottom left), and Adalbert Farkas (bottom right), are perched on scaffolding outside the Kaiser Wilhelm Institute for Physical and Electrochemistry in Berlin. It is the pose of young men full of purpose and hope who are unaware their world is about to fall apart. The Farkas brothers will be forced out of Germany because they are Jewish. Harteck will become a leader of Hitler's atomic bomb project. Bonhoeffer will be caught in between, attempting desperately to hold everything together even as World War II overwhelms them all.

List of Figures

Fig. 2.1	Karl Ludwig Bonhoeffer. Paul Fearn/Alamy Stock Photo	9
Fig. 3.1	Spartacist uprising, Berlin. United States Holocaust Memorial Museum, courtesy of Dottie Bennett	44
Fig. 4.1	Albert Einstein. Pictorial press/Alamy Stock Photo	57
Fig. 4.2	Ladislaus Farkas and Fritz Haber. United States Holocaust Memorial Museum, courtesy of Leorah Kroyanker	69
Fig. 6.1	Adolf Hitler in Nuremberg in 1929. United States Holocaust Memorial Museum, courtesy of William O. McWorkman	90
Fig. 6.2	Marinus van der Lubbe on trial for Reichstag fire. United States Holocaust Memorial Museum, courtesy of National Archives and Records Administration, College Park	101
Fig. 7.1	Nazi propaganda poster on forced sterilization. United States Holocaust Memorial Museum, courtesy of G. Howard Tellier	121
Fig. 7.2	Max Delbrück (left) and Salvador Luria at Cold Spring Harbor Laboratory in 1941. World History Archive/Alamy Stock Photo	128
Fig. 8.1	Dietrich Bonhoeffer. Sueddeutsche Zeitung Photo/Alamy Stock Photo	135
Fig. 9.1	Norsk Hydro, Vemork, Norway. Pictorial Press Ltd./Alamy Stock Photo	153
Fig. 10.1	Werner Heisenberg. Alpha Historica/Alamy Stock Photo	172
Fig. 11.1	Carl Friedrich von Weizsäcker in 1971. Interfoto/Alamy Stock Photo	193
Fig. 12.1	Hans Dohnanyi. Interfoto/Alamy Stock Photo	219

Fig. 12.2 Flossenbürg concentration camp in May 1945 after liberation. United States Holocaust Memorial Museum, courtesy of National Archives and Records Administration, College Park 233
Fig. 13.1 Martin Niemöller after his release from concentration camp. United States Holocaust Memorial Museum, courtesy of Joseph Eaton 241
Fig. 13.2 Soviet tanks roll through Leipzig in March 1945. United States Holocaust Memorial Museum, courtesy of National Archives and Records Administration, College Park United States 247
Fig. 13.3 Robert Havemann was accused of being a Soviet spy. PJF Military Collection/Alamy Stock Photo 256

Notes on Sources

The seventeen-volume *Dietrich Bonhoeffer Works* (DBW), published by Fortress Press, is magisterial. I could not have written *The Scientific World of Karl-Friedrich Bonhoeffer: The Entanglement of Science, Religion, and Politics in Nazi Germany* without the series because it includes the correspondence between the brothers. It took great dedication on the part of many editors and translators to bring the series into existence—retranslated, unexpurgated, with extensive footnotes and commentary. Prior to the publication of the series, some of Dietrich's books were shortened with sections reordered, inadvertently misleading readers.

Eberhard Bethge's *Dietrich Bonhoeffer: A Biography* (2000) is ancillary to *Dietrich Bonhoeffer Works.* It is the primary source for information on Dietrich's life and is destined to remain so because Bethge was a close friend who married into the Bonhoeffer family and whose correspondence with Dietrich was critically important in helping him shape his theology. Bethge dedicated his life to keeping Dietrich's legacy alive. Other biographies have been written, but all are built on the foundation Bethge laid down.

Primary material is always of great importance. Karl-Friedrich Bonhoeffer's papers are in the Max-Planck-Gesellschaft Archives in Berlin. Paul Harteck's papers are in the Institute Archives and Special Collections, Folsom Library, Rensselaer Polytechnic Institute. The correspondence between Paul Rosbaud and Samuel Goudsmit is in the Alsos Mission Papers, Niels Bohr Library & Archives with the Center for History and Physics, American Institute of Physics, and is available online.

Because of the large number of people who are mentioned in *The Scientific World of Karl-Friedrich Bonhoeffer*, I have included in the appendix *Biographical Profiles*, which include birth/death dates, academic affiliations, major events, and what occurred in the lives of these people following World War II.

Many excellent books and biographies have been written about science and scientists during the Nazi era and the German atomic bomb project. However, I found most helpful the books written by an elite group of physicists who elected, usually at points well along in their scientific careers, to bring their knowledge to bear on history. Among this group was Per F. Dahl whose book *Heavy Water and the Wartime Race for Nuclear Energy* (1999) is an excellent summation of research during this period. Dahl was a physicist who worked on superconductors at Brookhaven National Laboratory and Lawrence Berkeley National Laboratory. Because of his expertise, coupled with his being multilingual, he was able to delve into German and Norwegian archives in a way that few other researchers could. As a result, he brought scientific clarity to the history of heavy water and the connections between the Norsk Hydro plant in Vemork, Norway, and some members of the Uranium Club. I could not have written about Karl-Friedrich's work on heavy water without Dahl's exemplary scholarship, which underscored the interplay between pure science and atomic bomb research.

Stalin's Captive: Nikolaus Riehl and the Soviet Race for the Bomb (1996) by Nikolaus Riehl and Frederick Seitz is another valuable book written from the scientific perspective. It is mainly the memoir of Riehl whose job was to supply purified uranium and who went to the USSR following the war. However, the memoir needed to be set in historical context so as to be more understandable, including explanations of both science and espionage. A physicist, Seitz admirably provided these components. Working as a technical intelligence expert assigned to General Dwight Eisenhower's headquarters in Versailles, Seitz personally knew many German scientists. He also knew the members of the Alsos team, including the physicist Samuel Goudsmit whose mission was to learn everything about German nuclear research. After the war, Seitz led a distinguished career as director of the Oak Ridge National Laboratory, chair of physics at the University of Illinois, and president of The Rockefeller University.

Trained in physics and history, David C. Cassidy has written two biographies of Werner Heisenberg who was Karl-Friedrich Bonhoeffer's colleague at the University of Leipzig. The second book *Beyond Uncertainty: Heisenberg, Quantum Physics, and the Bomb* (2009) builds on the first *Uncertainty: The Life and Science of Werner Heisenberg* (1992). However, the second book was written after the declassification in 1992 of the secret tapes made during the incarceration of ten German scientists at Farm Hall in England immediately following the war. The tapes shed light on the workings of the Uranium Club, the personalities of some of its members, and Heisenberg's efforts to justify himself.

Klaus Hentschel is also a physicist and a prolific historian of science. In his important book *Physics and National Socialism: An Anthology of Primary Sources*, Hentschel provides an in-depth introduction as well as copious scholarly footnotes, making vital connections between people and projects, both well known and obscure. This was not a book I read once and put aside; it became a reliable reference work always near at hand.

Arnold Kramish was a nuclear physicist who worked on the Manhattan Project. His book on Paul Rosbaud *The Griffin: The Greatest Untold Espionage Story of World War II* (1986) is inaccurate in some respects because certain documents were still classified at the time he wrote the book. Otherwise, it is well researched, buttressed by facts gleaned from extensive interviews. The information about the relationship between Karl-Friedrich and Rosbaud is corroborated by Rosbaud's letters to Samuel Goudsmit in the Niels Bohr Library & Archives with the Center for History and Physics, American Institute of Physics.

Although a historian, not a scientist, Fritz Stern is worthy of mention not only because of his exemplary scholarship on Germany but also because his family history overlapped with the Bonhoeffers at several critical points. The Sterns and their relatives lived in Breslau, Germany (now Wroclaw, Poland), from the early nineteenth century to the 1930s when they were forced into exile because of their Jewish heritage, even though many of them had converted to Christianity. Fritz's father and grandfather were renowned physicians. During the first decade of the twentieth century, they and their families moved in the same medical and cultural spheres as the Bonhoeffer family with whom they were friends. Although Fritz was not born until 1926, fourteen years after the Bonhoeffers moved from Breslau to Berlin, he knew of the close connection through letters and stories about which he wrote in his memoir *Five Germanys*

I Have Known (2006). Yet, another critical juncture is that Fritz was the godson of Fritz Haber, the physical chemist and Nobel laureate who figured prominently in Karl-Friedrich Bonhoeffer's life. Fritz Stern was also a relative of Otto Stern, the renowned physical chemist who was driven from his position at the University of Hamburg after Hitler came to power. In 1943, Otto was awarded the Nobel Prize in Physics. Fritz Stern's branch of the family came to the USA in 1938 when he was twelve years old. Following the war, the Sterns sent care packages to the Bonhoeffers to help alleviate their suffering. Unfortunately, Fritz Stern died in 2016. Had I gotten the chance to interview him, I would have thanked him for his scholarship. For me, Stern was cautionary: He considered oversimplification a danger, but so also was the piling up of historical details to the point that the reader became numb to the terror embedded therein. Stern tried not to lose sight of the individual human being struggling to live a worthy life in a world that labeled entire races, religions, and nations as worthless.

CHAPTER 1

Introduction

Karl-Friedrich Bonhoeffer seemed to stand apart from his family's intense resolve to bring the brutality of Nazism to an end. A distinguished scientist who was nominated five times for the Nobel Prize, he was chair of the department of physical chemistry at the University of Leipzig from 1934 until 1946. Besides being physically distant from the center of the resistance movement in Berlin, he was intellectually distant from his family due to the complexity of his research, combining the apparent insubstantiality of quantum mechanics with the substantiality of classical physics and chemistry. Few people had the requisite brilliance to understand his work other than his colleagues. What Karl-Friedrich shared with all the Bonhoeffers was the quality of being steadfast. After the war, using what strength he had left, he helped rebuild the shattered field of physical chemistry in Germany, brick by brick, element by element.

In quantum mechanics, the word *entanglement* means that while the states of particles are uncertain, nonetheless, there is a correlation even though the particles can be far apart. As Albert Einstein put it with incredulity, there is "spooky action at a distance." Entanglement appears literally in Karl-Friedrich's work on spin and anti-spin in hydrogen isotopes. Entanglement is also an apt metaphor for his life. No matter how remote he appeared from his family's involvement in the resistance, he was not remote. Throughout the evidentiary record, there are photon flashes of his compassionate awareness. However, these can be easily missed in the

K. L. Housley, *The Scientific World of Karl-Friedrich Bonhoeffer*, Palgrave Studies in the History of Science and Technology, https://doi.org/10.1007/978-3-319-95801-9_1

steady glow of the light cast by his brother, Dietrich, whose powerful theological writings and death by hanging in Flossenbürg concentration camp led to his being proclaimed a Christian martyr.

For Karl-Friedrich, there was the world of pure science and then there was the rest of the world, but that does not mean the rest of the world was the lesser. In fact, he was very close to all his siblings with whom he corresponded frequently. An agnostic, he sometimes took issue with Dietrich's theological views but never to the point of snapping their brotherly bond. During Dietrich's incarceration in Tegel prison in Berlin, Karl-Friedrich wrote him warm letters, often accompanied by carefully selected books and food packages to supplement his sparse prison diet. In a letter dated May 30, 1943, less than two months after Dietrich's arrest, Karl-Friedrich wrote from Leipzig that he had not yet told his children what had occurred, adding on a light note, "I think they consider me somewhat peculiar since I always ask for homemade cookies and candies when I travel to Berlin."

In the catastrophic closing months of the war, when it was clear that his two brothers and two brothers-in-law would be executed along with many other friends and relatives, Karl-Friedrich tried desperately to visit them all in prison. At the same time, he was attempting to take care of his family, his brothers' families, and his elderly parents—some of them burned out and bombed out, some scattered, all traumatized, especially the children. It is no surprise that Karl-Friedrich suffered a severe heart attack not long after the war ended. On meeting him after many years of separation, his sister Sabine, who was Dietrich's twin, was astonished at the change in his appearance. "He had become very haggard, and as he looked at me, he had my mother's eyes."[1]

Karl-Friedrich did not speak publicly about what had happened. He had no choice but to remain silent because in postwar Germany the Bonhoeffers were not considered heroes but traitors for attempting to kill Hitler, who was still venerated by many people. Eventually, Germany would grapple with the truth, but Karl-Friedrich would not live to see it. He died of a second heart attack in 1957.

Karl-Friedrich was aware of his family's role in the resistance, although he did not know the degree. Besides Dietrich, his sister Christel and brother Klaus were involved as were his two brothers-in-law Hans Dohnanyi (married to Christel) and Rüdiger Schleicher (married to Ursula). In a letter to Sabine, who had gone into exile in England in

1938 because her husband was of Jewish descent, Karl-Friedrich wrote on August 3, 1945, that the family had received word of the executions of Dietrich, Klaus, and Rüdiger. The fate of Hans was still unknown, but Karl-Friedrich strongly suspected that he was dead as well. Then, Karl-Friedrich wrote movingly about the family's involvement:

> You can not imagine how much courage, prudence, caution and endurance was necessary, how often we all expected the imminent breakdown of this criminal tyranny. (I have since heard that no less than five attempts were made on Hitler's life), or how often we were disappointed. Our parents were aware of what they were doing, approved of it, and gave them assistance. I believe there can have been very few families in Germany during the past twelve years in which there was such complete agreement on political matters, and there is no doubt that this spirit gave them strength to continue their plans.[2]

Karl-Friedrich was also aware that some of his close scientific friends, including Werner Heisenberg, winner of the Nobel Prize for his seminal work on quantum mechanics, and Paul Harteck, who had worked with Karl-Friedrich on molecular hydrogen, were researching how to build an atomic bomb for Hitler. In fact, Karl-Friedrich's research on heavy water during the 1930s had been a crucial precursor step to the work being carried out by the Uranium Club—a euphemism evoking a group of colleagues sharing schnapps after a productive day spent in the laboratory.

Eberhard Bethge, Dietrich Bonhoeffer's friend and biographer, wrote that Karl-Friedrich "abandoned his earlier field of research into nuclear hydrogen, parahydrogen and heavy hydrogen in favour of electro-chemistry and the kinetics of electrode processes, in order to clarify the question of biological processes and basic electro-physical principles in nervous stimulation. He made this change so that he could avoid having to cooperate in the development of nuclear armaments."[3] That is only partially correct. Bethge did not know that during the war Karl-Friedrich was consulted several times by members of the Uranium Club about issues pertaining to heavy water on which their bomb research was reliant. Karl-Friedrich tried to distance himself by shifting the focus of his research and deflecting questions. However, his position at the University of Leipzig as well as his friendships with the Uranium Club members meant that he was privy to top secret information.

It would appear that Karl-Friedrich found himself in a monstrous ethical dilemma, caught between men trying to kill Hitler and men trying to provide him with a weapon of mass destruction that would guarantee Nazi domination of the world, or what was left of it. Tottering between them, Karl-Friedrich chose a third way: What he knew about nuclear research, he passed on to Paul Rosbaud, a scientific advisor to the publisher Springer Verlag. Rosbaud was also a master spy for the British known by the code name *The Griffin*. Following the war, Rosbaud wrote in a letter to the American intelligence expert Samuel A. Goudsmit that Karl-Friedrich was not only his "ally" but also one of his "best friends."[4] Lending credence to Karl-Friedrich's passing on information to the British is a report written after the war by Franz (Francis) Simon to Michael Perrin, one of the leaders of the British atomic bomb program, in which he related a conversation with Karl-Friedrich on heavy water that undercut protestations of innocence by the members of the Uranium Club.[5] There are also letters written after the war to Karl-Friedrich from several exiled Jewish scientists. This was not a group disposed to compassion for their "Aryan" colleagues in Germany, no matter how dire their circumstances. Yet, they offered Karl-Friedrich help as the Soviet Union took over East Germany and the University of Leipzig disappeared behind the iron curtain. Despite the evidence that will be presented in this book, there is no certainty of Karl-Friedrich's link to British espionage, one reason being that Britain's Secret Intelligence Service (commonly known as MI6) has never released the World War II files of Paul Rosbaud. Another reason is that much was destroyed, pages were ripped out of diaries, codes were lost, material in archives disappeared.

To this must be added the significant problem of long-standing obfuscation. Many German scientists tried very hard for decades to obscure their roles in research that benefited the Nazis. They devised numerous stratagems to present themselves and their work in a positive light, for example, Werner Heizenberg's oft-repeated claim that he was building an atomic reactor for civilian purposes, not a bomb. Probably, most notorious was the disingenuous argument that German scientists were more ethical than American scientists because they knew how to build an atomic bomb but did not do it, intentionally sabotaging the research, whereas the Americans pushed ahead, built the bombs, and dropped them on Japan. This would be a devastating argument if true. However, the Germans knew full well what they were doing and were committed to it. The reasons they failed have to do with the extreme difficulty of the science, lack of organizational unity, and insufficient financial resources—resources that were available in the USA but not in Germany.[6]

Yet, another difficulty in unearthing the truth is almost too obvious to mention: People who were passing on secrets or who were part of the resistance had to appear other than who they were, even taking on the guise of Nazis when it was essential. Dietrich Bonhoeffer had to do that repeatedly and in different ways, for example, when he told Bethge to give the Heil Hitler salute when they were sitting in the garden of a cafe and the surrender of France was announced over the loudspeaker. As people cheered wildly, jumping up on the chairs, Dietrich raised his arm in the salute while Bethge sat there dazed. "Raise your arm! Are you crazy?" he whispered to Bethge, later telling him, "We shall have to run risks for very different things now, but not for that salute."[7]

The title *The Scientific World of Karl-Friedrich Bonhoeffer: the Entanglement of Science, Religion, and Politics in Nazi Germany* points to the book's scope. This is not a straightforward biography of a single man. For example, Karl-Friedrich's belief in pure science was shared by almost all his colleagues. Einstein described pure science as a temple unsullied by connections to home, culture, and politics, dedicated solely to the pursuit of objective knowledge. Karl Bonhoeffer, the patriarch of the family and one of the leading physicians in Germany, also believed in pure science, which led him to remain detached from the political problems swirling around him. To understand that belief, it is essential to look at the extreme changes in German science from its glory days at the beginning of the century to its utter degradation after 1933. By holding fast to pure science, German scientists became dangerously vulnerable to manipulation. Yet, some men were beginning to scrutinize the nature of scientific autonomy and the effects of governmental and ideological control—among them was Michael Polanyi, Karl-Friedrich's colleague and close friend.

On one level, *The Scientific World of Karl-Friedrich Bonhoeffer* is the story of a compassionate man in love with chemistry, his family, and his nation, trying in the midst of chaos to do right by all of them. On another level, it raises ethical issues about the interaction of science, religion, and politics that are still relevant today. To tell that multi-level story requires me to step away from quantum mechanical entanglement and return to the cause and effect of linear history, beginning with Karl-Friedrich's childhood and the major influence of his father.

However, I do so holding in mind an illuminating observation made by Sir George Thomson, Nobel laureate in Physics, in a letter to Ronald W. Clark, the biographer of Albert Einstein: "Whenever a system is really complicated, as in the brain or in an organized community, indeterminacy comes in, not necessarily because of *h* [Planck's constant] but because to make a prediction so *many* things must be known that the stray consequences of studying them will disturb the *status quo*, which can never therefore be discovered. History is not and cannot be determinate. The supposed causes only *may* produce the consequences we expect."[8]

Notes

1. Sabine Leibholz-Bonhoeffer, *The Bonhoeffers: Portrait of a Family* (London: Sidgwick & Jackson, 1971), 197.
2. Ibid., 195.
3. Eberhard Bethge, *Dietrich Bonhoeffer: A Biography* (Minneapolis: Fortress Press, 2000), 562.
4. Letter from Rosbaud to Goudsmit, February 17, 1947. Samuel A. Goudsmit Papers, 1921–1979. Series IV, Alsos Missions. Subseries B. Paul Rosbaud correspondence and manuscripts 1946–1949 (Box 28, Folder 43). Niels Bohr Library & Archives, American Institute of Physics. www.aip.org/history/nbl/collections/goudsmit/colls/print/Box_28/Goudsmit28_43/Goudmit28_430006.html.
5. Kenneth D. McRae, *Nuclear Dawn: F. E. Simon and the Race for Atomic Weapons in World War II* (Oxford: Oxford University Press, 2014), appendix.
6. There are several books on the German atomic bomb project. See David C. Cassidy, *Beyond Uncertainty: Heisenberg, Quantum Physics, and the Bomb* (New York: Bellevue Literary Press, 2009); John Cornwell, *Hitler's Scientists: Science, War and the Devil's Pact* (New York: Viking, 2003); Paul Lawrence Rose, *Heisenberg and the Nazi Atomic Bomb Project* (Berkeley: University of California Press, 1998); and Mark Walker, *Nazi Science: Myth, Truth, and the German Atomic Bomb* (New York and London: Plenum Press, 1995).
7. Bethge, *Bonhoeffer*, 681.
8. Ronald W. Clark, *Einstein: The Life and Times* (New York: World Publishing, 1971), x.

CHAPTER 2

The Father's Scientific World

Karl Ludwig Bonhoeffer is presented in several biographies of his son, Dietrich, as kindly but slightly remote. He seemed to play only a marginal role in shaping the direction of Dietrich's life as evidenced by his bemused tolerance of his youngest son's budding interest in religion, an interest he did not share. However, in the life of his oldest son, Karl-Friedrich, Karl played a central role, bequeathing to him his scientific worldview and his agnosticism.

In 1904 when Karl was thirty-six-years old, he was appointed chair of psychiatry at the University of Breslau and director of the prestigious Breslau Psychiatric Clinic. Of primary importance to him in his profession was the establishment of a solid neurobiological foundation to psychiatry. This had been true for Carl Wernicke, his predecessor in Breslau, who studied aphasia caused by brain damage. It would also be true for Karl's successor, Alois Alzheimer, well known for his research into dementia. In Germany, where academic lineage was nearly as important as family lineage, this was an exemplary succession of scientists. In 1912, Karl became head of the famed Charité Clinic and chair of psychiatry at the University of Berlin where he kept the succession going by training many other doctors who would become famous, among them Hans Creutzfeldt and Karen Horney.

Until his death in December 1948, Karl Bonhoeffer was one of the leading neurologists and psychiatrists in Germany, whose research papers were widely read in Europe and America. At the age of eighty,

K. L. Housley, *The Scientific World of Karl-Friedrich Bonhoeffer*,
Palgrave Studies in the History of Science and Technology,
https://doi.org/10.1007/978-3-319-95801-9_2

he was visiting patients on a routine hospital round in Berlin when he suddenly developed a severe headache and dizziness. His obituary read, "He calmly announced his diagnosis and gave precise emergency instructions before he lapsed into deep unconsciousness from which he never awoke." Such was the nature of the man—the way he approached his death was in keeping with the way he approached his life.[1]

The preceding three paragraphs have compressed Karl Bonhoeffer's entire career, providing an uplifting but pallid image of the man: that of a respected physician so committed to his work that he died doing what he loved. The truth about his death is larger, encompassing the painful extremity of life in postwar Germany. Only by looking more broadly is it possible to comprehend why an elderly physician, retired for many years, was making rounds on a wintery day in the bombed-out ruins of a city under a Soviet blockade.

Karl had returned to clinical work at the age of seventy-seven immediately after the war because his world had collapsed, leaving him grief-stricken and appalled. In the 1930s, he had been cautiously silent about his anti-Nazi views, attempting to placate Nazi authorities while protecting the rights of his mentally ill patients. In a plan to overthrow Hitler in 1938, Karl's intended role was to perform a psychiatric examination following Hitler's arrest to determine his sanity. When the overthrow attempt failed, Karl continued to support the resistance, sometimes in strange ways that ran counter to everything he as a conscientious physician embodied, such as obtaining diphtheria bacillus for his daughter to smuggle to her husband in prison so that he would become too ill to be interrogated by the Gestapo.[2] For all Karl's study of the human condition, nothing had prepared him for this. Furthermore, as a German civil servant (as were all academics), he no longer had a pension after the war.[3] The USSR was the occupier of the section of Berlin where the clinic was located, so in late 1945, Karl contacted the Soviet authorities for permission to return to work, donned his white coat, and walked back into the clinic, seeing patients for three more years before he suffered a stroke in the hospital corridor (Fig. 2.1).

This entanglement of personal conscience, medical prowess, and historical events began early in Karl's career—well before the rise of Hitler—when he decided to turn his research attention to the

Fig. 2.1 Karl Ludwig Bonhoeffer. Paul Feam/Alamy Stock Photo

connection between trauma and mental disorders. This decision would involve him in some of the major medical issues of the century, including shell shock suffered by soldiers in the trenches of the First World War.

Karl's education had begun in the 1880s at a time of an enormous leap forward in medicine linked to a revolutionary new way of comprehending disease causation. The Frenchman Louis Pasteur had provided strong evidence for the germ theory of disease through his work on deadly puerperal fever, known as childbed fever because it struck women after they had delivered. He showed that it was caused by bacterial contamination, not by the spontaneous generation of germs that moved via miasma from patient to patient, as commonly thought. The fever was spread by the dirty hands of doctors and nurses and by the use of unsterilized surgical equipment. The way to stop the spread was strict sanitary procedures. While he was not the first to make the connection between dirty hands and disease, Pasteur backed up his ideas with hard evidence that helped convince the skeptical medical profession. He then turned to preventing, but not curing, the scourge of rabies. In 1885, he successfully vaccinated a nine-year-old boy who had been mauled by a rabid dog. The boy's survival brought Pasteur worldwide acclaim. Pasteur himself was utterly fearless and often reckless in how he carried out his research, one of the most famous examples being his extraction of saliva from the jaws of a mad dog via a suction tube in Pasteur's own mouth while his assistants held the animal down.

In Germany, the less flamboyant, more meticulous, but equally brilliant Robert Koch was hard at work on tuberculosis. Commonly known as consumption, it had reached epidemic proportions, accounting for approximately twenty-five percent of all deaths in Europe. Koch also did groundbreaking research on anthrax and cholera, receiving the Nobel Prize in 1905 for his work. Meanwhile, Paul Ehrlich was opening up the field of immunology, putting forth the idea of a therapeutic "magic bullet" that would selectively target a pathogen in the body. In line with that idea, he successfully used Salvarsan (derived from arsenic) against syphilis, a sexually transmitted disease that caused physical deformations and madness before it finally killed. Ehrlich also won the Nobel Prize.

To describe these advances as enormous does not begin to convey the upheaval in worldview, accompanied by a surge in hopefulness, they caused. Suddenly it appeared possible to cure and prevent all diseases. Pasteur, Koch, and Ehrlich were lionized even when they made serious mistakes or overestimated the impact of their discoveries. A new era in human history had begun, and it had been brought into being by the ingenious hands of men, not the hand of God, at least according to many scientists and physicians.

Karl's two fields of neurology and psychiatry were also undergoing vast change. Both had roots in the ancient world, but not until Karl's era were they set on a medical foundation. By the middle of the nineteenth century, the Frenchman Jean-Martin Charcot had developed clearer clinical pictures of many diseases, including multiple sclerosis, amyotrophic lateral sclerosis (called Charcot disease), and shaking palsy, which he renamed Parkinson's disease after the English doctor who had first identified it. Many physicians studied with Charcot, among them Sigmund Freud who was particularly interested in the etiology of hysteria.

Charcot's insistence on the use of the systematic neurological examination of patients was also significant. Doctors should not jump to a diagnostic conclusion on the basis of a superficial look. By the time that Karl entered practice, this was becoming standard procedure for neurologists. Karl himself insisted on it. Yet, unlike rabies and tuberculosis, the causes of neurological and psychological disorders were far less clear. Nowhere on the horizon was there the prospect of a magic bullet to cure Parkinson's, multiple sclerosis, epilepsy, or the strange psychosis the German psychiatrist Emil Kraepelin called *dementia praecox*, eventually renamed schizophrenia meaning a split mind. In his book *Clinical Psychiatry in Imperial Germany: A History of Psychiatric Practice*, Eric J. Engstrom makes the point that the simple model of expanding medical knowledge "is one that rather poorly describes the development of psychiatry." He quotes Karl Bonhoeffer as follows: "the development of psychiatry as a clinical field took a path different from other specialized disciplines. Unlike ophthalmology, otiatics, and orthopedics, it did not gradually specialize and split off from surgery or, like pediatrics, from internal medicine. Its path was just the opposite, it had to be brought laboriously from outside into the framework of the medical disciplines."[4]

In his book *Dreams and Delusions: The Drama of German History*, the historian Fritz Stern pondered what it was that Germany, as a new empire, sought at the end of the nineteenth century that drove it forward with such fervor, particularly in the sciences and medicine. Unified under Otto von Bismarck, its politically astute chancellor, in a way that preserved the privileges of the ruling class while denying those privileges to the middle and working classes, Germany exuded nationalist and militarist strength. Fearful on the one hand, aggressive on the other,

Germany longed for the greatness and glory it felt had been wrongfully denied by other European nations. One place Germany found that greatness was in science and medicine. According to Stern, those burgeoning fields were also the ones more open to Jewish scholars, a fact that would have serious repercussions in the 1930s:

> In Germany as elsewhere, the generation before 1914 was prodigious in talent and achievement; it was then that Germany attained a preeminent place in the natural sciences. German universities, which had long thought scientific studies a secondary concern, suddenly discovered that their scientists had won world renown. Universities had not been inhospitable to talent—provided it came in politically respectable, male, preferably Protestant guise. Any deviation from the norm had to be paid for by a superabundance of talent and, on the whole, was admissible only in the newer and politically neutral fields, such as medicine and the natural sciences. In those fields, achievement was more easily measurable and more immediately useful.[5]

The word *measurable* implies exactitude. The word *useful* implies pragmatism. The combination of the two means that objectivity was valued over subjectivity and that achievement was verifiable.

No doubt the prospect for success made it an exciting era to be a physician or a scientist, but for Karl personally it took some time to earn professional recognition. When Karl became Wernicke's doctoral assistant in 1893 at the age of 25, he equated the move to Breslau (now Wroclaw, Poland) as going into exile, writing, "I would have done anything to turn back."[6] Breslau was a far cry from the hilly Swabian towns of Neresheim, where he was born, and Tübingen where his university education began (followed by study at the Universities of Berlin and Munich). With a population of approximately 450,000, Breslau was the third largest German city behind Berlin and Hamburg. It was located on the Oder River in the lowlands of Silesia and was a major industrial and trading center, its principal products being locomotives, textiles, and beer. There were coal mines to the southeast to provide fuel for its expanding factories. There were also extensive agricultural lands to feed its burgeoning population. Over the centuries, Breslau had flown the flags of several conquering countries and empires, including Poland (when the city was called Vretslaw), the Hapsburg Empire, and Prussia. Since 1871, it had flown the flag of a united Germany. From a religious standpoint, its population was 60% Protestant, 35% Catholic, and about 5% Jews, of whom many were culturally assimilated. Adding to Breslau's sense of both separateness and inclusion, some of its citizens spoke a Silesian dialect, which distinguished them when they traveled to other German cities. Despite its geographical borderline status (which would lead to its near total destruction by Germans, Poles, and Soviets in the last few months of World War II), Breslau was industrially ambitious and proud of its artistic and scientific culture, best exemplified by its university.[7]

At first, Karl found Breslau's sprawling industry oppressive, a feeling apparently shared by many of its citizens as evidenced by the fact that Karl had no difficulty finding participants for one of his early research projects on alcoholic psychosis. It was not until 1896 when Karl met Paula von Hase at a social gathering that Breslau began looking much more pleasant. With flaxen hair and blue eyes, there was an appealing gentleness to Paula's face and an openness to her personality. Karl was her opposite. He had dark brown hair, a fashionable goatee, and an intense gaze that mirrored his intense mind. No matter how committed he was to scientific objectivity, Karl was immediately smitten by Paula in a way he described as mystical. Because she liked to ice skate on winter mornings on the canals, Karl took to doing the same so that they could meet. Paula was the daughter of Clara and Karl Alfred von Hase, a professor of practical theology at the university and a church official. Fortunately, Karl Bonhoeffer's agnosticism did not put a damper on the relationship; they were married on March 5, 1898. It turned out to be in every respect a very strong marriage. Sabine Leibholz-Bonhoeffer wrote about her parents that, "Each was only 'half a person' without the other. In a marriage lasting fifty years they managed to restrict their time of separation to a few weeks."[8]

For a two-year period from 1902 to 1904, the Bonhoeffer family was in a state of flux as Karl attempted to establish his career independently from his mentor Carl Wernicke, under whom he had habilitated in 1897, a process that required him to complete a second thesis to qualify for university teaching.[9] First, Karl was appointed professor of psychiatry at Albertina University in Königsberg (now Kaliningrad), a city with medieval roots located on the extreme northeastern edge of Prussia near the Baltic Sea. Then, the family moved to Heidelberg, where Karl replaced the renowned psychiatrist Emil Kraepelin who had moved to the University of Munich. At that time, he became a lifelong friend of the psychiatrist Robert Gaupp, a protégé of Kraepelin.

Hardly had the family unpacked their boxes in Heidelberg when Karl was offered the chair back at the University of Breslau as well as the directorship of the psychiatric clinic. Part of the upheaval was undoubtedly due to a major disagreement between Wernicke and the university administration, which led to the near collapse of psychiatric training at Breslau. Subsequently, Wernicke left Breslau and went to the University of Halle, only to be killed in a bicycle accident while on vacation.[10]

After having moved so frequently in such a short period, the family was grateful to return to Breslau and the warm, supportive circle of the large von Hase family. By that time, there were five children: Karl-Friedrich (1899), Walter (1899), Klaus (1901), Ursula (1902), and Christel (1903). The twins Dietrich and Sabine were born in 1906 and Susanne, known as Susi, in 1909. Thus, there was a natural division between the children: first the "three big boys" followed by the "two girls," and finally the twins, grouped (to their consternation) with Susi, the youngest. In the big boy group, Karl-Friedrich and Walter were very compatible, with the same fair coloring as their mother, while Klaus, though not standoffish, was his own individual self, with dark eyes and brown hair like his father. He balanced a temper with a sense of humor. Ursula, known as Ursel, was good-looking and serious, while Christel loved the outdoors and was known in the family as quick-witted. The twins, Dietrich and Sabine, were a team, forming their own dyad within the family, sharing their secrets with each other and concocting their own games. Susi was the energetic baby who, according to Sabine, turned everything upside down, was reluctant to pick up after herself but was very sociable.

At the Breslau clinic and the university, Karl had to undertake some repairs following Wernicke's tumultuous departure. Under his direction, a major building project was begun in 1906, bringing the clinic to an even higher level of esteem throughout Europe and attracting foreigners for treatment. Different from asylums that were often situated in the countryside to provide a calm environment away from the turmoil of the world, clinics were usually located in cities, were often connected to universities, and tended to put a greater focus on research than on patient care, although the latter was not slighted. According to his colleague Robert Gaupp, Karl was naturally empathetic, which in a psychiatrist was a gift of great therapeutic value.[11] He had the capability of winning the trust of severely disturbed patients, enabling them to talk with him. He could also calm down agitated, potentially dangerous patients, much to the admiration of his students who did not yet possess that essential skill. He was also considered a good administrator and teacher whose lectures were long, detailed, occasionally dull, but always worthwhile.

Because Karl was both a neurologist and a psychiatrist, he believed that the two specialties should be combined in the clinic, which would then be dedicated to treating nervous disorders. Comprehensive job titles, such as neuroscientist or neuropsychiatrist, were not yet in use. It was common for physicians to be trained in both fields, as were Wernicke, Alzheimer, and Freud. However, Karl did not agree with Freudian psychoanalysis, judging it to be unscientific, meaning unprovable, subjective, and muddled. Edith Jacobson, who was one of Karl's students in the 1920s, recalled that under his leadership the department of psychiatry and psychotherapy at the University of Berlin was a superior place to learn about clinical psychology, but it was unwise to mention psychoanalysis there.[12] It is worthy of note that Karl served as the Ph.D. advisor to Karen Horney in 1913–14. Although Horney would become a world-famous psychoanalyst, her dissertation was on the development of psychosis following a blow to the head. This was a topic totally within Karl's area of research expertise, which was on external causes, which he termed *exogenous*, of mental disorders. External causes could be pathogens, traumatic injury, the insidious damage done by ingesting alcohol, and breathing carbon monoxide. An internal cause, termed *endogenous*, was heredity.

Karl was not the first to focus his research on physical causes of mental illness. His contemporaries Emil Kraepelin and Hermann Oppenheim were involved in similar research. As a result of Oppenheim's early research, accident insurance for railroad workers was extended in 1889 to cover mental and nervous disorders caused by train accidents that in a rapidly industrializing nation were serious and all too frequent. According to the historian Paul Lerner in *Psychiatry and Casualties of War in Germany, 1914–1918*, Oppenheim's traumatic neurosis diagnosis was "a kind of precursor to today's PTSD (Post-Traumatic Stress Disorder), which treated post-traumatic nervous symptoms as a distinct entity resulting from the direct—anatomical or psychic—effects of traumatic events." This led to much controversy, with some psychiatrists insisting psychoses were hysterical reactions.[13]

Karl played a key role in this long-running controversy, which would reverberate throughout his professional life as three overwhelming historical events overtook him and the entire field of psychiatry, forcing psychiatrists into positions of ethical ambiguity in response to shifting cultural perceptions as well as economic and political pressures. The first event was shell shock, also known as male hysteria. The second was the 1918 worldwide influenza epidemic and the serious psychoses that developed in some survivors, which underscored Karl's premise that a pathogen (the influenza virus) could cause not only life-threatening physical illness but also severe mental illness. The third was the eugenics movement—the twisted offspring of Darwin's evolutionary theory that offered as its mission ridding the human race of hereditary diseases, thereby creating stronger, healthier, and happier people. Given what little was known about genetics at the time, eugenics seemed to be a sensible, even moral, way to use science to improve society. Besides politicians, many doctors believed in it, including Karl, albeit with caution. Eugenics was extremely popular all across Europe and in the USA. In fact, in 1927, the United States Supreme Court ruled in favor of sterilization of "imbeciles," although no one had taken the time to define the word diagnostically or to determine whether it was caused by faulty genes. However, it was in Germany that eugenics would meld horribly with Hitler's plan to create a master race. Beginning in 1933, he demanded that people with disorders such as schizophrenia go through "voluntary" sterilization so that their bad genes were not passed to future generations. But what if these mental disorders had multiple causes and were not solely genetic? And what did voluntary consent mean in the context of a disease in which a person did not know what reality was? These were the questions that Karl raised repeatedly. All these events were in the future, but by focusing on physical causation early in his career at the University of Breslau, Karl had set himself on a path that would lead inevitably to his involvement.

During the first decade of the twentieth century, Karl's research was on acute organic psychosis (sometimes called the Bonhoeffer reaction), motor aphasia, and asocial behavior. At one point, he studied "tramps and vagabonds" living in the congested streets of Breslau, finding that only 15% were mentally normal, while the other 85% showed mental derangement. He also studied prisoners in Breslau's crowded jails to assess the difference between those who took up a life of crime early and those who took it up later, arguing that it was an inborn psychiatric deficiency that played a part in young offenders, whereas it was the negative influence of the urban environment on congenitally normal adults that was the principal cause among those who became criminals when they were older.[14] These studies tied into the growing concern among psychiatrists about the steady increase in mental illness, particularly in the cities. Speculation was that the increase was caused by rapid urbanization and industrialization, which appeared to strengthen some people while enfeebling others, turning them into social misfits.

The other important aspect of Karl's early work was his insistence that only a psychiatrist was qualified to diagnose a mental disorder, a point he would make repeatedly throughout his career. Because it required specialized knowledge as well as the time to gather and assess a patient's history, psychiatric diagnosis should not be left in the hands of police or inadequately trained medical personnel. In cases involving the law, Karl argued that psychiatric examination should be obligatory. Those ideas found their way into the thinking of "alienists" (the word used in the early twentieth century for psychiatrists), criminologists, and welfare experts in Europe and the USA. For example, Olof Kinberg, a Swedish psychiatrist, presented a paper titled *Obligatory Psychiatric Examination of Certain Classes of Accused Persons* at the Seventh International Congress for Criminal Anthropology in Cologne in October 1911. In the paper, he referred to the work of Karl Bonhoeffer and his colleagues when he stated, "The occasion here for the attention of the alienist is based on one hand on the investigation made in late years by Bonhoeffer, Wilmans, Mönkemoller and others as to the mental qualities of tramps and beggars and the criminalist relations between this group of social incapables and recidivists."[15]

Significantly in regard to dubious mental diagnoses, Karl wrote in 1912 that, "The variety of the causative diseases is in contrast to a great uniformity of the psychic pictures." For example, delirium could be caused by many things acting alone or together, including severe brain injury, chronic intoxication, or strangulation. "At present, decisive information and etiological criteria cannot be derived from the psychological findings," he concluded, "but nearly exclusively from the somatic and neurological ones. In my opinion it will remain so."[16] This statement shows the hallmarks of a careful researcher who refused to push his findings too far, cautiously warning scientists and physicians not to jump to single-cause and single-effect conclusions, a warning that is still relevant.

Comparing the approach of Sigmund Freud, Alfred Adler, and Carl Jung to that of Karl Bonhoeffer, Robert Gaupp wrote many years later:

> In intuitive psychology and scrupulous observation Bonhoeffer had no superior. But he came from Wernicke's school, which was concerned solely with the brain and permitted no departure from thinking in terms of cerebral pathology. He practiced the psychology of empathy and understanding on the basis of his natural aptitudes and gifts. The theoretical interpretation of what was *behind* that being observed, of what occurred unconsciously and might project itself into consciousness, was foreign to his approach. Intuition was not alien to him, as his entire life's work shows. But he had no urge to advance into the realm of dark, undemonstrable, bold and imaginative interpretation, where so much has to be assumed and so little can be proven.[17]

It is doubtful that these issues were discussed around the Bonhoeffer dinner table as the children were growing up, although somehow they came to understand their father's point of view. Obviously, Dietrich's eventual dislike of Freudianism—one of several *isms* that captivated many of his young theological contemporaries in the 1920s and 1930s—shows his father's influence. It is also possible that it affected Karl-Friedrich's decision to study physics, which was nearer the ideal of pure science than psychiatry, being less swayed by cultural intrusion. The other subject that Karl would not have discussed with his children was the mentally ill people whose care filled his days—the tramps and vagabonds, the alcoholics, the pathological liars, the palsied, the hysterical—but he could not keep his children from realizing that beyond the safe circle of the family, there was much madness in the world. Such dark awareness was the common heritage of children of physicians, despite their privileged upbringing. Heightening that awareness for the Bonhoeffer children was the cemetery that was located across from their home; from the windows, the children often watched the solemn passage of horse-drawn hearses. Dietrich, in particular, had a fascination with death as a child, thinking about what eternity would be like and imagining himself on his deathbed speaking his last words to his gathered family.[18]

However, there were two aspects of Karl's character that he no doubt instilled overtly in his children. The first was the unwavering belief in the superiority of objective science and the transformative advances in the well-being of humanity that it achieved. The term used for science was *Wissenschaft*, and it was dedicated to verifiable truth. Taking on a quasi-religious cast, *Wissenschaft* demanded absolute dedication to education and hard work, which Fritz Stern called "the austere ethos." This ethos went hand and glove with the idea of *Bildung*, which had to do with self-actualization by means of cultural education (music, poetry, arts, philosophy, literature) in a way that ultimately benefitted not only the individual but also society. *Wissenschaft* and *Bildung* had many tangled roots that had been well-watered by Goethe who wrote, "He who possesses art and science has religion; he who does not possess them needs religion."[19]

The second aspect that Karl instilled in his children was the importance of self-control. In his memoir written in the last years of his life, Karl wrote, "I have always believed that the qualification for a psychiatrist must include not only a capacity for understanding those who think differently, but a high-degree of emotional control." Paula was in complete agreement with Karl about teaching the children the value of self-control as opposed to the imposition of external control. She provided the early education of the children herself precisely because she and Karl disliked the severely autocratic form of education in German schools, exemplified by the common saying that boys "had their backs broken twice in life; first in school and then in the military." Unfortunately, it was a saying that had much truth to it. For example, Albert Einstein's early education at Luitpold Gymnasium in Munich so embittered him that he renounced his German citizenship when he was only a teenager and set for himself a course of education that meant he was in some respects as much self-taught as taught.[20]

Karl did not place many restrictions on the children so as not to influence them too much. There is no doubt he was the authoritarian in the family whose study was sacrosanct and whose decisions were final, which was the cultural norm for German fathers. However, he was not a dictator, nor was he ever cruel. Sabine wrote, "He took it for granted that we would try to do what was right and expected much from us, but we could always count on his kindness and the fairness of his judgement."[21] Throughout his life, he was proud of his children and supportive of their decisions, even when those decisions went against the grain, for example, Dietrich's decision to study theology. Yet, his influence on all his children, including Dietrich, was substantial. In a letter Dietrich wrote from prison to his friend Eberhard Bethge on April 22, 1944, he reflected on the development of his thinking from his childhood: "I don't think

I have changed very much, except perhaps at the time of my first impressions abroad, and under the first conscious influence of Papa's personality. Then a turning away from the phraseological to the real ensued." His juxtaposition of the words *phraseological* and *real* has relevance to the development of his theology, particularly in light of the year he spent at Union Theological Seminary in New York City. However, it is just as likely Dietrich was referring to his father's insistence that all the children learn to separate "the inessential from the essential," saying things directly. "His dislike of clichés did at times make some of us inarticulate and uncertain of ourselves," Sabine wrote. "But it had the effect that as adults we no longer had any taste for catchwords, gossip, commonplaces or loquacity."[22]

Fortunately, Paula had received her teaching certification from the Royal Provincial School College in Breslau prior to marrying Karl. Not common practice for young women of her class, her certification revealed both her intelligence and a certain degree of independence from *Kinder, Küche, Kirche*—children, kitchen, church—that defined and constrained the lives of most German women. The family had servants and a governess, but it was Paula who took on the responsibility of teaching the children during their early years. Sabine remembered her as a very good teacher, not at all "school mistressy." As a result, the children liked to learn. The sprawling Breslau villa was perfect for a large family. Built of brick, it had multiple fireplaces and chimneys and ornate wood trim inside and out. Rose bushes climbed trellises on the exterior walls. There was a schoolroom, a hobby room, and, best of all, a slightly overgrown garden filled with old trees where the children could play and explore. There was even a tennis court that was flooded in the winter for ice-skating. The carriage house provided ample space for a combination of zoo and natural history museum. It was filled with live snakes, lizards, squirrels, and birds, with display cases of pinned and labeled insects and butterflies. This was the favorite domain of Walter, who of all the children was the most interested in the natural world. Sabine remembered admiring "his collection of butterflies, his stones and bird's eggs, the squirrels which he had caught himself, his doves, and the animals he kept in aquaria and cages. He loved them all." Once one of his non-venomous snakes fastened itself firmly to his thumb and would not let go even when Walter directed a jet of water onto its head. "In the end my mother had my father fetched from his consulting room, a quite unusual occurrence, so that he might teach the snake its manners. I believe that it was given ether."[23]

Paula's mother, Clara, had been a pianist as a young woman, studying with Clara Schumann and Franz Liszt. Paula herself loved to sing, so she made sure that every child studied music: Karl-Friedrich and Dietrich played piano, Walter and Sabine the violin, Klaus the cello, Ursula the lute. They took dancing lessons with the children of the Stern family, whose father, Richard, was a renowned physician and a friend of Karl's.[24] They learned about musical ensembles in the home of Toni and Albert Neisser, a medical colleague of Karl's at the university, whose beautiful villa served as a musical salon for the cognoscenti of Breslau. Eventually, Dietrich showed such musical promise (surpassing his siblings) that at one point a career as a pianist was seriously considered. As a teenager, he even wrote a trio based on Schubert's song *Gute Ruh*, which he played with Klaus on cello and Sabine on violin.[25]

Family festivities were enlivened by musical offerings, puppet shows, and readings of poetry sometimes composed by the children. To encourage their creativity, Paula made sure that an old chest was stuffed with wigs and costumes. Once Karl-Friedrich wrote a play based on the story *The Cold Heart* by William Hauff, a nineteenth-century writer and collector of folktales. In the story, a poor man receives three wishes from a good spirit only to squander them. He then turns to a bad spirit who offers to make him wealthy in return for his warm heart. The man agrees but finally comes to his senses after killing his wife in a fit of rage. With the help of the good spirit, the man receives his heart back and returns gratefully to his humble life, with moral order (and wife) restored. The play had a large cast of characters, including a delicate little spirit made of glass, a large evil spirit who chopped down trees and unleashed storms, a beautiful young woman, and a grieving mother, so there were more than enough parts for everyone. Those who did not want an onstage part could always make the sounds of wind, falling trees, and tinkling glass. As a result of their parents' solicitude, the years in Breslau were happy for the children. "Dishonesty and fibbing were severely punished; in comparison, broken windows and torn clothes hardly counted," Bethge wrote.[26]

The years were also scientifically productive for Karl, who did not have to waste time in commuting, because the psychiatric clinic was not far from the villa. He was a busy man, seeing patients, conducting research, and lecturing at the university, yet when he was home he was attentive to his children and encouraged their curiosity. As 1909 drew to a close, he looked back on a year in which his youngest child, Susi, had been born, and his oldest child, Karl-Friedrich, was about to turn eleven. He wrote in his diary on December 31, "In spite of having eight children – which seems an enormous number in times like these – we have the impression that there are not too many of them! The house is big, the children develop normally, we parents are not too old, and so we endeavor not to spoil them, and to make their young years enjoyable."[27]

The presence of friends and family living nearby made for a rich environment for the Bonhoeffers. So too did the fact that many of their friends were by background Jewish, although some had converted to Christianity, including the Sterns. The main reason was that medicine as a profession was very attractive to Jews who were hindered in advancement in civil service positions, which included academia. According to Fritz Stern, "In imperial Germany, the physician's white coat was the one uniform of dignity to which Jews could aspire and in which they could feel a measure of authority and grateful acceptance." As to their reasons for conversion, Stern wrote:

> My grandparents, two doctors and their wives, and their circle of friends, colleagues and assistants exemplified the wide range of responses to a world at once enticing and hostile. In a culture in which *Bildung* and *Wissenschaft* had profoundly altered the traditional outlook of many religions, Jews faced special quandaries. How could one reconcile ancient rituals and proscriptions with the prevailing post-Darwinian secular-scientific outlook? Like so many educated Protestants, my grandparents arrived at a worldview that fused a distillation of Christian ethics with rational precepts and national sentiment, the whole enveloped with a proper sense of awe.

Furthermore, conversion was not difficult because Protestantism had become theologically undemanding, as Stern phrased it, "having made its peace with modern life, the spirit of capitalism, and modern science." There was also the desperate hope that as Christianity became a convenient nicety rather than a spiritual necessity, anti-Semitism would disappear. Stern's insights help to explain decisions eventually made by Dietrich for whom the church's making peace with modern life, which meant acquiescence to Nazism, was a monumental issue.[28]

For the Bonhoeffers and Sterns, the friendships formed in Breslau were lifelong, not even broken by bitter exile and war. However, the sharing of daily life ended in 1912 when the Bonhoeffers moved to Berlin where Karl had been appointed as head of neurology and psychiatry at the prestigious Charité Clinic in Berlin and chairman of psychiatry at the University of Berlin. The Charité had been established over two hundred years before and had a fine reputation for both patient care and research. At the time of the move, Karl-Friedrich was thirteen years old, and Susi (the last in line) was three. The big change for the older children was that they began attending school outside the home.

Berlin was a bustling city that had grown very rapidly since German unification. It had over four million people and was expanding outward with the development of new suburbs and independent towns, such as Charlottenburg. Many of its citizens were outsiders who came from other parts of Germany. Berlin was the third largest city in Europe after London and Paris, but it wanted to be first and it wanted to be first fast. If it couldn't beat out London and Paris, then it would set its sights on beating out an American city, such as New York or Chicago. The comparison to American cities was apt because Berlin did not have medieval charm (it had torn down many of its old buildings to make room for new) and was, as one journalist wrote, "Condemned to a state of eternal becoming." As a result, it had an inferiority complex for which it overcompensated. If Munich was nationalistic, then Berlin was more so. If Hamburg liked the Kaiser, then Berlin adored him. After all, Berlin was the seat of royal power; Kaiser Wilhelm was theirs. Arrogant and proud of it, Berlin was at least never dull.

Berlin quickly became the hometown of the Bonhoeffers. For four years, they rented a house on the Brückenallee near the Tiergarten, a lush urban park where the children liked to play. It had broad sweeping lawns, flower gardens, ponds and lakes dotted with small islands, meandering streams, and wide paths. Throughout the park, there were sculptures of German political and cultural figures, including Wagner, Goethe, Beethoven, Haydn, and Mozart. In keeping with the name of the park (originally a hunting preserve, Tiergarten means animal garden), there were sculptures of animals including a bison for the pleasure of the children.

In 1913, the Bonhoeffers purchased a forester's lodge in the village of Friedrichsbrunn in the Harz Mountains to replace the much-loved retreat near the Bohemian border where they had spent their vacations while living in Breslau. Friedrichsbrunn provided relaxation for the adults and fun for the children, with hiking, swimming and mushroom hunting in the forests in the summer, and cross-country skiing in the winter.

It soon became everyone's favorite place, one reason being that its isolation enabled Karl (relieved of the necessity of seeing patients) to give the children his undivided attention. Eventually, Friedrichsbrunn would play a more important role, providing emergency shelter during World War II for children and grandchildren.

In 1916, when Karl-Friedrich was seventeen years old, the family moved to a large three-story villa on the Wangenheimstrasse in the area of Berlin known as Grunewald, an upper-class suburb populated by professors, scientists, physicians, and musicians. The families of Hans Delbrück (professor of history), Adolf von Harnack (professor of theology), and Max Planck (professor of physics and a Nobel laureate) all lived nearby. It was a group destined to be linked ever tighter, first by marriage, second by their work in the resistance movement, and finally by tragedy. According to Max Delbrück, whose older sister Emilie (called Emmi) would marry Klaus Bonhoeffer, Grunewald was a place where "Professors with large families intermingled with moderately successful businessmen. Some of the houses were quite palatial, but the houses that the Harnacks and the Delbrücks and the Bonhoeffers built were straight-forward accommodations for large families, nothing very fancy about them."[29] About the only thing distinctive about the Bonhoeffers's villa was that it was painted yellow.

All the children grew up together, playing in the gardens, making music, and concocting numerous games. "We are building an underground cave and tunnel [in the garden]," wrote ten-year-old Dietrich to a cousin. "It's there so that in case we fight with Klaus again, we can either bring reinforcements from the cave . . . or attack the enemy from behind."[30] Emmi Delbrück was fond of a more cerebral game in which each person wrote down a question on a slip of paper and threw it into a bowl. Once a question was selected, each child had to come up with an answer, and a winner was chosen. Sometimes the questions dealt with human relationships. Emmi remembered one question in particular because of the prescient winning answer. The question was, "What is the quickest way of getting to know another person?" One child answered that the person should be put on his hobby horse. Another child suggested that they go on a journey with the person. Suddenly, the game became serious when one of the boys said, "I get to know someone best when I offend him unfairly." The boy was Gerhard Leibholz who would marry Sabine Bonhoeffer, and who would be forced to leave Germany in 1938 because of his Jewish background.[31]

Throughout his life, Karl-Friedrich took his role as eldest son seriously, watching over his siblings, occasionally intervening on his worried mother's behalf when one of them, specifically Dietrich in the 1930s, began to push into dangerous territory. Sabine, who was seven years younger than Karl-Friedrich, wrote: "My mother used to say that if he had not come at the head of the list none of us would have come to any good. It was plain that soon he was as reliable as a nursemaid. His selflessness was striking. Invariably he took the smallest portion of food, and this trait lasted throughout his life."[32]

When in 1896 Karl had first met Paula, he had been struck by "her free natural manner, her open uninhibited gaze." Karl-Friedrich inherited both those characteristics. Whereas Dietrich looked like the idealized model of a German boy—light blond hair cut in bangs, square jaw, serious expression—Karl-Friedrich was softer in appearance. He had darker blond hair that he parted to the side, a slender build, and a quick smile that tended to light up his eyes. But while he took after his mother's side of the family physically, he took after his father's side mentally, being intrigued by the search to gain new knowledge coupled with the ability to speak as precisely as possible about research. Karl-Friedrich's paternal grandfather had been a lawyer and a judge; his uncle was an industrial chemist with Bayer. Karl-Friedrich was not interested in how the human mind worked but in how the universe worked, revealed in his early interest in astronomy. Sabine considered him very smart. "He was always at the top of his form without doing much work. He seemed to have plenty of free time and enjoyed painting and going to museums."[33]

To Max Delbrück, Karl-Friedrich was one of the few people in his boyhood world who encouraged his interest in science and mathematics. Not long after the Bonhoeffers settled in the Grunewald just around the corner from the Delbrücks, ten-year-old Max, who was the same age as Dietrich and Sabine, came under the benign influence of seventeen-year-old Karl-Friedrich, who had already decided on a career in science. As a distinguished professor of history at the University of Berlin, Max's father had no objection to his youngest son's interest in science and even encouraged him, although he did not share that interest. Max had acquired a telescope and would set a very loud alarm clock to wake him in the middle of the night so he could get up to watch the stars from a balcony of his parents' bedroom. Max was such a sound sleeper that the alarm clock invariably woke everyone else in the family before it roused him. Then, he would tiptoe through his parents' room, staying outside on the balcony for an hour before going back to bed. "I'm sure my mother was worrying herself stiff that I would freeze to death out there." Eventually, Karl-Friedrich learned of Max's interest and asked him some questions. By the answers Max gave, "it was obvious that I had really very little insight about planetary motions or parallaxes or stellar properties and so on. So he quickly found out that I really didn't know much, and he told me a fair amount, and from that developed our friendship. He took a great liking to me and I, of course, admired him." To Max, Karl-Friedrich was not only a role model but also a kind of hero. Following in Karl-Friedrich's scientific footsteps, Max studied at the Universities of Tübingen and Göttingen, earning his Ph.D. in physics. Eventually, he would win the Nobel Prize.[34]

Notes

1. "Obituary: Professor Karl Bonhoeffer," *Journal of Nervous and Mental Diseases* 109, no. 2 (1949): 194–197.
2. Karl Bonhoeffer had access to various microbes for research purposes at the Charité Clinic, including malaria. Elizabeth Sifton and Fritz Stern, *No Ordinary Men: Dietrich Bonhoeffer and Hans von Dolnanyi, Resisters Against Hitler in Church and State* (New York: New York Review Books, 2013), 118–119.
3. In a letter to Paul Harteck dated April 24, 1946, Karl-Friedrich wrote that his parents had no pension or survivors' benefits. "The only consolation in this catastrophe is that all the grandchildren remain healthy." *Paul Harteck Papers*, Box 1, Folder 8, Institute Archives and Special Collections, Folsom Library. Rensselaer Polytechnic Institute.
4. Eric J. Engstrom, *Clinical Psychiatry in Imperial Germany: A History of Psychiatric Practice* (Ithaca: Cornell University Press, 2004), 24–25.
5. Fritz Stern, *Dreams and Delusions: The Drama of German History* (New Haven: Yale University Press, 1987), 31.
6. Bethge, *Bonhoeffer*, 13.
7. Fritz Stern, *Five Germanys I Have Known* (New York: Farrar, Straus and Giroux, 2006), 16.
8. Leibholz-Bonhoeffer, *Bonhoeffers*, 13.
9. Karl Bonhoeffer's thesis was on unilateral chorea, a movement disorder thought to be psychiatric in origin but which he proved was caused by a brain lesion.
10. *Dietrich Bonhoeffer Works* (DBW), Vol. 17 (Minneapolis: Fortress Press, 2014), 279. See also Engstrom, *Clinical Psychiatry*, 223.

11. Bethge, *Bonhoeffer*, 22.
12. Bethge wrote that after the move to Berlin, Karl "turned the city into a bastion against the invasion of Freudian and Jungian psychoanalysis. His mind was not closed to unorthodox theories, nor did he deny the validity of efforts to investigate unexplored areas of the mind; but he refused to ever have any contact with Freud." However, this appears to be an overstatement. Ibid., 21. For additional information on the Bonhoeffers and Freud, see Clifford Green, "Two Bonhoeffers on Psychoanalysis," in *A Bonhoeffer Legacy*, ed. A. J. Klassen (Grand Rapids: Eerdmans, 1981), 58–75. Edith Jacobson was Jewish and politically active. She escaped from Germany and came to the USA where she became a leading psychoanalyst. Rainer Herrn, "Edith Jacobson," Institute of the History of Medicine of the Charité. http://gedenkort.charite.de.en/people/edith_jacobson.
13. Paul Lerner, "Psychiatry and Casualties of War in Germany, 1914–1918," *Journal of Contemporary History* 35, no. 1 (2000): 14.
14. Paul Lerner, *Hysterical Men: War, Psychiatry, and the Politics of Trauma in Germany, 1890–1930* (Ithaca: Cornell University Press, 2011), 21.
15. Olof Kinberg, "Obligatory Psychiatric Examination of Certain Classes of Accused Persons," *Journal of the American Institute of Criminal Law and Criminology* 2, no. 6 (1912): 858.
16. P. Propping, "Genetic Disorders Presenting as 'Schizophrenia': Karl Bonhoeffer's Early View of the Psychoses in the Light of Medical Genetics," *Human Genetics* 65, no. 1 (1983): 2.
17. Bethge, *Bonhoeffer*, 22.
18. Ibid., 14, 38.
19. Stern, *Five Germanys*, 17.
20. Ibid. See also Ronald W. Clark, *Einstein: The Life and Times* (New York: World Publishing, 1971), 20.
21. Leibholz-Bonhoeffer, *Bonhoeffers*, 12.
22. DBW, Vol. 8, 358; Leibholz-Bonhoeffer, *Bonhoeffers*, 12; Hans Christoph von Hase, "Turning Away from the Phraseological to the Real," in DBW, Vol. 10, 591–604; Bonhoeffer's essay fragment "What Does It Mean to Tell the Truth," DBW, Vol. 16, 603.
23. Leibholz-Bonhoeffer, *Bonhoeffers*, 20.
24. Richard Stern was the father of Rudolf and the grandfather of Fritz.
25. Bethge, *Bonhoeffer*, 25. See also Stern, *Five Germanys*, 31.
26. Bethge, *Bonhoeffer*, 19.
27. Ibid., 16.
28. Born in 1926, Fritz Stern was baptized as a child. He did not know of his family's Jewish background until he was seven years old when he hurled an anti-Semitic epithet at his sister and was summoned to his father's study. Stern, *Five Germanys*, 21–22, 98.

29. Max Delbrück. Interview by Carolyn Harding, July–Sept 1978, Archives California Institute of Technology.
30. Ferdinand Schlingensiepen, *Dietrich Bonhoeffer, 1906–1945* (London: T & T Clark, 2010), 12.
31. Dorothee von Meding, *Courageous Hearts: Women and the Anti-Hitler Plot of 1944* (Providence and Oxford: Berghahn Books, 1997), 5.
32. Leibholz-Bonhoeffer, *Bonhoeffers*, 17.
33. Ibid.
34. Max Delbrück and Karl-Friedrich Bonhoeffer remained lifelong friends. Max came to the USA in 1937.

CHAPTER 3

The First World War

At the time the Bonhoeffers moved to the Grunewald district of Berlin, the First World War battle lines were locked grimly in place along thousands of kilometers of trenches, with the two sides divided by barbwire and cratered no-man's-land. The 1914 dream of a short and glorious war had rapidly turned into a nightmare of industrialized slaughter in which powerful long-range artillery mixed with clouds of poison gas and machine-gun fire. Hundreds of thousands of lives were lost to gain a few meters of ground, only to have hundreds of thousands more lives lost in losing the same few meters. In one of the most egregious examples, in the Battle of the Somme, on a single day—July 1, 1916—the British suffered nearly 60,000 casualties. By the battle's miserable and inconclusive end four months later, after heavy rains had turned the ground into impassable mud, the British had suffered 420,000 casualties, the French 195,000, and the Germans 650,000. In total, five kilometers of land had been gained by the Allies. The *Encyclopedia Britannica* summed it up succinctly: "Sir Winston Churchill once described the battles of the Somme and Verdun, which were typical of trench warfare in their futile and indiscriminate slaughter, as being waged between double or triple walls of cannon fed by mountains of shells. In an open space surrounded by masses of these guns large numbers of infantry divisions collided. They fought in this dangerous position until battered into a state of uselessness. Then they were replaced by other divisions."[1]

K. L. Housley, *The Scientific World of Karl-Friedrich Bonhoeffer*,
Palgrave Studies in the History of Science and Technology,
https://doi.org/10.1007/978-3-319-95801-9_3

If anything good can be said about the First World War in comparison with World War II it is that civilian casualties as a direct result of bombing and artillery fire were low. Massive bombing of cities, such as Berlin and London, was a military strategy of World War II made possible by advances in aviation and munitions. What that meant for the Bonhoeffers was that Berlin was safe from direct attack. However, starvation—a killer predating bombs—grew worse and worse in the cities as the war lengthened. Food was strictly rationed. When potatoes became scarce in the winter of 1916–17, swede turnips were substituted. Derisively called Prussian pineapples, they were bitter, low in nutrition, and caused gastric distress. Many other substitutes, called "ersatz," were also common, such as extending pepper by mixing it with ash, adulterating milk, and using chicory for coffee.[2] Long after the war, Dietrich recalled eating bread in which sawdust had been mixed with flour. Because Kaiser Wilhelm had been convinced the war would be quick and victorious for Germany, he and his high command had not adequately taken into account the country's inability, even during peacetime, to feed itself without massive imports of food and raw materials. Great Britain's naval blockade effectively halted those imports, despite Germany's use of submarines to strike back by disrupting supply lines to the British Isles.[3]

The Bonhoeffers were fortunate to have a garden in which they grew vegetables. They also kept poultry and goats that provided eggs and milk. Toward the end of the war, as coal supplies ran short, they heated only the kitchen and the living room.[4] In terms of day-to-day sustenance, the family was better off than most, but in terms of losses on the battlefield, they were hit hard. Early in the war, Paula's youngest brother, Benedikt (known as Uncle Bubi to the children), became a prisoner of war. Karl's brother, Otto Bonhoeffer, lost two sons, Fritz and Werner, and his third son, Lothar, was severely wounded.[5]

Eight years old when the war started, Dietrich and Sabine felt these losses viscerally: "We were told that three of our cousins had fallen in the West, that our cousin Lothar Bonhoeffer had had his leg crushed and an eye shot out, and that another cousin had had to have his leg amputated," wrote Sabine. "A lasting impression was made upon me by my wounded cousin Lothar, half blinded and limping on crutches, head and leg swathed in bandages, yet quite unbroken in spirit and still continuing to sing his soldier songs and songs by Hermann Loös."[6]

At the start, the overwhelming majority of Germans were in favor of the war to the point of nationalistic frenzy. However, in Karl's reminiscences written thirty years later in the last few years of his life, he implied that he was ambivalent. "We did not really believe that there was going to be a war, even though the encirclement by the Triple Entente [England, France, and Russia] was felt to be alarming. All the same, I remember having been given the disquieting feeling in the company of some staff officers in the winter of 1913–14 that the possibility of an early armed clash must be taken into account."[7] Given that the war did not start until August 1914, Karl was referring to the significant arms buildup that had been going on for several years in anticipation of a conflict. He himself played a role in those preparations from the standpoint of psychiatric readiness. In a paper titled "Psychiatry and War" published earlier in 1914, Karl recommended that to prevent "war psychosis," men should be weeded out of the military who were vulnerable to mental illness in combat. Those soldiers who were "not up to the demands which discipline, the physical and mental exhaustion of military service places upon them, and are thus disposed as a consequence of their mental instability, to be thrown off kilter," were to be separated from their units and discharged.[8] What Karl failed to comprehend—what everyone failed to comprehend, holding in their minds an idealized image of the way prior wars were fought—was the level of carnage that humankind was about to achieve. Nothing, least of all psychiatry, was a match for what was coming.

German military leaders had long prepared to fight against Russia and France. Curiously, many of them had discounted the possibility of Great Britain's swift entry into the war, having convinced themselves that Great Britain had no stake in the Balkan turmoil, and that it would not stand by Belgium, its ally. Furthermore, despite Great Britain being a formidable sea power with superiority over Germany, it had only a small volunteer army. Against Germany's 98 regular army divisions, Great Britain had six.[9] On August 1, 1914, Kaiser Wilhelm II declared war against Russia. Standing on the balcony of the Royal Palace with the late-afternoon sunlight glinting on his helmet, he told the rapturous crowd packing the avenue, Unter den Linden, that there were no more factions in the nation, only united Germans who stood together as brothers, wielding the sword of victory with God's blessing. People cheered wildly all along the avenue, bursting into patriotic songs and the Prussian national anthem.

Two days later on August 3, Germany declared war on France and invaded neutral Belgium, which geographically blocked the way.

Then on August 4, Great Britain entered the war in defense of Belgium. Germany had spent years significantly building up its naval power; nevertheless, it was focused on fighting a land war. Great Britain's entry changed everything. That evening Karl went for a walk with his three oldest sons, Karl-Friedrich (fifteen years old), Walter (fourteen), and Klaus (thirteen). They found the atmosphere on the Unter den Linden markedly different. Karl wrote: "The enthusiasm of the previous days of the crowds in the streets in front of the castle and the government buildings had succumbed to a gloomy silence, which made the scene extraordinarily oppressive."[10] However, the drop in the nation's confidence was temporary. Germans remained convinced that the war would be short, victory was guaranteed, and as the Kaiser had promised, a glorious time was coming.

The total lack of a free press for the next four and a half years helped keep those beliefs alive. Patriotic propaganda became the only news available, leading people to believe that the German army was "undefeated in the field" right up to the day the armistice was signed on November 11, 1918. As Karl wrote in his diary entry at the end of 1916, "It's hard to form one's own opinion, given the censored press, and what happens politically is so frequently just poorly organized necessity that we are very disturbed. The constant appeal to the trust of [the Kaiser's] subjects, which only allows for retrospective observations, offers nothing very edifying."[11]

A personal incident that Sabine recalled provides credence to her parents' ambivalent response to the war. Once she was given a little brooch by her teacher on which were the words, "now we'll thrash them." It was pinned proudly to the collar of her white blouse when she went home for lunch. She was surprised by her parents' reaction. Her father asked, "What have you got there? Just give it to me." With that, the brooch disappeared into his pocket, while her mother promised to get her a prettier one.[12]

When Germany brought in its 420-millimeter siege guns to destroy the forts of Liège in Belgium, there was an immediate outcry from various scholarly groups in England, claiming that Germany had broken international law. In response, ninety-three German academics signed the "Manifesto to the Civilized World" in October 1914. Among the signatories were many famous scientists, including Max Planck, Fritz Haber, Paul Ehrlich, and Wilhelm Roentgen, the discoverer of X-rays. The manifesto was an overblown justification of Germany's invasion of

Belgium, stating that the claims of brutality were lies and that any killing of Belgian citizens by soldiers had been in self-defense. The manifesto also defended the ideal of German militarism, seeing it as the upholder of German civilization. Some of the language eerily foreshadowed what was to come in the 1930s, with mention of the "Russian hordes" and threats to the "white race."[13]

Karl did not sign the manifesto, but he did sign a petition called the "Reply of the Cultural Federation of German Scholars and Artists to the Declaration of the Professors of Great Britain and the English Academic Circle," which was similar in that it condemned Great Britain for disrupting the peaceful existence of the German Empire and denied that Germany had violated Belgian neutrality. Several physicians signed the petition.[14]

One signature was conspicuously missing on both the manifesto and the petition—Albert Einstein's. Having recently returned to Germany from Switzerland to head the Kaiser Wilhelm Institute of Physics, Einstein was vociferously opposed to the war to the point of helping to draft a counter-manifesto that called for international relations and the formation of a European league, upbraiding scientists and artists alike for speaking "in a hostile spirit" against peace. Only a handful of men signed it at the University of Berlin where it was circulated. Undeterred, Einstein continued to be an outspoken critic of the war. A few weeks after it started, he wrote in disgust to the Dutch theoretical physicist Paul Ehrenfest that, "Europe, in her insanity, has started something unbelievable. In such times one realizes to what a sad species of animal one belongs." Einstein believed that the war was fomented in part by Germans being psychologically submissive to the nation.[15]

Many psychiatrists, Bonhoeffer among them, saw the outbreak of war as improving the mental health of German citizens. Bonhoeffer asserted that the increase in community feelings brought on by the war caused citizens to disregard minor complaints and not seek medical advice as often. Furthermore, he noted a drop in alcoholism. Another psychiatrist noted a drop in suicides.[16] There were even arguments made that from the standpoint of eugenics the war would make a stronger race of Germans. The soft indifference and lax morality of the young men would be hardened to steel in the forges of battle. This argument was based on a complete misunderstanding of how genetics worked. By the end of the war, psychiatrists had realized ruefully their error: killing off healthy young men did nothing to improve the quality of the German gene pool.

Besides being influenced by the mystique of the defense of the fatherland, there was another factor at work among academic professors and physicians employed in public clinics. They were civil servants whose salaries were paid by the government; therefore, they could be called upon to help with the war effort. For example, Karl remained at the Charité Clinic and the University of Berlin during the war, but he was also a medical consultant to the Corp of Guards. In the First World War, the effect of being in government employment was that everyone was under pressure to come together in unity, suppressing their doubts and silencing their criticism.

In December 1914 at the first wartime meeting of the Berlin Society for Psychiatry and Nervous Illness held at the Charité, Karl addressed the gathering on a new psychosis seen in alarming numbers among soldiers. It came to be known as shell shock in Great Britain, while in Germany it was often referred to as male hysteria. Only four months after the war had begun, the armies were already exhausted and deadlocked, with the trenches stretching from the borders of Switzerland to the North Sea. And in those rat-infested, cold, and damp trenches, which also functioned as makeshift graves for fallen soldiers, something strange was occurring. The list of symptoms was long: amnesia, numbness, morbid agitation, tremors, paralysis, delirium, dizziness, loss of speech and hearing, severe headaches, hypersensitivity to sound, irregular heartbeat, and high blood pressure. Outwardly, there was no physical injury. There was contentious disagreement among psychiatrists and neurologists as to whether these symptoms were caused by physical trauma, such as the intense shock waves from incessant cannon fire over the confines of the trenches. Another view was that it was nothing but cowardice, whereas the cure was to return the victims to the front or court-martial them, the penalty being imprisonment or execution. Part of the diagnostic dilemma was whether the shell-shocked soldier had a pre-existing condition or a predisposed personality, neither of which could be easily determined in the crisis atmosphere of military hospitals with insufficient staff and supplies. In a system of psychiatric medicine that was rapidly becoming mechanized, there was no time for the subtleties required to make an accurate diagnosis. Even if there had been time, there did not yet exist any of the scanning diagnostic tools that might reveal physical brain damage. Everyone was working in the dark.[17]

One of the main issues at that first meeting in December 1914 was the use of the term hysteria, which had a long history of being associated with weak females and effeminate men. The year before in 1913, Emil Kraepelin went so far as to dismiss hysteria as the atavistic reaction of "women, young persons, and psychopaths whose will-power is not properly formed."[18] This was not the first time that psychiatrists proved they were vulnerable to the torque of language, whether they admitted it or not. In every respect, hysteria was a negatively freighted term.

Karl opened the meeting by presenting several patients with symptoms that he asserted were hysterical—some trembling, some with contractions of their limbs, some with speech disorders, some who could not walk. According to Paul Lerner in his book *Hysterical Men: War, Psychiatry, and the Politics of Trauma in Germany, 1890–1930*:

> Bonhoeffer drew a crucial distinction between the cause of the neurotic symptoms, or their "moment," and the symptoms' subsequent "fixation." He distinguished between the fear, terror, or shock at the root of a hysterical reaction and the ideas or desires that could cause these reactions to develop into full-blown, persistent neuroses. Under normal circumstances, Bonhoeffer added, these cases presented no great therapeutic difficulty; the symptoms could be easily removed when the diagnosis had been made accurately and promptly. But when these psychogenic cases were "misdiagnosed" by untrained army medics and erroneously labeled organic they could become "fixed" and often proved tenacious.[19]

It was Karl's opinion that when the cause was determined to be organic, the potential for a cure was diminished. The hidden danger in Karl's argument was that it left open the possibility that an hysterical soldier was a malingerer. Later visits to overflowing field hospitals and prisoner-of-war camps near Verdun in 1916 reinforced Karl's belief that hysteria was a way for German soldiers to avoid being sent to the front. He could find no other explanation for the fact that hysteria was not seen in French prisoners of war for whom the threat of returning to battle had been lifted.[20] However, Karl's assessment was based on behind-the-lines observations. Had he actually spent an extended period of time in the trenches, while under constant bombardment or during a gas attack, he might have drawn a different conclusion.

In September 1916, Karl was one of the participants in the annual conference of the Society of German Neurologists and the German Psychiatric Association in Munich, called the War Conference. By this time, shell shock had become an enormous problem among soldiers fighting on both sides. For the years 1914–18, according to the official report of the German Army Medical Service, 613,047 soldiers were treated in military hospitals for psychoses. In England and France, the statistics were also extreme. At the Battle of the Somme, which was still raging at the time of the Munich conference, of the approximately one million casualties, an estimated forty percent suffered shell shock.[21]

Deforming the entire debate about whether shell shock was really a form of hysteria was the economic cost of treatment and compensation, the shortage of hospital beds, and the political and military pressure to triage injured soldiers so they could return to the front as fast as possible. For many doctors, the imperative was not "first, do no harm" but "first, win the war." According to Paul Lerner, "By 1916, with Germany deeply in the grip of *Materialschlacht* [mechanized warfare] the last opposition of male hysteria had been silenced and soon the diagnosis turned into a rallying-cry for a psychiatric establishment eager to prove its value to the national cause. Karl's colleague and friend Robert Gaupp wrote: 'I hold it for the most important duty of the neurologist and psychiatrist to protect the Reich from a proliferation of mental invalids and war pension recipients.'"[22]

It was the view of the psychiatrist Hermann Oppenheim that shell shock was caused by physical trauma. He defended this approach strongly, indeed eloquently, at the Munich conference. Unfortunately, the consensus was that it was hysteria and involved no organic change. His opponents made their case by presenting evidence that shell-shocked soldiers responded positively to hypnosis, one of the chief proponents of this dubious method during the war being the psychiatrist Max Nonne who at the conference hypnotized—and supposedly healed—a group of soldiers. Only later did it become clear that there was a high rate of relapse.

Some psychiatrists presented examples of hysteria that were not caused by external physical trauma. Gaupp reported on one such case: "I saw one of the most serious forms of spastic paresis with tremors [partial paralysis] in a young officer who had to drive a munitions wagon over wounded German soldiers. When he heard the cracking and moaning, he lost consciousness, fell from the wagon, and showed serious symptoms. Such cases are numerous." This was not considered to be an ethical reaction of an individual soldier to prevent his continued participation in an atrocity, but as psychopathic. To Gaupp, such hysteria was egotistic, anti-social, and showed "ethical defects." He stated that the war neurotics were "hot-tempered grousers and malicious agitators." Blind to his own deeply embedded prejudices, Gaupp judged Poles and Jews to be more susceptible; so also were lower-class soldiers more than upper-class officers, and southern Germans more than northern Germans.[23]

Infected by the aberrant zeitgeist of extreme patriotism, many psychiatrists and neurologists saw Oppenheim's ideas as anti-German. Their criticism verged on ad hominem attacks, which may have been spurred by subtle anti-Semitism, Oppenheim being Jewish. Oppenheim was dismayed, saying it was inconceivable "that doctors trained in neurology and psychiatry could disregard to such an extent the effect of the war's violent mental and psychical traumas, and assume that they left only a fleeting impression on the nervous system."[24] Karl was not on Oppenheim's side. Nor was anyone else. The overwhelming opposition nearly broke Oppenheim, who, despite the ostracism, continued to uphold his belief in trauma as the cause of shell shock and that physicians must not ignore the very real suffering of the victims.

Karl was more diagnostically cautious and less strident than some other psychiatrists. He maintained that the principal cause was the predisposed personality, which made a soldier vulnerable to external trauma. Following the meeting, he published several papers on war psychosis. A paper in 1917 was titled "Experiences of the war on the etiology of psychopathological states with special consideration of the exhaustion and emotion." Another published in 1920, two years after the war, was titled "About the importance of the war experience for general psychopathology and etiology of mental illness." In it, Karl stated that war psychosis was caused by an inner conflict of the individual when faced with maximum expectation of death, anxiety, pain, and grief. "On the one hand the inescapable military duty and the threatening fatal machinery of war; on the other, affirmation of life and the desire to be away from the firing, outside the area of danger. . . . That war hysteria originated in this combination can be regarded as proven by the war."[25] Actually nothing was proven. In a way, Karl had gone against his own frequently stated belief that there could be many causes for the same set of neurological symptoms. No doubt some soldiers were malingerers who only wanted to get out of the war and receive a pension. No doubt some were predisposed to hysteria. But there is also no doubt that many had been seriously damaged by physical trauma.

By the mid-1920s, as Germany continued to suffer from severe postwar economic problems and was unable to meet its onerous reparations payments set by the Versailles Treaty, Karl and his colleagues maintained that monetary compensation, such as pensions, could turn a traumatic neurosis caused by inherent weakness into a permanent condition, the cure being not to compensate, thereby forcing the individual to return to restorative work.[26]

For Karl, what seemed to be uppermost was whether the diagnosis undercut the efficacy of therapy. He himself was a compassionate man who was opposed to the extreme forms of treatment that some psychiatrists tried during the war, including electroshock therapy that was nothing short of torture. He was acutely aware of the depredations of the war on all levels of society. Exactly how terrible it had been for institutionalized mental patients is evidenced by a statement Karl made at the annual convention of the German Association of Psychiatry in 1920, in which he voiced his concern that patients had been allowed to starve to death. What had occurred was so awful that, in his words, it was leading to a negative "transformation in the concept of humanity."

> Under the difficult experiences of war, we were forced to value the individual human life differently than before, and that in the years of hunger during the war we had to come to terms with watching as our institutionalized sick died en masse of malnutrition, and nearly to sanction this in the thought that this sacrifice may have saved the lives of the healthy. This emphasis of the right of the healthy to self-preservation, as is entailed in a time of need, conceals a risk of exaggeration.[27]

That careful statement is so ethically fraught, and yet so powerful, that it needs to be taken apart, oblique phrase by oblique phrase, in order to grasp the full meaning of *coming to terms* with patients starving to death. First, there was capitulation on the part of physicians. Second, there was an effort to justify the capitulation—*nearly to sanction* the *sacrifice*—as if the mentally ill patients had bravely and knowingly willed their own starvation so that the healthy could live. Furthermore, justification was predicated on *the right of the healthy to self-preservation*, implying that the right of mentally ill patients to self-preservation was the lesser. Finally, there are the words *conceals a risk of exaggeration*, meaning that perhaps in *time of need*—of war—such a massive ethical failure could be countenanced, but not in time of peace.

Karl's troubled statement that the concept of humanity was being negatively transformed is of great importance, because what occurred in the hospitals during the First World War can be seen as a precursor to Hitler's decision in 1939 to murder institutionalized mental patients. Under Nazism, the belief was that mentally ill people led "lives that were not worth living." In the book *The Kaiser Wilhelm Institute for Anthropology, Human Heredity and Eugenics*, the historian Hans-Walter Schmuhl writes that Karl Bonhoeffer was in the minority following the First World War as to the treatment of the mentally ill. The majority, being race hygienists, believed people with mental defects should in some way be eliminated or controlled so that a bright future of perfected humanity could be achieved. "Voices exhorting caution like [Bonhoeffer's] found little resonance, however. Under the impression of the immense loss of human life in the 'storms of steel' of World War I, the argumentation of the race hygienists appeared compelling."[28]

Paul Lerner presents a different view with respect to war psychosis and how it was conceptualized after the war. He contends that the denial by psychiatrists of the reality of mental trauma and the pathogenic potential of war was part of the process by Germans during the Weimar period to work out a collective memory of the war. By "undermining any links between war service and mental illness, they denied nervous veterans their status as the war's victims, ultimately condemning many as back-stabbers, who along with Jews and Marxists, had sabotaged the war from within. In such a way, the conservative psychiatric profession joined forces with the German right; contemptuous of the new republic and longing for the restoration of an imagined status quo ante bellum, the medical science of memory was complicit in the fateful act of forgetting."[29]

Lerner makes a solid point, but the fact remains that Karl was troubled by what he had witnessed during and after the war. In the 1920s, he strongly opposed forced sterilization of the mentally ill.[30] His stress in his research remained on correct diagnosis and therapeutic efficacy. In relation to shell shock, he believed that the diagnosis alone was a factor in fixating the disease, hindering the return to health. Looking back on the war, he considered it to be degenerative, increasing the number of "inferiors" within German society, including hysterics and psychopaths.[31] It was not difficult to reach that conclusion, regardless of the influence of eugenics. However, there were two other enormous events that weighed heavily on Karl—in the last year of the war, his son, Walter, was killed by a shell, and his son, Karl-Friedrich, was wounded.

In 1917, when Karl-Friedrich was eighteen years old and preparing to attend the University of Tübingen to study physics, he was called up for military service. Because the USA had entered the war on April 6, 1917, the German army needed another infusion of soldiers. Motivated by strong loyalty to the fatherland and full of the fervor of young men, Karl-Friedrich went first, taking his physics textbooks with him, followed by Walter in early 1918.[32] Before they left, the brothers showed Dietrich how to keep track of German military progress on a map with colored pins. Filled with pride, Dietrich wrote to his grandmother, Julie Bonhoeffer, on January 23, 1918: "Karl-Friedrich is writing us more often. Recently he wrote that he had won the first prize in a race in which all the junior officers of his company competed. The prize was 5 marks."[33]

Walter was known in the family as a lover of nature with a gift for languages. He was also Paula's favorite son, perhaps because he was born premature and needed more care as an infant. Only eleven months younger than Karl-Friedrich, he was very close to his brother. Like most young men, he was keen on joining the army and doing his part to make Germany victorious. Sabine wrote that Walter, "spent a short time in the Boy Scouts but then wanted to subject himself to harder training on his own and on his expeditions he constantly increased the number of sandbags he carried in his rucksack."[34] He matriculated from school early and enlisted as an officer cadet. On April 22, 1918, Walter was severely wounded when he was marching in close order on a road that was under artillery fire. Without antibiotics, which had not yet been discovered, it was nearly impossible to fight infection caused by shrapnel. Karl wrote, "A shell fell into the column, killing several and wounding the rest. Walter received several shell fragments in both legs. In the judgment of the doctors this did not at first seem dangerous, but an inflammation developed from which he died six days later in the field hospital on the 28th April 1918." Just hours before his death, Walter wrote hopefully to his parents:

> I am using my technique of thinking of other things so as not to think of the pain. There are more interesting things in the world just now than my wounds. The Kemmelberg affair with its possible consequences, and the fact that news has been brought to us that we have taken Ypres, give us much hope. I cannot bring myself to think of my poor regiment at all. The last few days have been so hard for it. How can things be going for

> the other officer cadets? I am full of longing as I think of you minute by minute throughout the long days and nights. Your loving Walter, still so far away from you.[35]

Sabine called his death a "terrible shadow" that fell over the family on "one bright morning in May" and which did not lift for many years. "My father was just in the act of leaving the house to drive to his clinic, and I was on the point of going through the door on my way to school. But when a messenger brought us two telegrams I remained standing in the hall. I saw my father hastily open the envelopes, turn terribly white, go into his study and sink into the chair at his desk where he sat bowed over it with his head resting on both his arms, his face hidden in his hands. I had never seen my father like that." It was also devastating to Paula who went into severe depression, made worse by Karl-Friedrich's being wounded six months later in October 1918. There were no more family vacations for a long while, and the entries in Karl's New Year's Eve journal stopped for many years. When Walter's name was mentioned, Karl would quietly leave the room. Prior to the start of each Advent season, the family would visit his grave.

Sabine wrote: "For almost an entire year I prayed that it might not be Walter whom we had buried but a strange soldier, and that Walter might return. It seemed to me the only way of bringing back any happiness to my parents. For never a day passed without us feeling Walter's loss. After Walter's death it was a long time before my mother recovered her broken strength. Only her faith and the loving care of our father were able to restore her to health."[36]

Like everyone in the family, Dietrich, who was twelve years old, was overcome by Walter's death. It was around this time that he made the firm decision to study theology, although his family did not regularly attend church. When Dietrich was confirmed a few years later, his mother presented him with the Bible that Walter had received on his confirmation. For the rest of his life, Dietrich used it for worship services and for his personal meditations.[37]

The severity of Karl-Friedrich's injury is not known. What is known is that he spent two months in the Charité Clinic, during which time the war came to an end on November 11, 1918. There was immediate revolutionary upheaval. Two days earlier on November 9, the day that Kaiser Wilhelm II abdicated, Karl had walked to the Charité amidst chaos, writing, years later, his impressions: "Raggedly marching parties of troops, mingled with civilians carrying posters with the inscription 'Don't shoot;' pale, famished-looking creatures covered with sweat, gesticulating excitedly and pleading with the soldiers and grabbing rifles from their hands, in so far as they put up any resistance." On reaching the safety of the clinic, he was accosted at the door by the porter but he took no notice and entered without incident.[38]

Many Germans blamed the defeat on betrayal, what became known as the "stab in the back," by Jews and some government leaders. The men who signed the capitulation papers were labeled as the November criminals. This was a dangerous and delusional mind-set that was totally at odds with the fact that early in the fall of 1918 the German army had completely run out of resources to continue fighting. Its military successes in the spring and early summer of 1918 were Pyrrhic victories achieved at unbearable cost. By August, the victories were replaced by decisive defeats. In September and early October, General Erich Ludendorff and Field Marshal von Hindenburg had pleaded with the Kaiser for an armistice, with Hindenburg stating emphatically that the army could not wait another two days. Fed by glowing reports printed in the censored press of victories over the summer, German citizens refused to accept the truth even after it became indisputable. This self-delusion prepared the way for the rise of Hitler who believed so totally in "the stab in the back" that he used it with great effectiveness to stoke the glowing coals of anti-Semitism. The Versailles Treaty signed on June 28, 1919, made the situation worse. It was so vituperative—its intent being to break the economic back of Germany—that even some non-Germans were appalled. One of the clauses forced Germany to accept all guilt for the war.

Years later, Dietrich wrote that the things he saw at the end of the war shook him profoundly: "The number of suicides increased in a terrifying way. I remember very well, I had on the way to my school to pass by a bridge and in the winters of 1917 to 1919 almost every morning when I came to the bridge I saw a group of people standing on the river and everybody, who passed by, knew what had happened. These impressions were hard for young boys."[39]

To add to the upheaval, the influenza epidemic broke out in 1918, raging around the world for the next three years, eventually claiming from 50 to 100 million lives. Because of living in close quarters in unsterile conditions, soldiers were particularly hard hit. Calling it the *Grippe* in a lecture he delivered in the USA while at Union Theological

Seminary in 1930–31, Dietrich wrote that more than 100,000 Germans succumbed because they "were physically too enervated and could not resist."[40]

Unfortunately, for some people the disease led to severe after-effects and syndromes, including forms of psychoses. A few of the psychoses were short-lived, but others, such as encephalitis lethargica, known as the "sleepy" sickness, were incurable. Just as the question had been asked of shell shock whether it was a moral failure of the patient, so similar questions were asked about these worrisome and inexplicable epidemics. It is doubtful that there had ever been an outbreak of disease in the history of humankind that could more clearly prove Karl's point that some forms of mental illness could be caused by physical sickness. However, as has been pointed out already, Karl Bonhoeffer was a cautious scientist and physician. He was quoted in an article written by the American psychiatrist Karl Menninger titled, "Psychoses Associated with Influenza" published on January 25, 1919, in *The Journal of the American Medical Association*. Bonhoeffer warned that the diagnoses of psychoses and the communications concerning the diagnoses "proceed from the pens of others than psychiatrists." He meant that general practitioners were not sufficiently familiar with the "frequency and nature of the mental illnesses" to make an accurate diagnosis. Menninger concurred and recommended that the best place for accurate diagnosis was in a psychiatric hospital where the staff was trained and the patient could be observed over a five-day period, during which time clinical tests would be run and an effort would be made to determine whether the patient or any member of his family had a history of psychosis. Bonhoeffer and Menninger, who was then practicing in Boston and teaching at Harvard Medical School, agreed that there was a strong correlation between influenza and post-influenza psychoses, but they did not want the data skewed by misdiagnoses. The influenza epidemic did not end until 1921, although smaller outbreaks continued to occur over the next few years.[41] One of the victims in 1922 was the mother of Gerhard Leibholz, who would become engaged to Sabine Bonhoeffer in 1924.

Karl-Friedrich's wounds were not disabling, but his wretched wartime experience and the loss of his brother turned him against politics as practiced by the German empire, confirmed his agnosticism, and made him very bitter. When he was given a badge for being wounded, he threw it out in disgust when he learned the badge had also been awarded to a drunk officer who had been injured in a fall

Fig. 3.1 Spartacist uprising, Berlin. United States Holocaust Memorial Museum, courtesy of Dottie Bennett

from his horse.[42] Adding to his anger was the chaos in Berlin immediately after the war as left-wing and right-wing factions battled to form a new government. While he was still recuperating in January 1919, the Spartacist uprising took place in Berlin. It began with the calling of a strike by Rosa Luxemburg and Karl Liebknecht who wanted to establish a communist form of government. On January 7, approximately 500,000 people took to the streets in Berlin, with protests in other cities as well. The Bonhoeffers could hear gunfire at Halensee train station not far from their home (Fig. 3.1).

In a letter from Dietrich to his grandmother in Tübingen, he wrote about the gunfire and billeting of soldiers with the family: "Karl-Friedrich has at last been discharged from the Charité. He would like to be part of this somehow, but mama and papa do not yet agree."[43] By the time the uprisings were suppressed, over 150 people had been killed in Berlin by war veterans ordered into the fray. Luxemburg and Liebknecht were beaten and shot to death. It was hardly the end of the political turmoil, which would continue into the 1920s, but it was the end of it for Karl-Friedrich. He became a socialist, which according to Eberhard Bethge "led to conflicts with other branches of the family outside Berlin," and made him "something of an outsider" in their home.[44] The branch of the family to which Bethge referred was Otto Bonhoeffer's, Karl's brother. Having lost two sons, Otto did not want to hear of Germany being the victimizer; all he wanted to hear was that Germany was the victim that had been "stabbed in the back." Karl-Friedrich's battle experience had taught him otherwise.

There was yet one more aspect of the postwar world that exacerbated Karl-Friedrich's disillusionment. Just as shell-shocked victims were blamed for their condition and were accused of trying to get undeserved pensions, these same accusations were sometimes directed toward physically wounded soldiers. Many were treated very poorly, as if their existence was an unwelcome reminder of Germany's defeat, and their wounds were their own fault. When Dietrich was in prison during World War II, he worked on a play (unfinished) that was partially autobiographical. In one scene, he depicted a disgraceful incident that was probably based on fact. The main character is an older brother named Christoph who has returned from the war with a fatal but hidden wound about which he refuses to speak, but his father, who is a doctor, is aware of the wound and the prognosis. One day the father returns from the hospital in angry frustration and tells his wife: "Anna, I saw things today . . . I'd give anything never to have seen them. A wounded young soldier about Christoph's age hauls himself across the street on crutches. Behind him walk two young men about his age, fit as a fiddle and all dressed up. I see one of them pointing his finger at the wounded man, putting on a repulsive grin, and saying in a voice for all to hear: 'That's another one of those eternal fools; there are still too many of them around.'" The father tells his wife how he had grabbed the young men by their collars and shaken them violently, too angry to speak. Afterward, he says to himself, "Better dead than this," when suddenly he realizes Christoph is standing beside him. "Yes, Father, you're right," Christoph says, "better dead than this."[45]

Sabine had a compassionate view of her brother's disillusionment and depression. To her, Karl-Friedrich's rejection of some of the middle-class values with which he had been raised arose from his empathetic interactions as an officer with the soldiers under his command. He had grown up in a society with a strict class structure in which he and his brothers (but not sisters) had attended the Friedrich Werder Gymnasium in preparation for university study. Because he had stepped directly from that elitist environment into the military, Karl-Friedrich had little experience of what were then considered lower classes, which historically had little to no political power and were poorly educated. Once in the military, Karl-Friedrich had his eyes opened. "He was full of admiration for the manifold gifts, including artistic ones, which he had discovered among the simple soldiers, gifts which he then believed, had never been brought to light because these men had had no opportunity to develop them." Sabine's reflection aligns with another family recollection in which one of Karl-Friedrich's comrades declared, "What do you mean fatherland? My father has no land!"[46]

For Dietrich, his brothers' service in the First World War created a kind of divide between himself and Karl-Friedrich. In his powerful speech "The Führer and the Individual in the Younger Generation," delivered as a radio broadcast on February 1, 1933, Dietrich wrote:

> There is an invisible but impenetrable barrier between those who fought in the war and those who are only a little younger, who reached maturity during the period of collapse. This is felt even more greatly among the younger ones than among the older. For the first postwar generation there is nothing more impressive than the fact that among them live these men who have escaped from the world of death. There is something frightening, worrying, terrifying in this knowledge.[47]

It was Bethge's opinion that the barrier came down between the brothers in 1933, although Dietrich continued to feel compelled to explain himself to Karl-Friedrich in the years ahead. When Dietrich was running great personal risk in 1937, he received a letter from Karl-Friedrich on behalf of their mother who was very worried about him. Dietrich responded: "We will not be able to persevere on behalf of the church without making sacrifices. After all, you yourselves risked significantly more during the war. Why should we not do the same for the church? And why should anyone want to dissuade us from doing so? Certainly none of us is particularly keen on going to prison, but if it does happen, then—at least I would hope—we will do so with joy precisely because the cause is worthwhile."[48]

There was also the divide between the living and the dead. For Germany, military fatalities in the First World War were approximately two million. For all countries, upward of 8,500,000 soldiers died, although an accurate number has never been determined. Even less accurate is the number of civilian casualties, estimated at 13,000,000. Among the men whose physics and chemistry lectures Karl-Friedrich would attend in the next few years were Max Planck who had lost one son and Walter Nernst who had lost two sons. Entering the 1920s, no one remained undamaged, including Karl-Friedrich.

Notes

1. "The World Wars," *Encyclopedia Britannica*, 15th ed., Vol. 29, 987.
2. Belinda J. Davis, *Home Fires Burning: Food, Politics, and Everyday Life in World War I Berlin* (Chapel Hill: University of North Carolina Press, 2000), 180, 205.
3. The U-boat campaign began February 1, 1917, and was directly related to the economy. See Avner Offer, *The First World War: An Agrarian Interpretation* (Oxford: Clarendon Press, 1989), 354–367.
4. DBW, Vol. 10, 414.
5. DBW, Vol. 17, 280, 349.
6. Leibholz-Bonhoeffer, *Bonhoeffers*, 4.
7. Bethge, *Bonhoeffer*, 25.
8. Lerner, *Hysterical Men*, 55.
9. When the figure for Germany is combined with that for Austria-Hungary, the total of regular divisions at the start of the war was 146. German manpower was 1,900,000. Austria-Hungary manpower was 450,000. British manpower was 120,000. "The World Wars," *Encyclopedia Britannica*, 15th ed., Vol. 29, 963.
10. Bethge, *Bonhoeffer*, 26.
11. Ibid., 27.
12. Leibholz-Bonhoeffer, *Bonhoeffers*, 5.
13. Clark, *Einstein*, 180.
14. Lerner, *Hysterical Men*, 42.
15. Clark, *Einstein*, 181–183.
16. Ibid., 49.
17. Not until 2016 was there hard evidence linking PTSD to blasts in combat. Alan Schwarz, "Research Links PTSD to Blasts in Combat: Evidence Found of Tissue Damage That Did Not Come From Injuries," *The New York Times* (June 10, 2016): A19.

18. Doris Kaufmann, "Psychiatry in the First World War and Weimar Germany," *Journal of Contemporary History* 34, no. 1 (1999): 131.
19. Lerner, *Hysterical Men*, 66.
20. Ibid., 68.
21. Kaufmann, *Psychiatry*, 1.
22. Lerner, *Hysterical Men*, 17.
23. According to Kaufmann, the definition of the war neurotic as a psychopathic personality was the result of the state's handing the problem to the psychiatric community, which in turn shifted the responsibility to the victims. This meant that "war neuroses are not caused by war, but by supposedly inferior human materials." Kaufmann, *Psychiatry*, 136–138.
24. Ibid., 134.
25. Ibid., 134–135.
26. Matthew J. Friedman et al., eds., *Handbook of PTSD* (New York: Guilford Press, 2010), 22.
27. Hans-Walter Schmuhl, *The Kaiser Wilhelm Institute for Anthropology, Human Heredity and Eugenics, 1927–1945* (Berlin: Springer, 2008), 10–11. See also footnote 17 Breicht über die Verhandlungen psychiatricher Vercine, 598.
28. Ibid., 10–11.
29. Lerner, *Hysterical Men*, 29.
30. Voluntary sterilization consent and mental illness became a major issue for Karl Bonhoeffer after Hitler's rise to power in 1933. Lawrence A. Zeidman, "Neuroscience in Nazi Europe Part II: Resistance against the Third Reich," *Canadian Journal of Neurological Science* 38, no. 6 (2011): 834.
31. Lerner, *Hysterical Men*, 216.
32. In *All Quiet on the Western Front* by Erich Maria Remarque, the character Müller "still carries his school textbooks with him, dreams of examinations, and during a bombardment mutters propositions in physics." (Boston: Little, Brown, 1929), 5.
33. DBW, Vol. 9, 18.
34. Leibholz-Bonhoeffer, *Bonhoeffers*, 21.
35. A fierce battle had taken place at Kemmelberg in Belgium on April 25, 1918. Ibid., 21.
36. Ibid., 24. See also DBW, Vol. 9, 55.
37. Bethge, *Bonhoeffer*, 28.
38. Ibid., 30.
39. DBW, Vol. 10, 414.
40. Ibid., 414.
41. Karl A. Menninger, "Psychoses Associated with Influenza," *The Journal of the American Medical Association* 72, no. 4 (1919): 1.

42. DBW, Vol. 7, 7.
43. Bethge, *Bonhoeffer*, 30.
44. Ibid., 30.
45. DBW, Vol. 7, 31–32.
46. Leibholz-Bonhoeffer, *Bonhoeffers*, 17–18. DBW, Vol. 7, 69, see footnote 64.
47. Another title for the radio version of the speech, which was also a lecture, was "The Younger Generation's Altered View of the Concept of Führer." DBW, Vol. 12, 269–270.
48. DBW, Vol. 7, 21. See also DBW, Vol. 14, 320–321.

CHAPTER 4

The Promise of Pure Science

At Max Planck's sixtieth birthday celebration in 1918 in Berlin, Albert Einstein addressed the members of the Physical Society in Berlin, using as a metaphor the "temple of science." Were an angel to drive from the temple all those scientists who were there out of personal ambition or for utilitarian reasons, only a few men would be left, one of them being Planck. Einstein then ruminated about what had brought those few men into the temple in the first place.

> To begin with, I believe with Schopenhauer that one of the strongest motives that leads men to art and science is escape from everyday life with its painful crudity and hopeless dreariness, from the fetters of one's own ever shifting desires. A finely tempered nature longs to escape from personal life into the world of objective perception and thought; this desire may be compared with the townsman's irresistible longing to escape from his noisy, cramped surroundings into the silence of high mountains, where the eye ranges freely through the still, pure air and fondly traces out the restful contours apparently built for eternity.[1]

It was the world of pure science—a world apart from the cruelties of the battlefield and the economic chaos of a defeated nation—that Karl-Friedrich now entered. The belief that science should remain unsullied by any connections to home life, cultural life, and political life, dedicated solely to the pursuit of knowledge, was so common it amounted

K. L. Housley, *The Scientific World of Karl-Friedrich Bonhoeffer*, Palgrave Studies in the History of Science and Technology, https://doi.org/10.1007/978-3-319-95801-9_4

to dogma among German physicists and chemists. With his gift for language, Einstein caught the feelings attendant to it with great verve; he was also prescient about its grave dangers even though he did not speak of them at Planck's birthday party.

Pure science was not an unobtainable ideal but a day-to-day reality, especially in the vibrant intellectual environment of the 1920s when Karl-Friedrich would do some of his most important work. To walk down the halls of the various Kaiser Wilhelm institutes was to rub shoulders with Nobel laureates who were picking the locks to the secrets of the universe, not by using the old skeleton keys of classical physics bequeathed to them by Isaac Newton and his descendants, but by using space-time keys of their own devising. However, even among the pure science believers, there was a rigid stratification, with the theoretician (the untainted thinker) at the top and the experimentalist (the tainted doer) far down in the pecking order, at least according to the theoretician.

Physics and chemistry shared equal stature, with physicists and chemists working together on problems of mutual interest, including the structure of the atom. These disciplines were also united in the use of the scientific method, comprised of rigorous observation and data collection, elegantly designed experiments that were repeatable, and unbiased intellectual analysis, with the results disseminated so that other scientists could prove or disprove them. That was the ideal pursued since the days of Galileo, although quantum mechanics was in the process of unraveling the scientific method by attempting to cut the thread between cause and effect. (Because the biological sciences did not adhere as tightly to abstraction and mathematical analysis as did chemistry and physics, they were not considered to be on the same level.)

The pure science mind-set existed in other countries as well, but in Germany it was uniquely supported by the compartmentalized and hierarchical structure of society in which academic scientists were paid by the government, providing them with a degree of financial security. Like a velvet trap, this was a system that was relatively benign as long as the political sphere was benign. However, it was vulnerable to manipulation when pressure was applied, as had occurred with research on the cause of shell shock during the First World War. In the 1930s, that pressure increased exponentially. Until then, Berlin was the place, the 1920s the time, when for a twinkling moment the world of pure science did indeed exist, at least in the minds of its acolytes. In 1919, Karl-Friedrich picked up where he had left off before marching off to war and entered that world. He began to study at the University of Tübingen, his father's alma mater, but then changed to the University of Berlin (the Friedrich Wilhelm University), renowned for science, earning his Ph.D. in physical chemistry in 1922.

There is only one known incident worthy of mention that occurred during Karl-Friedrich's year at Tübingen: He refused to join the Hedgehogs, the upper-class Swabian fraternity that his father had joined when he was a student more than thirty years before. As a wounded veteran, Karl-Friedrich objected to the nationalist pressure the Hedgehogs exerted on its members, including the call to participate in the suppression of uprisings in Stuttgart and Munich in 1919. He was offended by its play-soldier mentality, which had a darker anti-Semitic side. His brother Klaus, who had served in the German army as an aid in the closing months of the war and who matriculated at Tübingen, also refused to have anything to do with the Hedgehogs. Klaus was perturbed by the anti-democratic attitude of many of his fellow students, not just the Hedgehogs, who were walking backward into the twentieth century, their eyes turned to the early days of German unification under Chancellor Otto von Bismarck. In May 1922, Klaus wrote to his friend, Hans Dohnanyi, about a lecture he had attended in which a pro-democratic professor had been verbally assaulted by the students and another professor. "The students bawled at him, stamped their feet, and shouted personal insults," he wrote. Klaus's indignation was so apparent that he got into an argument with a man sitting behind him at the lecture, "an idiotic former lieutenant wearing a monocle." Presciently, Klaus wrote, "Hans, only think of the trouble we shall later have with these people. If only I were a bit older!"[2]

However, in 1923 when Dietrich began to attend Tübingen at the age of seventeen, he readily joined the Hedgehogs, perhaps because it was a way for him to come out from the long wartime shadows of his older brothers. It was not until 1933 that Dietrich officially resigned; by then, the Hedgehogs had openly aligned themselves with the Nazis.[3]

It is not known whether the political climate at Tübingen had anything to do with Karl-Friedrich's decision to transfer to the University of Berlin. Matriculating at one university but graduating from another was common practice in Germany as a student progressed from generalized to specialized learning. All the German universities were public, and while getting into them was difficult, moving among them was relatively easy. In the sciences, it was especially important for a student to learn firsthand from being in the presence of Noble laureates, such as Fritz Haber, Max Planck, Walter Nernst, and Max von Laue, each of whom had fully equipped laboratories to conduct experiments. Because scientific institutes within universities were rather like fiefdoms with a single scientist (Geheimrat) at the head and subordinates beneath, a student had to go to that institute to learn.[4] In the new relativistic world of quanta, there were no textbooks or set curriculum. A student learned as much by osmosis and hands-on doing as by attending class. In Berlin, there were always impromptu talks, spirited hallway discussions, standing-room-only colloquia, and scientific demonstrations. Listening to the ideas of older students and professors, while sitting around wooden trestle tables drinking liters of coffee, could be more enlightening than attending a lecture. Isaac Newton had said, "If I have seen farther, it is by standing on the shoulders of giants." In Berlin, a student could stand *next* to them.

To save money during a period of hyperinflation and economic ruin, Karl-Friedrich lived at home while attending the University of Berlin, which was normal even in more bounteous times because the university was basically lecture halls with no dormitories. It was a place to learn, not to live. An advantage for Karl-Friedrich's family was that they could entertain some of his scientific friends, among them Hanfried Ludloff who was working on his degree in theoretical physics. Ludloff was a childhood friend from Breslau. This resulted in rich family discussions and the further exposure of the younger children to science. About Ludloff, Dietrich wrote to his grandmother, Julie Bonhoeffer, on May 3, 1921, "H. Ludloff is now attending our university. He visits us every Sunday but was not able to live with us because he had already rented a room."[5]

Rudolf Stern was another friend from Breslau who visited the Bonhoeffer home during this period. Like Karl-Friedrich, he had served in the war and had emerged from it injured, sick from influenza, and deeply disillusioned with the mendacity and class egotism of the German high command. He was more than a little lost, wanting to make up for all the wasted time but uncertain how to do so. Qualified for a medical residency or assistantship, Rudolf came to Berlin to seek the advice of Karl Bonhoeffer and Fritz Haber. The upshot was that Haber gave him a position at his research institute in the area of biochemistry. However, Rudolf really yearned to be a practicing physician. His discussions with

Karl about his future in medicine must have been influential because he decided to leave Haber's institute and return to Breslau. There, he finished his doctorate and began to focus on the effect of trauma on internal diseases, making correlations between high stress and heart disease, paralleling Karl's research on trauma and mental disorders.[6]

The sixty years from 1870 to 1930 were known as the period of the "Great Berlin Physics" for good reason. Physical chemistry, the close cousin of physics, achieved world-renown slightly later in the period. The three major centers of research of the discipline were the University of Berlin, the Kaiser Wilhelm Institute for Physical and Electrochemistry, and the Charlottenburg Institute of Technology. Prior to the preeminence of Berlin, Leipzig had been the nucleus of the physical chemistry world under the guidance of Friedrich Wilhelm Ostwald, who became chair in 1887, the same year that he founded the scientific journal *Zeitschrift für Physikalische Chemie*. Having won the Nobel Prize in Chemistry in 1909, Ostwald inculcated in his students the importance of excellence. Several would become exceptional scientists, including Max Bodenstein, Herbert Freundlich, and Walter Nernst who would himself win the Nobel Prize in Chemistry. Ironically, after Ostwald's retirement, Bodenstein, Freundlich, and Nernst stole the glory from Leipzig and took it to Berlin. There, all three men would play major roles in Karl-Friedrich's life, with Nernst as his mentor during his doctoral studies, Freundlich as the organizer of the physical chemistry colloquia that Karl-Friedrich attended, and Bodenstein as a source of encouragement for Karl-Friedrich's research on chemical kinetics and photochemistry. In 1934, Karl-Friedrich would assume Ostwald's position as chair of physical chemistry, thereby returning some of the luster to Leipzig, although by then the glory days of German science were over.[7]

At that time, the various branches of science were not as differentiated as they were to become after World War II. Because physical chemistry dealt with intermolecular forces, such as thermodynamics and kinetics, it was common for physical chemists and physicists to work on the same problems. The hybridization of physics and chemistry was encouraged by Nernst. Famous for formulating the Third Law of Thermodynamics, which he called his heat law, Nernst saw himself more as a physicist than a chemist. However, he stressed that both sciences dealt with the same fundamental problem, which was to reduce "to the simplest rules the complicated phenomena of the world." Therefore, it was not "advantageous to separate the two." Physicists and chemists needed to work in concert to explore "the boundary region."[8]

Nernst stood out because he was a highly intuitive thinker with a practical bent, inventing a type of electric light and even an electric piano. A polymath, he loved Shakespeare, liked to hunt, and worked to improve both farming and fish hatcheries, which he considered, "A thermodynamically sound enterprise because fish, unlike cows, do not waste energy to heat the universe."[9] According to Wilhelm Jost (who was later to become a friend and colleague of Karl-Friedrich at the University of Leipzig), Nernst knew how to improvise and take risks. In the essay, "The First 45 Years of Physical Chemistry in Germany," Jost wrote, "He always tried to reach his goal by the simplest possible means which, however, did not prevent him from experimenting under very extreme conditions. He always tried to find relations of general validity, and he tried to avoid any unnecessary details, to find conditions favorable for an experimental proof."[10]

It was Nernst, along with Planck and Haber, who in 1914 was responsible for the appointment of Albert Einstein as the first director of the newly established Kaiser Wilhelm Institute for Physics as well as professor at the University of Berlin, coaxing him away from Zurich, Switzerland, and back to Germany just before the First World War broke out. Einstein's presence in Berlin was galvanizing. No scientist wanted to miss a colloquium at which he would be lecturing or simply asking astute questions. It wasn't just his ideas about relativity, it was the man himself. He was a mixture of egotism and humility, capable of injecting levity into the most serious discussion, sometimes at his own expense (Fig. 4.1). For example, in 1927 an astronomer lectured at Nernst's physics colloquium about several curious experiments, following which Nernst asked: "Herr Einstein, if all these experiments were correct, would they not disprove your theory of relativity?" Einstein replied, "One single experiment would suffice."[11]

Karl-Friedrich's doctoral thesis was titled "Photochemical Sensitivity and Einstein's Equivalence Principle." This was a subject critically important to the development of the field of photochemistry, which was at an impasse in part because of the photochemistry equivalence principle (or *law* as Einstein called it in 1912, but not yet with everyone's agreement), which stated that one quantum of light was absorbed per molecule.[12] On March 6, 1920, James Franck wrote to Karl-Friedrich from the University of Göttingen where he had just

Fig. 4.1 Albert Einstein. Pictorial press/Alamy Stock Photo

been appointed as chair of experimental physics. Franck was in the process of writing about photochemistry to be included in a book on excitation of quantum leaps and other related subjects. He wanted to incorporate the unpublished result of the absorbency of hydrogen that had been experimentally obtained by Karl-Friedrich. "Please do me the favor of reading a couple of pages I am sending to you and tell me if I possibly can use your result, with a friendly message," Franck asked, following up his request by recommending some scientific articles to Karl-Friedrich as well as a series of lectures he had recently delivered at Oxford University. He concluded, "I would love not only your answer in relation to the hydrogen spectrum, but also your assessment of the overall picture. At the moment I am writing a few pages about chemiluminescence." Given that Karl-Friedrich was only twenty-one years old, the letter from an esteemed senior scientist was clear affirmation of an exceptional student.[13]

For all the fervent research, there was no fixed university curriculum for physical chemistry, with little mathematics and virtually no theory. Instead, there was a great deal of laboratory work, "of the order of 40 hours a week for three to four years," according to Jost. Advanced training in physical chemistry occurred during thesis work. Jost wrote that Karl-Friedrich was free to choose what he wanted to research. "His subjects were far enough away from everyday problems to promise interesting results and were simple enough in the demands on technical tools to allow answers to questions at a reasonable cost within a reasonable time."[14] Foremost among Karl-Friedrich's subjects was the simplest chemical substance that was also bewilderingly complex—hydrogen.

Besides Nernst's guidance, the scientist who supervised Karl-Friedrich's thesis work was the chemist Walter Noddack, who was six years older than Karl-Friedrich. At the time, Noddack was also working with Ida Tacke (whom he would marry) and Otto Berg. In 1925, the three discovered rhenium (atomic number 75), one of the rarest elements in the earth's crust. They also claimed to have discovered an element they named masurium (later called technetium, atomic number 43); however, their claim was hotly disputed.[15] Later in life, Karl-Friedrich expressed gratitude to Noddack for his early guidance on photochemistry, but it was not a friendship that was to last due to Noddack's and Tacke's eventual support of Nazism.[16]

At that time, Berlin was a strange brew of class antagonisms and right and left ideologies, some carried by refugees from other European conflicts, including exiles from the Bolshevik Revolution in Russia, and people of Jewish background attempting to escape from anti-Semitism and vicious pogroms in what was left of the Austro-Hungarian Empire. Among them was a remarkable group of young Hungarian scientists who would make their mark first in Germany, and then after 1933—exiled again and desperate to find employment—in the UK, the USA, and elsewhere. Karl-Friedrich came to know them all well, including Michael Polanyi, Eugene Wigner, Leo Szilard, John von Neumann, and George Hevesy.[17] They discovered in Germany a level of science so high it was intoxicating. Writing in 1929, ten years after leaving Hungary, Polanyi (who had trained first as a physician and then as a chemist) found Germany to be a place where "the professors grab the student's hands, if he is supposed to be gifted. They are like art collectors whose obsession is discovering talent. They educated me and gave me a position where I could address myself to my abilities. They gave me everything and demanded nothing of me." Polanyi was extremely grateful for the opportunities German professors offered him, writing, "They trust that who gets to know the joy of scientific work, will never leave it as long as he lives." However, he also grieved for the Jewish scholars of promise who, having been thrown out of Hungarian universities, never had a chance. "Looking back I see other Michael Polanyis bogged half-way down and disappearing," he wrote.[18]

Anti-Semitism was a problem in Germany as it was everywhere in Europe, but Jewish scientists could still find work and could excel on the merits of their research. Many of them converted to Christianity, among them Haber and Polanyi; others considered themselves atheists who were culturally non-Jewish. In the book *One Hundred Years at the Intersection of Chemistry and Physics: The Fritz Haber Institute of the Max Planck Society*, Jeremiah James et al. state, "In any given year after 1923 roughly half of the scientists officially employed by the Institute were Jewish in this sense, and the physical chemist Robert Havemann would remark privately in 1932 that Jews also made up roughly half of the much larger number of total researchers at the Institute."[19]

There was an edgy quality to life in the Weimar Republic. In January 1923, after Germany defaulted on its reparations payments, the French and Belgian armies occupied the Ruhr Valley. On November 8, 1923, Hitler and his brown-shirted thugs attempted a coup in Munich called the Beer Hall Putsch. The date was chosen because five years before in November 1918, Germany had capitulated. While in prison, Hitler began to write *Mein Kampf*, setting forth his obsession that Germans (Aryans) were a master race that must keep themselves pure by destroying all Jews and subjugating all Slavs. Aryans were entitled to "living space" (Lebensraum) to the east, which encompassed Poland and Russia.

In the Bonhoeffer family, all this was cause for mounting concern. On June 14, 1922, near Dietrich's school, the German statesman and businessman, Walther Rathenau was murdered not only because he had negotiated a treaty between Germany and its former enemy, the Soviet Union, but also because he was Jewish. Dietrich heard the gunfire from his classroom. In a letter to her fiancé, Hans Dohnanyi, written not long after Rathenau's murder, Christel Bonhoeffer observed that it was the worst "of all the hideous acts of these accused swastika people."[20] Of further concern in the family was that Sabine became engaged to Gerhard (Gert) Leibholz, a young legal scholar of great promise, who had been a classmate of Klaus. Although he had converted to Protestantism, his parents were Jewish. Karl and Paula were not opposed to the engagement, but they told Sabine straight out, and with prescience, that "now things will be difficult for you. It is doubtful whether Gert will advance as far in his profession as he deserves, and that is an experience which it is very sad for a wife to have to watch."[21]

Everything was made much worse by hyperinflation, culminating in 1923 when the value of the mark plunged. Even the wealthy Bonhoeffer family struggled to make ends meet. In his first year at the University of Tübingen, Dietrich traveled back and forth from Berlin in fourth class on the cheaper local train, a trip that took two days. In October, he wrote to his parents that "every meal costs 1,000 million marks."[22] Fortunately, by the end of the year a new currency was introduced and the economy began to stabilize. The withdrawal of French and Belgian troops from the Ruhr Valley in the spring of 1924 also helped to lower tensions.

Despite the unease and the steep cuts in funding, life went on as usual within the halls of science. With his doctorate in hand, Karl-Friedrich began to work as an assistant to Haber, one of the most famous scientists in Germany, at the Kaiser Wilhelm Institute for Physical and Electrochemistry in Dahlem—Berlin, commonly known as the Haber Institute.[23] It was a place crowded with Nobel laureates and future laureates, all competing vigorously, building on each other's ideas, and goading each other to go further. Polanyi headed the section on basic research in physical chemistry. Herbert Freundlich was head of colloid chemistry. James Franck, who would be awarded the Nobel Prize in 1925 with Gustav Hertz for their research on mercury vapor, was head of atomic physics until he went to the University of Göttingen. He was replaced by an equally capable man, the atomic physicist Rudolf Ladenburg. All of these men believed firmly in pure science and the universal truth of mathematics. In his Nobel acceptance speech, Franck sounded rather like Einstein at Planck's birthday party, substituting a "magnificent edifice" for Einstein's "temple of science" metaphor: "Is it not rather we, who have every reason to be thankful?—thankful for the opportunity to work, thankful also to destiny for permitting us, in an epoch so rich and vibrant in our science, to carry building blocks to the magnificent edifice of quantum and atomic theory that men like Planck and Niels Bohr, in particular, have erected."[24]

No classroom teaching was supposed to take place at the institute, that being done at the University of Berlin, but some classes and lectures were held there, which helped to establish a stronger link between the different centers of physics and physical chemistry. All seven of the Kaiser Wilhelm institutes were structured to be places where scientists could be free of the burden of teaching and administration. Their funding came partially from the Kaiser Wilhelm Society (Kaiser Wilhelm-Gesellschaft), patents, and the government. It also came from industry and philanthropists, including the John D. Rockefeller Foundation, established in 1913 to "promote the well-being of mankind." One of Haber's biggest and oddest undertakings at his institute during the 1920s was an attempt to extract gold from seawater. If he had succeeded, Germany could have easily paid off its crushing war debts. The Kaiser Wilhelm institutes were never far from the needs of the country and of industry even though the individual scientists working within them liked to think otherwise.

More important than classroom teaching was the Haber colloquium held every other Monday afternoon to a standing-room-only crowd. From its start in 1919, it was a magnet for scientists throughout Berlin. Colloquia had long been a cornerstone in German universities. Nernst's colloquia at the University of Berlin were famous worldwide, being routinely attended by Einstein, Max von Laue, Planck, and others. It was a must-attend event for scientists visiting from other countries who saw it as the seedbed of new ideas in physics. However, Nernst's colloquia were more formal (famous scientists sat in the front row, students in back, questions followed the lectures) and were devoted solely to physics. Haber's colloquia were more wide-ranging in both speakers and subject matter, "from the helium atom to the flea" as David Nachmansohn, a physical chemist and frequent attendee, put it. According to

Nachmansohn, who was at the Kaiser Wilhelm Institute for Biology in the 1920s, the colloquia provided "an ideal opportunity for Haber to realize his aim of breaking down the barriers between physics, chemistry, physical chemistry, and the biological sciences." Discussion was not only vigorous but also interdisciplinary, and "automatically stimulated interests beyond one's own field and frequently suggested ways for trying a new approach." Because controversy was encouraged, the young scientists in attendance "became immunized against authoritative and dogmatic thinking." Nachmansohn found it exhilarating to learn that at a time when all fields of science were rapidly expanding, "nobody could claim to have the right answers."[25]

Herbert Freundlich, who had studied under Wilhelm Ostwald, was in charge of organizing the colloquia until 1930 when the venue changed from the Haber Institute to the Harnack Haus. Newly opened, the Harnack Haus had been built to serve as the spacious and elegant conference center for all the Kaiser Wilhelm institutes where German proficiency in science could be showcased to an international audience. At that point, the administration of the colloquia was taken over by Polanyi until he and Freundlich were forced to leave Germany. Only younger scholars who had not yet habilitated were supposed to be invited so as to "free the discussion as much as possible from consideration of academic standing," but no one checked credentials at the door.[26] All this made for very lively sessions. Paul Harteck, who was to become a colleague of Karl-Friedrich, frequently attended Haber's colloquia. He recalled that presenters expected to be given a hard time, especially when their ideas were novel and as a result were more likely to be wrong.[27] Einstein could get away with interrupting a presenter. "Einstein was someone who was not embarrassed if he didn't know something," said Harteck. "He would say, 'I don't know that. I don't understand that at all. Can you explain that a little better for me?'" Usually, the attempt to "explain a little better" forced the presenter into a higher order of thinking or a retraction.[28]

Haber's approach was more Socratic in nature than Einstein's. To J.E. Coates, the British scientist who had been Haber's assistant in Karlsruhe and who frequently attended the colloquia when he was in Berlin, Einstein displayed in them one of his "most brilliant qualities, the capacity to grasp quickly the essentials of subjects not his own, to perceive their true bearings beyond the details, to discover errors, to indicate lines of advance." According to Fritz Stern in *Dreams and Delusions*, in the empyrean of scientific Berlin, "Einstein was the solitary master, the genius of theoretical creation," whereas Haber was "the impresario of collective greatness, the genius of practical achievement."[29]

Fritz Haber's life is a study in ethical confliction. No other scientist had as significant a role in directing Karl-Friedrich's early career. As will be seen, even Haber's death had a major impact on Karl-Friedrich, badly shaking his belief in the goodness of humankind under the onslaught of Nazi cruelty. An ardent supporter of the ideal of pure science, Haber's own life gave the lie to the separation of science from politics and industry. He had won the Nobel Prize in 1918 (received in 1920) for his research on nitrogen-fixing and the synthesis of ammonia. In 1909, his research had been advanced by Carl Bosch, a chemist at the large industrial company, BASF, who had substituted an impure iron catalyst for the expensive and rare osmium that Haber had used in his experiments. The substitution worked, to Haber's pleased surprise, making the process economical on an industrial scale. The Haber–Bosch process, as it came to be known, led to the production of artificial fertilizer, bringing about a revolution in agriculture that reduced hunger worldwide. By lessening Germany's dependence on imported natural fertilizers, such as guano shipped in from Peru, artificial fertilizers made the nation more capable of growing enough food to sustain its population. In his Nobel lecture delivered in Stockholm in June 1920, Haber said that until a different solution was found, "improved nitrogen fertilization of the soil brings new nutritive riches to mankind and that the chemical industry comes to the aid of the farmer who, in the good earth, changes stones into bread." The Haber–Bosch process also made a difference in the supply of saltpeter used in munitions, a fact Haber did not mention in his Nobel lecture out of deference to the memory of Alfred Nobel whose fortune had been made in explosives.[30]

Haber's work on nitrogen led directly to his involvement in the development of poison gases used in the First World War. At the beginning of the war, Haber was appointed head of the Chemistry Section in the Ministry of War. He was also given the rank of captain by the Kaiser. Convincing military leaders that poison gases could be used to clear enemy trenches, Haber personally directed their first use at the Battle of Ypres in 1915. From then on, his institute was given over to, as he put it, "poison instead of air." He proved as able an administrator as a chemist, and by 1916, his Kaiser Wilhelm Institute for Physical Chemistry and Electrochemistry had turned into the Military Institute, employing almost 150 academicians and approximately 1,300 non-commissioned officers, soldiers, and other employees. In his biography of Haber, Dietrich Stoltzenberg writes, "Reports of Haber's activities in this period give the impression that he was obsessed with the task he had set himself. Everything else in his life was pushed into the background."[31]

In a manner that presaged the involvement of German scientists in military research during World War II, including the atomic bomb and rocketry, Haber was a fierce patriot, proclaiming in regard to science, "for humanity in time of peace, for the fatherland in time of war."[32] This was far more than a platitude; it was a foundational belief he shared with most Germans. It was not linked to who was in power but to threats to the *fatherland*; that threat in the First World War was the triple entente of England, France, and Russia, joined in 1917 by the USA.

Haber was listed as a war criminal by the Allies after the war. Faced with death threats and a war crimes trial, he fled to Switzerland and did not return to Germany until the Allies had dropped their extradition request. In the years that followed, he never tried to distance himself from his involvement in chemical warfare, arguing that "gas weapons are definitely no more inhumane than flying bits of metal." He also claimed that France had used chemical weapons first, forcing Germany to respond. Secretly, he advised German leaders to continue supporting research on chemical weapons, because other countries were pursuing such research. Yet for all his pride in having served Germany so well during wartime, it haunted him for the rest of his life for a singularly terrible reason: Nine days after the first use of poison gas at Ypres, Haber's wife, Clara, had taken his service revolver and killed herself.[33]

With the same energy that he had given to poison gas, Haber returned to physical chemistry, doggedly rebuilding his institute. His role in the war was not held against him in Germany. When he received the Nobel Prize, he was lionized. Tragically, none of his accomplishments could save him after Hitler came to power. The great man would die an outcast. Yet during the decade of the 1920s, Haber helped bring German science to its glorious peak, not by dint of his own research but by his ability to catalyze the research of his colleagues and students. As James Franck succinctly put it, "I would see a keyhole somewhere, and Haber would open the door."[34]

One way Haber achieved this was by engendering an egalitarian atmosphere at his institute, which was in stark contrast to the autocratic approach at the other six Kaiser Wilhelm institutes. Karl-Friedrich, who considered it an honor to work under Haber, wrote that he was "free of all academic narrowness" and attempted to preserve the ties between scientific progress and practical life.[35] In a memoriam written following Karl-Friedrich's death in 1957, Paul Harteck reflected back on Haber's importance during the 1920s as well as on what he considered to be Karl-Friedrich's "most beautiful work" done while at Haber's institute:

> Since Berlin was one of the leading centers for physics and physical chemistry in the years following World War I, Bonhoeffer was influenced by such men as Walter Nernst, Max Planck, and Albert Einstein. However, the man who influenced him the most was Fritz Haber. It was in Haber's Institute, where he worked as a young man through the years 1923 to 1930, that Bonhoeffer made what was perhaps his most beautiful work. Some of his most important works are as follows: "The Behavior of Active Hydrogen"; "The Chemiluminescence of Active Hydrogen"; "Work of Active Nitrogen"; "Investigation of the Absorption Spectra of Hydrogen Iodide" and also of ammonia which led to the understanding of the predissociation; with Haber, a fundamental paper concerning band spectroscopy and flame processes, experimental evidence for the existence of SiO and the OH Radical; and with this writer, experiments on ortho- and para-hydrogen.[36]

According to Jeremiah James et al., one of the most prestigious strands of research to emerge at the institute began with Karl-Friedrich's arrival and his investigations of hydrogen. "Given that atomic hydrogen was the only substance for which physicists felt they might have an acceptable quantum theoretical model at the time, it was a topic of manifest, if nebulous, potential for bridging physics and chemistry. From the outset, Bonhoeffer took advantage of the expertise in low-pressure gas chemistry available at the Institute."[37]

The precursor to Karl-Friedrich's research on hydrogen occurred in 1900 when Max Planck set forth his quantum theory in which only discrete packages of energy—quanta—are exchanged, their size being proportionate to the frequency of radiation. In Zurich, Einstein picked up Planck's idea and took it much further, applying it in 1905 to the photoelectric emission of electrons from metals. Then eight years later in Copenhagen, Niels Bohr postulated an atomic model that ingeniously combined electrons circling the nucleus in orbits between which the electrons could transition, giving off light quanta in the process. According to Adalbert Farkas, who earned his doctorate under Karl-Friedrich, Bohr showed that by calculation on this model, hydrogen's atomic spectrum could be derived with great accuracy. "However, the quantum theory could not explain the reason for the intensity variation of spectral lines or the dualism in the interpretation of interaction of matter and radiation according to which radiation sometimes behaved as waves and sometimes as corpuscular quanta." The answer began to come in the mid-1920s, according to Farkas, "through the development of quantum mechanics by convergent attacks from different directions by several researchers." Among them were Louis de Broglie, Erwin Schrödinger, Werner Heisenberg, and Paul Dirac.[38]

Studying in Berlin, Max Delbrück often attended the physics colloquia, sometimes accompanied by Karl-Friedrich. He recalled vividly the one in which Werner Heisenberg, then only twenty-three years old, announced his breakthrough on quantum mechanics:

> I went to the seminar, didn't understand a word, but I remember as I walked into the building – the grimy old building, the physics institute in downtown Berlin, the lecture hall on the third floor, enormous staircases – as I walked in there, at the same time Einstein came in from one side and Nernst from the other side. And I heard Nernst ask Einstein whispering "Do you think, there's anything to this? Do you think there's anything to this?" And Einstein said, "Ja, Ja, I think it's a very good paper, very important." So they walked up there and the place was packed, standing room only.[39]

On December 3, 1924, Franck wrote to Karl-Friedrich from Göttingen: "I thank you very much for the transmission of your beautiful work which has interested me very much. Almost more than the content of the work, which I personally knew about, I was interested in the news that Mr. Frumkin told me about obtaining chemiluminescence from sodium vapor by means of molecular hydrogen." (Aleksander Frumkin was a Russian physicist.) Franck was working on the molecular dissociation of nitrogen, which he wanted to discuss with Karl-Friedrich on an upcoming visit to Berlin. A man with a reputation for kindness as great as his reputation for physics, Franck asked Karl-Friedrich to share his thinking on nitrogen. This was similar in tone to the letter he had sent in 1920 to Karl-Friedrich, then a student, when he asked for his thinking on photochemistry. Franck reassured Karl-Friedrich that he was not attempting to push ahead of him: "I find it important to speak to you sometime about your fine researches or to become acquainted with your views on whether you think like I do on the question of nitrogen.

Probably you have already seen this with nitrogen; in any case, it is clear because Frumkin gave me your results with hydrogen, so you do not need to eye me as some scoundrel trying to take away something from you."[40] Karl-Friedrich never eyed Franck that way but with great respect. They remained friends throughout their lives, even though Franck was forced to leave Germany in 1933.

Under Franck's influence as well as that of Max Born, the University of Göttingen was fast becoming a world center for physics rivaling Berlin, attracting not only Heisenberg but also scientists from other countries, among them: J. Robert Oppenheimer (USA), Edward Teller (Hungary), Paul Dirac (England), and Edward Condon (USA). Franck's grandson, Frank von Hippel, wrote that Franck's approach to understanding molecular processes was more visual than mathematical. It was Condon who mathematically fleshed out Franck's visual idea, bringing into focus what became known as the Franck–Condon principle.[41] Such a combination of the mathematician (who literally worked out the numbers) with the theoretician was commonplace. It played an important role in making the universe of German pure science so powerful. In 1907, well before the arrival of Franck and Heisenberg at Göttingen, the mathematician Hermann Minkowski had taken Einstein's special theory of relativity—so brilliant yet so amorphous—and blessed it with mathematical formalism, to Einstein's bemused gratitude. Einstein remarked in his humorous manner, "Since the mathematicians have attacked the relativity theory, I myself no longer understand it any more."[42]

In June 1925, there was a major fire in one of the laboratories at the Haber Institute. The surprise was not that the fire occurred but that there were not more of them, considering the lack of safety precautions in the laboratories, made exponentially more dangerous by the daring-do of the young scientists, heightened even more by Haber's interest in the chemical processes of flames and explosions. Subsequently, the laboratory was rebuilt and expanded. With the help of grants from the Rockefeller Foundation, Haber added new equipment, including high-voltage apparatus, new X-ray machines, and a glassblower's workshop—glassblowing being essential to the design of experimental apparatus. It was an ideal environment for Karl-Friedrich to work on hydrogen, building on the quantum mechanics of Heisenberg, the dissociation work of Franck, the adsorption research of Polanyi, as well as the work of many others.

For Karl-Friedrich, despite the seriousness of the research and the intensity of the interactions with colleagues, there was also time for travel, and not merely to give guest lectures at other universities, or to relax at the family's summer home in Friedrichsbrunn. Like most Germans, Karl-Friedrich loved to hike, often with his brothers or Max Delbrück, spending time hiking in the Harz Mountains and when the opportunity arose, in the Austrian Alps. Cultivating a childhood interest in painting, he also enjoyed visiting museums, among them the National Gallery in London in 1924. In August 1926, he, Dietrich, and their mutual friend, Robert Held, who was studying law at the University of Berlin, traveled through the northern part of Italy, stopping at Lake Garda, hiking in the Dolomites, and touring Venice. Dietrich wrote to his sister about a steamboat ride they took near the end of that trip: "Thus far we encountered lemon trees, forests of olive trees, and cypresses that circle the sea, which was often as blue as the beautiful evening sky and as clear as glass. You were able to see the fish and sea flora that were far beneath the surface. Yesterday evening we went sailing in beautiful stormy weather. The whole sea bore little crests of foam and tossed the ship around quite a bit."[43] The Germans had long had a work-hard-play-hard approach to life, which was true of the Bonhoeffers. The only component that was absent was women friends. Three of the Bonhoeffer sisters were married by 1926, Ursula to Rüdiger Schleicher; Sabine to Gerhard Leibholz; and Christel to Hans Dohnanyi—all lawyers. But none of the brothers were married. As Karl-Friedrich told a reporter in 1929, he simply did not have time.

In 1927, at the age of 28, Karl-Friedrich habilitated and began teaching courses in physical chemistry and atomic physics at the University of Berlin. Habilitation in German universities was a further step beyond the doctorate, requiring a second, more advanced, thesis. It was an arduous task that was obligatory for a teaching position. In 1929, Karl-Friedrich was promoted to associate professor at the University of Berlin while continuing his work at the Haber Institute. He soon made his mark, assisted by his student Ladislaus Farkas (the brother of Adalbert Farkas), whose doctoral thesis was on the quantum implications of photochemistry. In his essay "Reflections on Physical Chemistry: Science and Scientists," Joshua Jortner, a physical chemist, wrote about their work together: "In 1927, Bonhoeffer & Farkas observed that predissociation in the electronically excited ammonia molecule is manifested by spectral

Fig. 4.2 Ladislaus Farkas and Fritz Haber. United States Holocaust Memorial Museum, courtesy of Leorah Kroyanker

line broadening, establishing the first spectroscopic-dynamic relation, providing experimental verification of the Heisenberg energy-time uncertainty relation, and pioneering the field of intramolecular dynamics (Fig. 4.2)."[44]

In the same period, Karl-Friedrich began an important collaboration with Paul Harteck, an Austrian physical chemist with whom he would remain friends for the rest of his life. Three years younger than Karl-Friedrich, Harteck had studied chemistry at the University of Vienna and the University of Berlin, receiving his doctorate under Max Bodenstein, who had taken over Nernst's position upon his retirement in 1923. Bodenstein was famous for the kinetics of gas reactions. Harteck had also worked as an assistant to Arnold Eucken at the Technical University in Breslau, gaining valuable experience in low-temperature specific heat experiments. Adalbert Farkas remembered meeting Harteck for the first time in Karl-Friedrich's laboratory.

"Paul Harteck won my admiration immediately as one of those rare individuals who combines personal charm, experimental skill and mastery of theory."[45]

In 1928, Harteck became a staff scientist at the Haber Institute, working closely with Karl-Friedrich to discover ortho-hydrogen and para-hydrogen, the spin isomers of hydrogen. A few years earlier, the physicists Werner Heisenberg and Friedrich Hund had speculated about their existence, postulating that it was analogous to the helium atom in which two identical nuclei exist in two forms, with the ortho form having parallel nuclear spins and the para form having antiparallel nuclear spins. The American physicist, David Dennison, who was doing postdoctoral work with Bohr at the Institute of Theoretical Physics in Copenhagen (from 1924 to 1926), proposed a theory that incorporated the observed specific heat of hydrogen. At low temperatures, the para form would be favored over the ortho form. However, Dennison predicted that the transition between them would be slow. Testing the new theory was very difficult work for Harteck and Karl-Friedrich. The physicist Fritz London, an assistant to Erwin Schrödinger at the University of Berlin, suggested to them that to facilitate the transition, they should keep the hydrogen at a low temperature as long as possible. On October 22, 1928, Karl-Friedrich wrote to Dietrich, who was then serving as an assistant pastor for a German church in Barcelona, Spain, "We got the idea of experiments that should illustrate that the normal hydrogen . . . is a mixture, as some theoreticians assert. Isn't that very interesting? If it works, we've succeeded, since there would surely be some people interested in that. But at the moment it's not working, and in the course of the many failed efforts I've already lost half my hair."[46]

In "From Berlin to Jerusalem: Ladislaus Farkas and the Founding of Physical Chemistry in Israel," Michael Chayut explains that Karl-Friedrich and Harteck were able to isolate para-hydrogen and to show its stability at room temperature. "Although quantum theory forbade ortho-para conversion because of the different symmetries of their wave functions, Bonhoeffer and Harteck showed that para-hydrogen could be converted to ortho-hydrogen. They found that the conversion was accelerated by high temperatures and pressure, by electric discharge, and by platinized asbestos."[47]

Karl-Friedrich and Harteck announced their discovery at the Haber colloquium in February 1929. It was a very cold day in the coldest winter in a hundred years; even the rivers were frozen solid. But in the lecture hall, the announcement generated considerable heat because of the excitement in the room. Essentially, by means of a verifiable experiment, they had put legs under Heisenberg's theory of quantum mechanics. Polanyi, who was in attendance, wrote despondently in his diary, "Oh, if only I could claim such a solid discovery. My achievements are scattered. I am depressed." His depression was not due to his having been beaten in a race with his friends but to his blaming himself for not focusing his attention on one subject. For Polanyi, being a polymath—as interested in philosophy and economic theory as in physical chemistry—was both a blessing and a curse.[48]

Beating to publication a similar announcement by Arnold Eucken, Karl-Friedrich and Harteck reported their findings in two brief articles in March 1929 in the journal *Die Naturwissenschaften*, the prestigious scientific journal published by Springer Verlag. C.A. Gearhart in his essay "*Astonishing Successes* and *Bitter Disappointment*: The Specific Heat of Hydrogen in Quantum Theory" sheds more light on this discovery and the importance of catalysts in the experiments: "In perhaps their most striking result, they found that hydrogen absorbed on activated charcoal at liquid-hydrogen temperatures and then pumped off was transformed in a matter of minutes to the nearly pure parahydrogen expected at this low temperature. If stored carefully without impurities in a glass container, it remained stable for some weeks even at room temperature." Gearhart also points out that Harteck and Bonhoeffer had used a new technique in which they measured the thermal conductivity of the gas by recording the temperature of a platinum wire in a container filled with hydrogen. Measurements of thermal conductivity could be made more rapidly and conveniently than heat capacity. "Thus a higher thermal conductivity corresponded to the larger specific heat expected at most temperatures for parahydrogen."[49]

Eucken carped that the success of the work was due to Harteck who had been his assistant at the Technical University in Breslau. Having first published on the specific heat of hydrogen seventeen years earlier in 1912, Eucken may have had some justification for his carping. However, in an interview conducted years later, Harteck said that despite his experience in Eucken's laboratory, initially he had no idea how to set up the experiment, and neither did Karl-Friedrich. "I just had the idea that you couldn't measure specific heat with a calorimeter because that would be too much work. I have always done everything with the minimum of effort. That was one thing. The second was that we had the heat conductivity method." Harteck also told the interviewer that he had an instinctive feel for chemistry. "It didn't matter what it was, hydrogen or anything else, I had an instinctive understanding of it."[50]

This discovery made Bonhoeffer's name well-known to scientists in Europe and the USA, although the sudden fame did not go to his head. In a world of arrogant brilliance, personified by Nernst who was accustomed to always being right and did not take being wrong as an option, or Werner Heisenberg, whose faith in his intellectual abilities was boundless, Karl-Friedrich was brilliant but modest. He insisted on sharing full credit with Paul Harteck as well as recognizing the antecedents to their work.

Notes

1. Albert Einstein, *Ideas and Opinions*, trans. Sonja Bargmann (New York: Three Rivers Press, 1954), 224–225. www.scienceforums.com/topic/4842-einsteins-view-on-schopenhauer/.
2. Bethge, *Bonhoeffer*, 34.
3. Ibid., 49.
4. The Geheimrat structure in academia was in some respects stultifying, but it was counterbalanced by the ability of students to change universities and, therefore, Geheimrats.
5. Ludloff became a professor at the University of Breslau, now the University of Wroclaw, coming to the USA in 1939. His specialty was fluid mechanics. DBW, Vol. 9, 43.
6. On his return to Breslau, Rudolf Stern first worked for Oskar Minkowski, a specialist in diabetes. Rudolf's father Richard had done significant research on trauma and disease. In 1930, Rudolf Stern published a revised version of his father's work *The Traumatic Origins of Internal Diseases*. Stern, *Five Germanys*, 57, 76.
7. Berlin had at least one stellar physical chemist prior to Nernst and Bodenstein. The Dutchman Jacobus Van't Hoff, the first winner of the Nobel Prize in Chemistry in 1901, moved to Berlin in 1896. Ostwald considered Van't Hoff the best physical chemist of the age, the other being the Swede Svante Arrhenius.
8. Nernst made these points repeatedly including in his Silliman lecture at Yale University in 1906.
9. W. Karl Jost, "The First 45 Years of Physical Chemistry: Science and Scientists," *Annual Review Physical Chemistry* 57 (2006): 7.
10. Ibid., 7–8.
11. Ibid., 4.

12. The photochemical equivalence principle was not the same as the equivalence principle which stated there was an equivalence between gravitational and inertial mass.
13. The typed date on the letter is March 6, 1920. The letter is on University of Göttingen stationery; however, in 1920 Karl Friedrich would have been at the University of Berlin, not Haber's Institute, yet Franck mentions Haber. The date may be in error. Abtielung Rep 23, Personal Correspondence, Nachlass Karl Friedrich Bonhoeffer, Archiv der Max-Planck-Gesellschaft.
14. Jost, "Memoriam Karl-Friedrich Bonhoeffer," *Die Naturwissenschaften* 44, Jahrgang, Heft 24 (Zweites Dezemberheft, 1957): 6.
15. Element 43 was very controversial. It was confirmed by Carlo Perrier and Emilio Segrè at the University of Palermo in 1936. en.wikipedia.org/wiki/technetium.
16. During the war, the Noddacks were at the Nazified University of Strausbourg. So also was Weizsäcker.
17. Hevesy's name often appears with either *von* or *de*, but their use is inconsistent. This is true for other Hungarian names, including Dohnanyi. Von Neumann maintained its use.
18. István Hargittai, *The Martians of Science: Five Physicists Who Changed the Twentieth Century* (Oxford: Oxford University Press, 2006), 23. See endnote 52.
19. Jeremiah James et al., *One Hundred Years at the Intersection of Chemistry and Physics: The Fritz Haber Institute of the Max Planck Society: 1911–2011* (Berlin: Walter de Gruyter, 2011), 55. See footnote 32. Cf. Rürup Schüring, Schicksale, 98. Havemann was a physical chemist.
20. Sifton and Stern, *No Ordinary Men*, 15.
21. Leibholz-Bonhoeffer, *Bonhoeffers*, 63–64.
22. Bethge, *Bonhoeffer*, 47.
23. The Haber Institute is now known as The Fritz Haber Institute of the Max Planck Society.
24. Frank von Hippel, "James Franck: Science and Conscience," *Physics Today*, American Institute of Physics (June 2010): 43.
25. David Nachmansohn, *German-Jewish Pioneers in Science 1900–1933: Highlights in Atomic Physics, Chemistry and Biochemistry* (Berlin: Springer-Verlag, 1979), 184.
26. James, *One Hundred Years*, 43.
27. Paul Harteck interview with Joseph J. Ermenc, July 6, 1967, PHP, MC 17, Box 10, Folder 2, pg. 13, Institute Archives and Special Collections, Rensselaer Polytechnic Institute (RPI).
28. Harteck interview with Medikus, 1982, Box 10, Folder 4, RPI.
29. Fritz Stern, *Dreams and Delusions*, 70.

30. Dietrich Stoltzenberg, *Fritz Haber: Chemist, Nobel Laureate, German, Jew* (Philadelphia: Chemical Heritage Press, 2004), 90, 216–217.
31. Ibid., 138, 140.
32. Ibid., xxii.
33. Clara, who had her doctorate, was apparently suffering from depression. Her body was found by her son, Hermann, who was twelve years old. There are connections between Haber and the development of Zyklon, which was used solely for pest control. In WWII Zyklon B was used in the gas chambers. Larry Rasmussen, "Dietrich and Karl-Friedrich Bonhoeffer: The Brothers Bonhoeffer on Science, Morality, and Theology," *Zygon* 44, no. 1 (2009): 106.
34. Stoltzenberg, *Haber*, 238. See endnote 21.
35. Fritz Stern, *Dreams and Delusions*, 57.
36. Harteck, *In Memoriam Karl Friedrich Bonhoeffer*, May 15, 1957. www.ac.eis.com.
37. James, *One Hundred Years*, 71.
38. Adalbert Farkas, "Paul Harteck: The Triumphant Decade 1925–1934," *AMBIX* 36, part 2 (July 1989): 1. http://www.maneyonline.com/doi/abs/10.1179/amb.1989.36.2.91.
39. Delbrück interview, 29.
40. Franck to Bonhoeffer, December 3, 1924, Abtielung Rep 23, Personal Correspondence, Nachlass Karl Friedrich Bonhoeffer, Archiv der Max-Planck-Gesellschaft.
41. Von Hippel, *James Franck*, 43.
42. Clark, *Einstein*, 122.
43. DBW, Vol. 9, 164.
44. Joshua Jortner, "Reflections on Physical Chemistry: Science and Scientists," *Annual Review Physical Chemistry* (2006): 5.
45. Farkas, "Paul Harteck," 1.
46. Bethge, *Bonhoeffer*, 45.
47. Michael Chayut, "From Berlin to Jerusalem: Ladislaus Farkas and the Founding of Physical Chemistry in Israel," *Historical Studies in the Physical and Biological Science* 24, no. 2 (1994), 243.
48. William Taussig Scott and Martin X. Moleski, S.J., *Michael Polanyi: Scientist and Philosopher* (Oxford: Oxford University Press, 2005), 112.
49. C.A. Gearhart, "'Astonishing Successes' and 'Bitter Disappointment': The Specific Heat of Hydrogen in Quantum Theory," *Archives for History of Exact Sciences* 64, no. 26 (March 2010): 189–190.
50. Harteck interview, Box 10, Folder 4, pgs. 10, 17, RPI.

CHAPTER 5

Traveling with Polanyi

The publication of the paper on the spin isomers of hydrogen led immediately to Karl-Friedrich receiving an invitation from Irving Langmuir, president of the American Chemical Society, to address the society's annual meeting, scheduled for September 1929 in Minneapolis, Minnesota. One of the most distinguished scientists in the USA, Langmuir, who worked for General Electric, had earned his doctorate in physical chemistry at Göttingen in 1904. Influenced by Nernst's belief that the various branches of science should not be rigidly separated, Langmuir's research over the course of his long career included chemistry, physics, and engineering, with meteorology joining the list in the 1930s. To these fields, he mixed in a large dose of practicality, which began at GE with his work on the incandescent light bulb. Because he had kept in close touch with scientific advances in Germany, he had immediately recognized the significance of Karl-Friedrich's discovery.

Around the same time, a second invitation to visit the USA came from Hugh Taylor, head of the chemistry department at Princeton University as well as chair of the American Chemical Society's annual meeting. In May of that year, Taylor had been in Berlin, visiting his friend, Max Bodenstein, under whom he had studied prior to the First World War. A world expert on chemical reaction velocities, Bodenstein was a professor of physics at the University of Berlin and director of the Kaiser Wilhelm Physical Chemistry Institute.

K. L. Housley, *The Scientific World of Karl-Friedrich Bonhoeffer*, Palgrave Studies in the History of Science and Technology,
https://doi.org/10.1007/978-3-319-95801-9_5

While Taylor was visiting Bodenstein, work was progressing on the construction of the new Frick Chemical Laboratory at Princeton. Its exterior was of stone with gothic arched windows and doors, while its interior had wood-paneled halls, lecture rooms, and most importantly, well-equipped bench space to conduct experiments. Because it was intended to be a showpiece of chemistry in the USA—the intent being to make Princeton the scientific equivalent of the University of Berlin and the Kaiser Wilhelm institutes—Taylor wanted to invite the best kineticists in the world for the dedication in the fall. Besides Bodenstein, Taylor hoped that Karl-Friedrich would be able to attend because his discovery was creating a scientific stir. Furthermore, Taylor's own work on catalytic theory, published in the mid-1920s, had a connection to Karl-Friedrich's research. The third invitee was Michael Polanyi who was already famous for his research on adsorption and X-ray defraction, among other subjects. For all of these reasons, Karl-Friedrich, Polanyi, and Bodenstein were pleased to accept Taylor's invitation to the Frick dedication, which was felicitously timed to occur following the annual meeting of the American Chemical Society.

Polanyi and Karl-Friedrich sailed on July 24 from Bremerhaven on the liner *SS München,* landing in New York ten days later on August 4. They were met by their friends the physical chemists Henry Stewart and George Kistiakowsky, the latter having studied under Bodenstein and who had recently accepted a position at Harvard. The group lost no time in seeing the sights, their final stop being the Hotel Pennsylvania, located across the street from Penn Station, where Karl-Friedrich and Polanyi were staying on the seventeenth floor, giving them a grand view of the bustling city so different from Berlin.[1] The New York stock market was booming, the Roaring 20s were roaring despite Prohibition, and the German dirigible, the *Graf Zeppelin,* (with funding from the American newspaper magnate William Randolph Hearst) was about to begin its circumnavigation of the globe, taking off from the Naval Air Engineering Station in Lakehurst, New Jersey, on August 8. However, Karl-Friedrich and Polanyi did not have time to linger in New York to celebrate the triumph of German aeronautics because Polanyi was scheduled to give a series of lectures during the month of August at a symposium on chemical kinetics at the University of Minnesota.

Karl-Friedrich was aware that Polanyi had interests extending far beyond physical chemistry and physics. Handsome and affable with dark hair and a winning smile, Polanyi was the type of man whose mind never rested, always observing, analyzing, and asking questions of everything and everyone, sometimes to the dismay of his colleagues, who wished he would stay more focused on science in which he was obviously a gifted researcher. The difficulty was that he was gifted in other fields as well. Such broadness ran in the family; his older brother, Karl, was a lawyer, historian, socialist, and economist whose ideas on the emergence of market-based economies would become highly influential.

Michael Polanyi would later write, "To recognize a problem which can be solved and is worth solving is in fact a discovery in its own right."[2] Polanyi saw such problems everywhere in many fields, but they rarely overwhelmed him. Put in the context of entanglement, Polanyi had an entangled mind, which he considered both an *opportunity* and a form of *grace*. The opportunity for Polanyi was to untangle the thinking processes of scientists, beginning with his own. Among the questions that were to occupy him more and more were: what motivated scientists; what values did they share with their culture; how did they collaborate; what did they know tacitly that they could not express; how did they know what was a fact? As to entanglement being a form of grace, one colleague said it best: "What made him differ most from those around him was his reverence."[3]

Polanyi's ideas did not coalesce in the 1920s, but there is no doubt that they had their genesis during his time at the Haber Institute. Years later, Eugene Wigner, who was mentored for his doctoral dissertation by Polanyi, recalled with gratitude that it was Polanyi who had taught him that "science begins when a body of phenomena is available which shows some coherence and regularities, that science consists in assimilating these regularities and in creating concepts which permit expressing regularities in a natural way." Wigner appreciated Polanyi's enthusiasm, guidance, and active intervention on his behalf during the 1920s. It had been Polanyi who had convinced Wigner to give up a job in a tannery in Budapest and take an assistantship with a crystallographer in Berlin, thereby changing the direction of Wigner's life. Wigner won the Nobel Prize in Physics in 1963.[4]

Because Polanyi's interests were so broad, traveling with him across the USA meant that for Karl-Friedrich the trip took on far more than a see-the-sights agenda. For example, Polanyi spent time comparing and contrasting scientific working conditions, not finding them up to the standards of Berlin. He was disappointed with the University of Minnesota, concluding that the scientists there were spending too much time on research that had practical applications. As a result, those who were working in pure theoretical science—whom Polanyi considered precious assets—were neglected.[5]

Polanyi had visited the Soviet Union the year before to study the effects of Stalin's first five-year plan, coming away with a negative impression of Communism that a subsequent trip to the Soviet Union in 1935 only reinforced. Repelled by the brutal fanaticism he encountered, he disagreed sharply with his brother, Karl, (who was then living in Austria) as to the potential for good in the communist model. John Jewkes, an economist and a colleague of Polanyi during his years at Oxford, wrote, "Running through all his writings in economics, there seems to be one central strand: how best to reconcile the safeguarding of individual liberty with the controls upon the individual inseparable from a complex and organized society, or as Polanyi succinctly put it, the relation between spontaneous and social order. His analysis was pointing not only to the inhumanities of totalitarianism but also to the muddle and drift which so often result from overconfident planning in freer societies."[6] Jewkes's comment on Polanyi's economic perspective also pertains to his scientific perspective.

As they traveled to Minnesota, Polanyi took notes on prices, wages, and labor relations. In preparation for the trip, he had read a book that had just been published titled *Overview of Middletown: A Study in American Culture* by the sociologists Robert Staughton Lynd and Helen Merrell Lynd. Middletown was in actuality Muncie, Indiana. While on the trip, he acquired a document titled "Trends of Philanthropy," published by the US National Bureau of Economic Research. On his return to Germany, he used it as the discussion topic with a group of friends (among them, Karl-Friedrich, Leo Szilard, Fritz London, and John von Newmann) who met on Sunday evenings at the Harnack Haus to listen to presentations on economics, sociology, and science.[7]

One of the stops on the trip west was Oberlin College in Oberlin, Ohio, where they visited Polanyi's old friend, Oszkár Jászi, a progressive Hungarian who was a political scientist, social scientist, and historian. While in Hungary, Karl Polanyi had worked with Jászi to establish a political party in 1914 and had also contributed essays to his journal. Although interested in politics, Michael was less involved than his brother, his principal direction after the First World War being to pursue research in physical chemistry. Even so, the discussion among the men must have been scintillating, if for no other reason than Jászi's book, *The Dissolution of the Hapsburg Empire*, had been recently published by the University of Chicago. For a brief time after the First World War, Jászi had held a position in the new Hungarian government in the hope that by taking part he could uphold the democratic inclusion and cultural autonomy of the numerous ethnic groups in what was left of the Austro-Hungarian empire. Failing in his efforts, Jászi withdrew from politics and left Hungary, emigrating to the USA in 1925, where he accepted a professorship at Oberlin.[8]

It is not known how much Karl-Friedrich participated in these discussions, if at all, but in a letter he wrote to Dietrich a year later when Dietrich was attending Union Theological Seminary in New York City, he spoke about the appalling situation of black Americans in a way that indicated his observations about racial segregation were not superficial. It is likely that under the influence of Polanyi, he had studied closely what was going on in the USA. Just the trip from New York to Minneapolis, stopping to give lectures along the way (including at the University of Michigan in Ann Arbor), would have put Karl-Friedrich in contact with Negro porters and conductors, some of whom had their Ph.D.'s from reputable colleges, but who could not get any employment other than on the railroad. Certainly, the trip to Johns Hopkins University in Baltimore, Maryland, would have exposed Karl-Friedrich to the depredations of segregation, in which blacks were not even allowed to drink from the same water fountains as whites. Karl-Friedrich was so disturbed by what he saw that it was the principal reason he subsequently turned down an offer to teach at Harvard University.[9]

On August 16, 1929, Karl-Friedrich wrote to Harteck in Berlin that he had finally arrived in Minneapolis and was waiting impatiently for the package that Harteck was supposed to have sent. He then gave a description of Chicago: "I visited Chicago yesterday. The city in her own peculiarity is very exciting. The wonderful blue Michigan lake is like the ocean glittering in the sun. Along the lake, between lawns, are magnificent highways with joyride traffic, then treeless lawns with the most beautiful fountains I have ever seen. Then the skyscrapers (30–45 floors high) lined up like fortresses, many kilometers wide, and running along the front is Main Street with unbelievably heavy traffic." Three days later, he wrote again to tell Harteck that the package with the records and laboratory equipment to present demonstrations had finally arrived. "Unfortunately everything was broken in little pieces. Only a few records are useful. It looks like they played baseball with the package." He began to build from scratch a heat-transfer vessel, but he complained about the lack of "mechanical workers," such as glassblowers and locksmiths, who were always on hand to help at the Haber Institute. He told Harteck that he was waiting in Minneapolis for Bodenstein to arrive and then the two of them were going to travel to Yellowstone National Park. Meanwhile he was by himself because Polanyi had gone with Samuel C. Lind, a professor of chemistry at the University of Minnesota (who had studied in Germany under both Ostwald and Bodenstein), on a three-day fishing trip in the northern woods near Lake Superior. He concluded the letter by writing, "I don't know if the chances in America are better than in Germany. I believe the opportunities are better for everybody over here."[10]

In September, Bodenstein, Polanyi, and Karl-Friedrich attended the American Chemical Society's annual meeting on the campus of the University of Minnesota. There Karl-Friedrich demonstrated para-hydrogen to the enrapt crowd. In an article published in the *Montreal Gazette* with the misleading title "Hydrogen Shown not to be Element but only Mixture; Two Components Have Different Melting and Boiling Points: Demonstration Given," the reporter wrote that "the first experimental proof of the new wave mechanics theory of atomic structure" was "expected to do to chemistry what the Einstein theory did to physics." The reporter then described the demonstration in which Karl-Friedrich used a cylinder of ordinary liquefied hydrogen provided by the University of Minnesota:

> He drew off some of the liquid to be used as an outside bath around a glass tube filled with ordinary gas-mask charcoal. The liquid gas bubbled and froze and the charcoal rapidly approached the ultimate cold of physicists.
>
> When everything was cold enough, Mr. Bonhoeffer passed some ordinary hydrogen from the cylinder through the supercooled tube of charcoal and it came out the other end parahydrogen, from which it had been divided.
>
> "It is the greatest chemical achievement of 1929," said Dr. Hugh Taylor, chairman of the meeting. It is as important as the Royal Society's first expedition to observe whether light passing close to the sun was bent, as Einstein predicted.[11]

Irving Langmuir was equally impressed with Karl-Friedrich's demonstration, telling the reporter that the division of the hydrogen mixture "would have a compelling effect in spreading the new wave mechanics among chemists."

Another reporter at the annual meeting asked Karl-Friedrich about his scientific research and his personal life in a more lighthearted way, even querying him on marriage prospects. Karl-Friedrich's answers reveal how much he valued friendship and mutual cooperation. The reporter wrote that Karl-Friedrich "declared he would not talk for publication unless his pal and co-worker Dr. Paul Harteck, 27, was mentioned equally, jointly and with whatever other scientific terms of parity are feasible. His attitude revealed that the co-discoverers of the fact that hydrogen is twins, are themselves twins in friendship." According to the reporter, both men were unmarried, did not care about dancing, had quiet tastes and an aversion to talking about themselves. Besides insisting on Harteck getting equal recognition, Karl-Friedrich explained that the impetus for their groundbreaking experiment came from the research of the American scientist, David Dennison, described as "a master of the new wave mechanics," who had returned to the University of Michigan following his seminal work with Bohr in Copenhagen. The reporter concluded that the discovery of the isotopes of hydrogen was "one of the most important of all time," but that there was no practical application "other than to upset some fundamental scientific ideas and to start a host of workers looking for chemical discoveries in places previously unexplored."[12]

Years later, Polanyi drew on his early research to stress the relationship between chemistry and quantum mechanics. "For chemistry answers questions regarding the interaction of more or less stable chemical substances, and these questions cannot be raised without experience of these substances, and of the practical conditions in which they are to be handled. . . . While quantum mechanics can explain in principle all chemical reactions, it cannot replace, even in principle, our knowledge of chemistry."[13] Karl-Friedrich's demonstration at the American Chemical Society was an excellent example.

In late September as the leaves were turning color, the group traveled to Princeton University for the dedication of the Frick Chemical Laboratory. Other international scientists on hand were the Frenchman, Francis Perrin, representing his father, Jean Perrin, who had won the Nobel Prize in Physics in 1926, and the Englishman,

Cyril Hinshelwood, who would receive the Nobel Prize in Chemistry in 1956. Of course, Irving Langmuir was there, having come directly from the American Chemical Society meeting. Revered in America, he would receive the Nobel Prize in Chemistry in 1932 for his discoveries and investigation of surface chemistry, the first American scientist to win. In the years ahead, the names of three of the attendees at Princeton would be linked in what was known as the Langmuir-Bonhoeffer-Hinshelwood mechanism (also known as the Bonhoeffer-Farkas mechanism), which dealt with chemical adsorption and reflection processes in the interaction of hydrogen atoms. In summary, the scientists at the opening of the Frick Chemical Laboratory were not only illustrious but also competitive, especially Polanyi and Langmuir who had recently formulated strikingly differing theories on adsorption.[14] Papers were presented by several attendees, including Karl-Friedrich and Bodenstein. Polanyi spoke on "Atomic Reactions and Luminescence in Highly Dilute Flames." There was also a formal tea and the awarding of honorary degrees.[15]

The festivities were a highlight of a period at Princeton that Hugh Taylor described as "the golden age" of physicochemical research. Besides the dignitaries, in attendance at the dedication was Harry Julius Emeléus who had just arrived from England with his newly earned doctorate in hand. Emeléus had received a two-year fellowship to work at Princeton on photochemistry with Taylor. The dedication made a vivid impression on him. Noteworthy were the lectures on the theme "Rates of chemical reactions," including the one by Karl-Friedrich.[16]

It should also be mentioned who was not at the dedication. Henry Eyring, then twenty-eight years old, had gone to Berlin on a National Research Council Fellowship to study with Bodenstein or Polanyi, both of whom were, of course, not there. After the dedication, Bodenstein stayed on in the USA to lecture at Johns Hopkins, while Polanyi returned to Berlin. So it would be with Polanyi that Eyring would collaborate, following up mathematically on Karl-Friedrich's work, publishing together in 1930 and 1931 two important papers on chemical activation energy and the quantum mechanics of chemical bonds. Only after he finished his work with Polanyi did Eyring go to Princeton. Two other men who had just been hired by Princeton were Eugene Wigner and John von Neumann, whose recent lectures in Berlin on mathematics had astounded everyone with their brilliance. It is unknown whether they were in attendance at the Frick dedication, but what is known is that Princeton was becoming a formidable scientific institution and the visit of Bodenstein, Polanyi, and Bonhoeffer only added to the luster.

In fact, a shift had begun to occur, although the scientists themselves were not yet overtly aware of it. Up until this point, the kineticists had focused on macroscopic measurements; however, the way forward would rely more on the development of microscopic molecular-based tools. So also, the lines of communication between chemists and physicists were becoming attenuated in Germany, which ultimately meant that the blossoming of chemical crystallography and theoretical chemistry would occur elsewhere.[17] Only Polanyi seemed to have intuited the coming change.

Right after the Frick dedication, Polanyi and Karl-Friedrich returned to Germany on the elegant steamship *Bremen*, landing on October 10 at Cherbourg, France, where they were met by Polanyi's wife, Magda.[18] Then on October 29, 1929, the New York stock market crashed. It recovered somewhat in 1930, only to begin a steady decline, reaching its nadir on July 8, 1932. The shock waves rippled out worldwide. In 1930, German unemployment rose to approximately three million. Angry at the unwillingness of the Weimar government to prop up the economy, voters in the 1930 elections helped push the number of seats held by the National Socialists (the Nazi party) from twelve to 107 seats.

Karl-Friedrich received his first of five Nobel Prize nominations (the last one being in 1952) in 1930. The nomination was divided between himself and Arnold Eucken, and the nominator was Theodor Svedberg of Uppsala University in Sweden. Svedberg was a prominent physical chemist, a member of the Nobel Committee, as well as a Nobel laureate. Karl-Friedrich was nominated again in 1934 by Andreas von Antropoff, but the prize went to Harold Urey at Columbia University in New York City, who had discovered deuterium.[19] When Heisenberg received the Nobel Prize in 1932, the work of Bonhoeffer and Harteck was referenced as follows: ". . . for the creation of quantum mechanics, the application of which has, inter alia, led to the discovery of two allotropic forms of hydrogen."[20] However, they were not mentioned by name.

In 1930, Karl-Friedrich married Margarete (Grete) Dohnanyi, sister of Hans Dohnanyi who had attended school with Klaus and Dietrich. This was a double bond, because Hans was married to Karl-Friedrich's sister, Christel.[21]

At that point, it would be safe to assume that Karl-Friedrich was on the fast track to his own Nobel Prize and that any barriers to continuing to do stellar research had been removed. Buttressing this assumption is a photograph of Karl-Friedrich, Paul Harteck, Adalbert, and Ladislaus Farkas. They are perched on triangular metal scaffolding next to a hoist with dangling chains outside the Haber Institute. Highest up is Harteck wearing knickerbockers and a short jacket, his dark wavy hair standing up as if catching the breeze. Lower down behind the scaffolding, which he grasps with both arms, is Karl-Friedrich wearing a suit and tie. His knee is bent over one of the diagonal braces. The Farkas brothers, also in suits and ties, are at the bottom, with Adalbert reaching around behind him, while Ladislaus leans outward, holding on with one hand. It is the pose of young men full of purpose and hope; their facial expressions are serious, while their physical positions are jaunty. Their student days are behind them. Work of great importance lies ahead of them. The sun is shining. The photograph was taken around 1930. Almost half of the scientists at the Haber Institute were about to lose their jobs because of severe budget cuts. Support throughout Germany for Hitler was rising steadily. So also was vicious anti-Semitism. The golden years were over. For these young men, they would never come again.

Wilhelm Jost's brief history of physical chemistry in Germany ceases abruptly with the year 1933. Revolted by what he called the deluge that followed what had been a golden era, he was unable to go further with the history, writing at the end, "Unique opportunities for physical and theoretical chemistry lay open and were being pursued. They could be ruined only by a blow from the outside. And that blow fell."[22]

Notes

1. Scott and Moleski, *Polanyi*, 114.
2. Michael Polanyi, *Personal Knowledge: Towards a Post-critical Philosophy* (Chicago: University of Chicago Press, 1958), 120.
3. Paul Ignotus, "The Hungary of Michael Polanyi," *The Logic of Personal Knowledge: Essays Presented to Michael Polanyi on His Seventieth Birthday* (London: Routledge and Kegan Paul, 1961), n.p.
4. Hagittai, *Martians*, 52.
5. Ibid., 115.
6. E.P. Wigner and R.A. Hodgkin, "Michael Polanyi," *Biographical Memoirs of the Royal Fellows of the Royal Society* 23 (November 1, 1977): 424. http://rsbm.royalsocietypublishing.org.
7. Scott and Moleski, *Polanyi*, 121.

8. Bonhoeffer and Polanyi did not travel together all the time while in the USA. They occasionally went their separate ways, although they kept in touch even then. There is no extant evidence whether they both visited Jászi or whether Polanyi went alone. Mary Jo Nye, *Michael Polanyi and His Generation: Origins of the Social Construction of Science* (Chicago: University of Chicago Press, 2011), 280.
9. DBW, Vol. 10, 276.
10. Unfortunately, there are no letters describing the trip to Yellowstone National Park. Letter to Harteck, PHP, Box 1, Folder 1, RPI.
11. *Montreal Gazette*, September 11, 1929, PHP, Box 10, Folder 1, RPI.
12. Ibid., Box 10, Folder 1, n.d., RPI.
13. Nye, *Polanyi*, 264.
14. Hugh Taylor, "Fifty Years of Chemical Kineticists," Woodrow Wilson National Fellowship Foundation, Princeton, New Jersey. *Annual Review of Physical Chemistry* 13 (1962): 11.
15. It is worthy of note that while Michael Polanyi never received a Nobel Prize for chemical kinetics, his son, John Charles Polanyi, did in 1986. He shared the prize with Dudley Herschbach and Yuan T. Lee.
16. Norman N. Greenwood, "Harry Julius Emeléus," *Biographical Memoirs of the Royal Fellows of the Royal Society.* http://www.jstor.org/stable/770201.
17. Email from Roald Hoffmann to author, February 8, 2017. Hoffmann wrote that the situation in the early 1930s was complex: "German physical and theoretical chemistry was being overtaken by other trends. The lack of communication between chemists and physicists in Germany was a real problem; as a result, chemical crystallography and theoretical chemistry developed outside of Germany. I also feel that the German physical chemists had reached a kind of maximal sophistication of macroscopic measurements of electrochemistry, kinetics, and were not developing the microscopic molecular-based tools that became important with time."
18. Scott and Moleski, *Polanyi*, 116.
19. See Nobel Prize Nomination Database, http://www.nobelprize.org/nomination/archives.
20. James, *One Hundred Years*, 73.
21. Grete and Hans were the children of Ernö Dohnanyi, a famous Hungarian pianist and composer, and Elisabeth Kunwald, a pianist. Ernö had abandoned the family during World War I when he had an affair with a German actress, eventually marrying her in 1919. Without support from her philandering husband, Elisabeth Kunwald taught piano to provide for the family. Hans earned his law degree at the

University of Berlin before moving in 1924 to Hamburg where he was employed by the Institute of Foreign Affairs while also earning his doctorate from the University of Hamburg. He and Christel returned to Berlin in 1929 when Hans entered the Ministry of Justice. Sifton and Stern, *No Ordinary Men*, 44–45.

22. Jost, *The First 45 Years*, 14.

CHAPTER 6

Turbulence and Conformity

The depression put the Haber Institute—and everyone in it—under grinding pressure. From 1930 to 1932, the financial support provided by the Kaiser Wilhelm Society plummeted from 342,000 marks to 12,000 marks, forcing staff to be cut from 65 to 42 in 1931.[1] The Kaiser Wilhelm Society in turn tried to convince the German government not to cut subsidies, arguing that scientific development was essential to the nation. Michael Polanyi did his part by publishing an article in the May 1930 issue of *The German Economist*, pleading that government support of science was essential despite the fact that practical benefits might not be immediately obvious.[2] It was all to no avail: Equipment broke and was not replaced. Laboratory supplies ran short. Nor were there funds available for travel to scientific conferences, so when Karl-Friedrich, Polanyi, and Freundlich were invited to present papers on adsorption at the Faraday Society in Oxford, England, in February 1932, they sent their papers in advance but did not attend.[3]

As he had done several times before, Haber turned to the Rockefeller Foundation for help. The foundation funded some postdoctoral positions: in Ernest Rutherford's Cavendish laboratory for Paul Harteck; and in Niels Bohr's laboratory for Hans Kopfermann. It also made individual grants for equipment purchases, one going to Polanyi. However, the foundation was unable to stabilize the Haber Institute financially, the result being the curtailment of research.

K. L. Housley, *The Scientific World of Karl-Friedrich Bonhoeffer*, Palgrave Studies in the History of Science and Technology, https://doi.org/10.1007/978-3-319-95801-9_6

As director of theoretical physics at the University of Göttingen, Max Born remembered receiving the government order that extreme economic measures would have to be taken due to the collapse of the American financial system. Specifically, the universities were to dismiss a large number of the younger assistants and other paid staff:

> Many members of our scientific faculty found this shocking: firstly, because it was cruel to single out the young, struggling ones, many of whom were married, for dismissal, and to deprive them of their already meager incomes; furthermore, it would paralyze the activities of the institutes, practically bringing them to a standstill. We formed a committee and decided to propose to the faculty that we would pay most of the people affected out of our own pockets by offering to make a voluntary contribution which amounted to less than 10 per cent of our salaries. The battles this caused with the faculty still make me shudder. In the course of an interminable meeting we won with a considerable majority. But those who were outvoted displayed an animosity we had never before experienced; amongst them were some historians, but most of them were agriculturalists and forestry people. Six months later we knew what they really were: disguised Nazis, who considered solicitude for the individual just as superfluous as the existence of scientific institutes. The only bright spot occurred when I personally told the Curator of the University, Geheimrat Valentiner, the faculty's decision. He was so moved that tears came to his eyes; he said something like, "if all corporate bodies acted as unselfishly as your faculty, our country would soon be rid of its problems."[4]

In 1930, Karl-Friedrich accepted the position of chairman of the physical chemistry department at the University of Frankfurt am Main. It had been founded only sixteen years before in 1914 on the eve of the First World War, with construction financed largely by donations from prosperous Jewish businessmen and philanthropists in the Frankfurt area. As a result, it had a higher percentage of Jewish professors than other German universities. Consequently, their expulsion in 1933 would seriously damage not only the individuals who were hounded into exile but also the university itself. In 1932, its name was changed to Goethe University in honor of Frankfurt's most famous citizen; even so, the original name remained in use.

The University of Frankfurt was probably not where Karl-Friedrich wanted to be, given that its reputation for scientific research was not yet firmly established. However, with conditions deteriorating rapidly at the Haber Institute and Karl-Friedrich and Grete expecting their first child (Karl Walter Paul Bonhoeffer was born on January 10, 1931), he took what he could get. He wrote to the physicist, Gregor Wentzel, in Zurich that the position at Frankfurt was like having a bird "in hand," instead of a possible position in Zurich that was like a bird "in bush." Paradoxically before the economic situation had worsened, he had been offered positions at Charkow (in the Ukraine), Breslau, and in the USA at Harvard University. He had turned them all down.[5]

Ebullient with the joy of being a new father as well as with the prospect of building up the physical chemistry institute, Karl-Friedrich was initially hopeful about Frankfurt, writing to Dietrich, who was studying for a year in the USA, that as soon as Dietrich returned, he should visit them. "So come have a look at the boy, who it seems to me looks a little like you sometimes. Probably that was the reason he pleased you in the photograph. And have a look at where we are living and what our life is like. If you have a little time, you also have to see the institute which I've built myself in the meantime".[6]

Adalbert Farkas, who was still working on his doctorate, accompanied Karl-Friedrich to Frankfurt while his brother Ladislaus remained in Berlin as Haber's assistant. Paul Harteck also remained in Berlin. Despite the distance, the men continued to work on joint projects, publishing papers together. Karl-Friedrich and Harteck even collaborated on writing a textbook titled *The Fundamentals of Photochemistry*, published in 1933. However, for all of them, the intellectually exhilarating closeness that had existed under the sheltering roof of the Haber Institute was no more. Especially for the Farkas brothers, the situation was becoming desperate, forcing them to seek positions in other countries. Germany had become a cold, dangerous place for them. The very small window of opportunity that had existed for Jewish scientists under the Weimar Republic was rapidly closing. Einstein was one of the few who saw clearly what was happening. He left Germany in 1932 with the intent of returning periodically to Berlin to teach, but once Hitler came to power, Einstein strongly denounced him. For the rest of his life, the USA would be his home (Fig. 6.1).

Yet, at the start of the decade, no one associated with Haber wanted to admit that the golden days of science were over. Even the scientists who were not German remained loyal. Michael Polanyi, who was from

Fig. 6.1 Adolf Hitler in Nuremberg in 1929. United States Holocaust Memorial Museum, courtesy of William O. McWorkman

Hungary with Austrian citizenship, received an excellent offer in 1932 from the University of Manchester in England that he initially turned down even though Haber recommended he accept. Polanyi's reasons were that he felt rooted in Germany and that it would be very difficult to leave the country as it struggled through hard times.[7]

On February 29, 1932, James Franck wrote from the University of Göttingen to Karl-Friedrich, seeking a postdoctoral position for Eugene Rabinowitch, one of his students. Franck and his colleague, the chemist and mineralogist Victor Goldschmidt, believed Rabinowitch was of the highest caliber, but Franck already had four men in habilitation and, therefore, could not take on one more. "I turn to you today for advice and help," Franck wrote in a tone of desperation. "I do not know what to do to promote Dr. Rabinowitch who deserves all my support." He hoped that the research he and Rabinowitch were doing on photochemistry could serve as Rabinowitch's habilitation thesis at the University of Frankfurt, if only Karl-Friedrich could find a place for him there. "On this occasion, I would like to look forward to a vigorous effort to accommodate Rabinowitch somewhere for habilitation according to his abilities where he will not be rejected immediately because of his Russian-Jewish origin."[8] There is no extant response from Karl-Friedrich whose future at the University of Frankfurt was already uncertain. Furthermore, he was faced with the similar problem of finding a position for Adalbert Farkas. All that is known is that Rabinowitch left Germany, temporarily working with Bohr in Copenhagen before taking a position at University College in London. Bohr used his laboratory during this period as a kind of scientific way station for displaced Jewish scientists.[9] Not long after Rabinowitch was there, Franck himself was forced to seek shelter under Bohr's capacious scientific wing, before going to Johns Hopkins in the USA.

In August 1930, Dietrich Bonhoeffer, then twenty-four years old, had sailed from Germany to undertake postdoctoral studies at Union Theological Seminary in New York City. A month later on September 19, he wrote to his brother that he was exhilarated to be "walking here in your footsteps," referring to Karl-Friedrich's trip to the USA the year before.[10] While Dietrich was at Union, the brothers exchanged letters often, and Karl-Friedrich sent Dietrich books he thought would interest him, one being a copy of the novel *Nachkrieg* (After War) by Ludwig Renn. Like Karl-Friedrich, Renn had served in the military, keeping a diary on which he based his novels. "The book *Nachkrieg* is much discussed here," Karl-Friedrich wrote. "To the extent I'm familiar with it, it's not really to my taste. Perhaps to yours? Try it!" Dietrich replied that the book "did greatly interest me; and since by chance I received two copies of it, I passed one along to a professor here," the professor most likely being the theologian Reinhold Niebuhr who was teaching the course *Ethical Viewpoints in Modern Literature*, which Dietrich was taking. In his reading notes on war literature, Dietrich listed Renn's *Nachkrieg*, Erich Maria Remarque's *All Quiet on the Western Front*, Robert Graves's autobiography *Good-Bye to All That*, and Robert C. Sherriff's play *Journey's End*, which had been performed in London in 1928 with Laurence Olivier in the lead. Dietrich was most impressed with *Journey's End*, but he questioned whether by presenting war as literature "in an aesthetic shape," the horrors of war were being idealized and romanticized. "My experience at least in Berlin of talking with people who were not in the war and saw war shows and pictures make this question very serious," Dietrich jotted in his notes.[11]

Another book that the brothers discussed via correspondence was *Recovery of Germany* by the American economist James Angell. On November 30, 1930, Karl-Friedrich wrote that he was reading "a relatively optimistic book about Germany's reconstruction by an American named Angell, professor of economics at Columbia University." He asked Dietrich whether Angell was highly regarded. Dietrich responded that he had made Angell's acquaintance at the time that Hjalmar Schacht, the German economist and banker, was visiting the USA. Schacht had just written *The End of Reparations* in which he was highly critical of the extreme war reparations forced on Germany after the First World War, contending that they continued to ruin the economy. Although Schacht was not a Nazi, he was sympathetic to Hitler, who would appoint him president of the Reichsbank in 1933. By the end of the 1930s, Schacht's sympathy would turn to revulsion, and he would be fired.[12] Dietrich wrote to Karl-Friedrich, "I listened to the two arguing together, and what Schacht said was incredibly depressing and pessimistic, though Prof. Angell really didn't know what to say either."[13] A major problem was that as the depression worsened, it put Germany in an impossible position in terms of its war debt and access to international credit.

During his year in the USA, Dietrich often took a nationalistic approach in the speeches he delivered to church and school groups. He explained the destructiveness of reparations while insisting that it was unjust to hold Germany solely responsible for the First World War. He described the terrible conditions in Berlin during the war and the suffering inflicted on its citizens, but he did not speak of the suffering inflicted by Germany on citizens of other nations. He did, however, state his belief that Germany was guilty of complacency, lack of humility, and lack of faith in God.[14] Calling for international brotherhood, Dietrich was not yet an ecumenist or a pacifist, although his friendship with Jean Lasserre, a French Reformed pastor studying at Union, had begun to bend his thinking in that direction.[15]

In his social relationships with the other seminarians at Union, Dietrich sometimes came across as egotistical, considering himself superior because he had already earned his doctorate at the prestigious University of Berlin. In comparison, he considered theological education in the USA to be lacking in substance as were its practitioners. Dietrich came to regret his attitude, confessing to his friend, Elizabeth Zinn, in a letter dated January 27, 1936, that he had "plunged into my work in a very unchristian way, quite lacking in humility. I was terribly ambitious, as many people noticed, and that made my life difficult and kept me from the love and trust of people around me." What changed Dietrich's life was discovering the Bible. Throughout his childhood, he had certainly read the Bible, and he had studied it at the university, but he had not experienced its personal power. Dietrich admitted to Zinn that until his discovery he had been using Jesus Christ for his own advantage, which he prayed would never happen again.[16]

The place Dietrich heard the Bible preached vibrantly was not from the pulpits of the white churches in New York City. It was from the pulpit of the Abyssinian Baptist Church in Harlem under the charismatic leadership of the black minister Adam Clayton Powell, Sr. He went there with his friend and fellow student Albert F. (Frank) Fisher who was one of only a handful of black students at Union Theological Seminary. Fisher was a graduate of Howard University in Washington, D.C. Little is known about his friendship with Dietrich, but there is no doubt that in introducing him to the Abyssinian Baptist Church, and the teeming streets of Harlem—at that time packed with thousands of new arrivals fleeing from the Jim Crow south in a desperate search for employment, education, and a better life—Fisher began to change Dietrich's understanding of justice and human equality. Certainly, Dietrich's flaxen hair, blue eyes, and white skin made him stand out in the pew on Sunday morning. His singing of the gospel hymns in a heavy German accent would also have set him apart. However, because Fisher was standing beside him making introductions and smoothing his transition, Dietrich was welcomed warmly. Soon, he became so involved (working with the youth and teaching Sunday school) that he no longer felt like a foreigner but a member of the church community.

Dietrich also fell in love with spirituals, writing to Karl-Friedrich that "the spiritual songs of the southern Negroes represent some of the greatest artistic achievements in America." When he returned to Berlin, he had in his trunk several records and the *Book of American Negro Spirituals* compiled by the famous black writer, educator, and diplomat, James Weldon Johnson. Given to Dietrich as a gift by Fisher, it contained 120 songs, including *Deep River* and *Go Down Moses.* When, in 1935, Dietrich established Finkenwalde, the illegal seminary to train pastors who were opposed to the Nazi takeover of the churches, he often played the spirituals for his students, sitting at the grand piano in the music room. He also let the students play his records. Bethge recalled the rooms ringing with spirituals, such as *Swing Low, Sweet Chariot*. The somber beauty of its refrain "comin' for to carry me home" provided an uplifting power in a time of great danger.[17]

Until Dietrich came to know black Americans personally, he had only seen the world through the eyes of a privileged white man. For the first time, he was seeing the world through the eyes of an oppressed people. And that perspective would remain with him, finding expression in his writing over and over, most eloquently in his short essay "After Ten Years" written in late 1942: "It remains an experience of incomparable value that we have for once learned to see the great events of world history from below, from the perspective of the outcasts, the suspects, the maltreated, the powerless, the oppressed and reviled, in short from the perspective of suffering."[18]

At the beginning of December 1930, Dietrich encountered segregation firsthand while on a trip to Washington, D.C. with a white student, Fisher, and another black student. He attended a Federal Council of Churches meeting, toured the Lincoln Memorial and the Washington Monument, and visited Howard University, which had been founded after the Civil War to educate black students who were denied entry to all-white colleges. At that time, Howard had become an important center of the New Negro renaissance, which came to be known as the Harlem Renaissance, as promulgated by Dr. Alain Locke, professor of philosophy, in his book *The New Negro*. One of the Howard students Dietrich met was Lee Cato Phillip, a senior who would attend Union the following year after his graduation. According to Dietrich, while in Washington, D.C., he "was able to become acquainted with all the leading figures of the Negro movement." However, outside the homes of the black students, it was a different matter: "The conditions are really rather unbelievable," Dietrich wrote in dismay to his parents. "Not just separate railway cars, tramways, and buses south of Washington, but also, for example, when I wanted to eat in a small restaurant with a Negro, I was refused service."[19]

After his return to New York City, Dietrich finished the fall semester, and then he and fellow seminarian, Erwin Sutz, who was from Switzerland, spent the Christmas holiday in Cuba. During their stay, Dietrich's awareness of the depravity of racial prejudice became even more pronounced. In a letter to Karl-Friedrich written on shipboard in January 1931 while returning from Havana, Dietrich expressed his outrage at the treatment of blacks in the USA as compared to what he had observed in Cuba, finding it shameful and repugnant, especially in the south. For him, segregation and racial prejudice gave the lie to the oft-repeated American claim of "freedom and justice for all." He was also disturbed by the blindness of white Americans to the injustice and cruelty they were perpetrating. "It is a bit unnerving that in a country with so inordinately many slogans about brotherhood, peace, and so on, such things still continue completely uncorrected," he charged.[20]

This experience prepared Dietrich—in a way that few people in Germany were prepared—for the malignant idea that there could be a pure *master* race: meaning *white* in the USA, and *Aryan* in Germany. In the USA, a person was white if he or she had no black blood whatsoever; if a person were only one-eighth black, he or she was a non-white octoroon, the pejorative word used in the south. To Adolf Hitler, any Jewish blood, even in grandparents who had converted to Christianity, was enough to brand a person as a mixed-breed, half-Jew, or mongrel.[21]

Dietrich also followed the Scottsboro case closely in which, in a gross miscarriage of justice, nine black men were wrongly charged with the rape of two white women. He even asked a leading church official in Germany to join his voice with many others around the world in protest, but the official refused to do so, saying it was not his responsibility, much to Dietrich's disappointment. Under Hitler, not only Jews but also communists and social democrats faced trumped-up charges in a rigged judicial system. For the Scottsboro men, the guilty verdicts led to the penitentiary; for the Germans, they led to the concentration camp.[22]

Upon his return to Germany, it would take Dietrich considerable time to assimilate all that he had seen and heard in the USA. It was so utterly different from everything he knew and everything that was happening in Europe, upending much of the theology he had imbibed as a student at the University of Berlin, pitting the possibility of a singular religious truth (the one true way that made all other ways wrong) against the messy religious pluralism mixed with inequity he encountered in America. To his professors, Reinhold Niebuhr and John Baillie, it seemed that the onslaught of information forced Dietrich to withdraw for a while into his orthodox European shell, only to emerge later as "a very astute political analyst."[23]

Helping Dietrich to reach a new understanding of the word of God, centering it in the pain and injustice of the world, was his ongoing correspondence with Karl-Friedrich, who was willing to share his own concerns about racism. Karl-Friedrich replied to Dietrich on January 24,

1931, drawing on his experience as well as on the sharp observational skills of his traveling companion, Michael Polanyi, who had kept notes during their trip across the USA on everything from wages to the behavior of automobile drivers:

> I'm glad to hear you have the opportunity for studying the Negro problem so thoroughly. When I was there I had the impression it was really *the* question, at least for people with a conscience, and when I received the offer of the appointment at Harvard it was one of the primary reasons for my disinclination to move there completely, since I wanted neither to come into this inheritance myself not to pass it on to my hypothetical children. I really can't see how it can be corrected, and I think that in this case as in mathematics there really are some insoluble questions. (A bit stupid! I notice it myself.) In any event, our "Jewish question" is a joke compared to it; only a few would still claim they are repressed here. At least not in Frankfurt.[24]

It is significant that racism was the main reason Karl-Friedrich turned down the position in physical chemistry at Harvard. He wanted nothing to do with it, and looking ahead to the future, he wanted his children to have nothing to do with it either. In hindsight, Karl-Friedrich would have rued his use of the word *joke* in the above quotation to refer to the Jewish question in Germany. However, there was no way for him to foresee what was about to happen. Furthermore, the University of Frankfurt was considered among the most liberal schools in Germany. Compared to the University of Tübingen that had been strongly anti-Semitic for decades, the University of Frankfurt was a place where Jewish scholars could thrive, among them the theologian, Martin Buber. Besides his brother-in-law, Gert Leibholz, a respected law professor, Karl-Friedrich's circle of colleagues and close friends included many scientists of Jewish lineage, first and foremost being Haber himself, then Polanyi, the Farkas brothers, and Freundlich.

In the USA in 1929, Karl-Friedrich had been troubled by the prejudice he had seen against blacks on the basis of their skin color, but he had not seen the prejudice against Jews because it was less obvious. As in Europe, few American Christians saw anything wrong with Jewish quotas, which were so common in higher education that many outstanding colleges, including Yeshiva University in New York City and eventually Brandeis in Waltham, Massachusetts, were established to accommodate qualified Jewish students who could not gain entry elsewhere. Although Dietrich and Karl-Friedrich would not have known it, Columbia University had strict quotas.[25] Often enforced covertly, quotas also were widespread in business and industry and in numerous social institutions. It was lack of awareness that led Karl-Friedrich to consider the race question more serious than the religion question.

Within the Bonhoeffer family, there was a tendency in 1930–32 to downplay the increase in violent anti-Semitism and the surging enthusiasm for Nazism. Under Hitler's leadership, there had been an insidious change in anti-Semitism: A Jew was a Jew even if he or she were culturally assimilated as well as a Christian, Fritz Haber, Michael Polanyi, and Gert Leibholz being examples. It was race—blood—that mattered. Unaware of this shift in how a Jew was defined and what it portended, Paula Bonhoeffer wrote to Dietrich on February 23, 1931, about Leibholz: "For his own part, Gert fears the Nazis as far as his job is concerned: I think that's exaggerated, but Sabine is also tormenting herself because of it."[26] Unfortunately, their fear was not exaggerated. Yet Sabine and Gert put off making a decision whether to emigrate until almost too late, forcing themselves to believe that Nazism was just a short-term aberration that would pass and everything would return to normal.

It was Klaus, who had recently married Max Delbrück's sister, Emmi, who felt it was necessary for Dietrich to comprehend how serious the situation in Germany had become. Klaus was considered the most pessimistic, even moody, of the Bonhoeffer siblings, but he was also clear-eyed and had an international perspective because he was a corporate lawyer for Lufthansa. On November 2, 1930, he wrote to Dietrich in New York City: "Since your departure the political situation has changed greatly. The success of Nazism has convinced the widest circle that the democratic regime has failed in the past ten years. The consequences of the world economic crisis are explained in purely domestic terms. People are flirting with fascism. If this radical wave captures the educated classes, I am afraid it will be all over for this nation of poets and thinkers."[27] Perhaps because of his position with Lufthansa, Klaus and Emmi had a broad circle of friends, not all of the same political persuasion, giving them a unique perspective. Among their friends were Arvid Harnack, whom they had known since childhood, and his American wife, Mildred, who were academics, literati, and committed leftists.[28]

In his biography of Dietrich Bonhoeffer, Eberhard Bethge alluded to the muted response of Dietrich's father, Karl, to entrenched anti-Semitism. His response was in contrast to the strong stand that his neighbors, Adolf von Harnack (Arvid's uncle) and Hans Delbrück, had taken many years before against Adolf Stoecker, the court chaplain to Kaiser Wilhelm, who had pushed for the imposition of Jewish quotas:

> Both [Harnack and Delbrück] had been much more active and risked more politically than Karl Bonhoeffer. Years before they had fought together to keep Adolf Stoecker's anti-Semitic activity in check, and successfully blocked his attempt to discuss the Jewish question at the Protestant Social Congress, forcing him to leave the meeting. Likewise, during the 1920s they watched the political course taken by the Protestant church with concern, for royalism and anti-Semitism had not died with the disappearance of the monarch.[29]

One of the reasons why Delbrück and Harnark were more outspoken than Karl Bonhoeffer had to do with the nature of their professions, history and theology, respectively, whereas Karl was a physician who held sacrosanct the privacy of the doctor–patient relationship.

Delbrück died in 1929; Harnack died the following year in 1930 as he was preparing to open the annual meeting of the Kaiser Wilhelm Society, of which he had served as president since its founding in 1911. Dietrich Bonhoeffer delivered one of the eulogies at his memorial service, speaking on behalf of Harnack's former students who had learned that "truth is born only of freedom" by watching Harnack speak the truth when the majority of people were too fearful to do so. "Because we knew that with him we were in good and solicitous hands, we saw him as the bulwark against all trivialization and stagnation, against all the fossilization of intellectual life."[30]

Bethge believed that the deaths of Delbrück and Harnack "epitomized the demise of a great liberal and humanitarian tradition," leaving Germany dangerously vulnerable to extremism and silent compliance.[31] Max Delbrück's recollections of his father and uncle (Delbrück's mother and Harnack's wife were sisters) align with those of Bethge. Both men were in the public eye. However, according to Max, his father "was often very critical of what went on and several times he had run-ins with the imperial government." He never tried to remove himself from the fray; for example, in 1919 at the age of seventy, he was a member of the German delegation at Versailles, arguing against the language in the peace treaty that held Germany solely responsible for the outbreak of the First World War. For his part, Harnack was as politically perceptive as Delbrück, but more diplomatic. "He never expressed himself in public so as to incur the wrath, at least not the way my father did," said Max Delbrück. Nonetheless, the two men were in constant discussion on current issues, sharing ideas. Every Sunday evening after dinner, they would meet to talk.[32] Then, every Wednesday, they would get together with intellectuals who often had clashing points of view, among them the theologian and historian Ernst Troeltsch, the historian Friedrich Meinecke, the sociologist Heinrich Herkner, and the politician Theodor Heuss.[33]

Even though he participated in the discussions, Karl's work as a psychiatrist at the Charité Clinic did not dispose him to taking public positions on political issues, which left him unprepared for Hitler's rise to power. In his memoir written after the war, Karl wrote:

> From the start we regarded the victory of National Socialism in 1933 and Hitler's appointment as Reich chancellor as a misfortune—the entire family agreed on this. In my own case, I disliked and distrusted Hitler because of his demagogic propaganda speeches, his telegram of sympathy on the occasion of the Potempa murder, his journeys by car through the country, horsewhip in hand, his choice of colleagues, about whose qualities we here in Berlin may have had a more detailed knowledge than those elsewhere, and finally because of what I heard from my professional colleagues regarding his psychopathic symptoms.[34]

The telegram to which Karl referred was sent by Hitler to five Nazis who had beaten a communist to death in the Silesian village of Potempa. Pledging the murderers his "unconditional fidelity," Hitler pardoned them as soon as he took power.

As Karl noted, Hitler had "psychopathic symptoms," but it is of critical importance to recognize that he was also an astute politician who knew how to tap into the deep fears of many Germans and to play on those fears. Having served in the army during the First World War, Hitler himself felt the sharp stab of wounded pride that Germany had been treated unjustly at Versailles, carrying all the guilt for the First World War. He stridently proclaimed to enrapt crowds that France remained an implacable enemy. Building on this paranoia, he then summoned up past glory going back to Norse mythology, promising a return to that glory, while firing up a storm of hatred for the people he was convinced had caused the nation to fall from grace, first and foremost the Jews and then the communists. His speeches had a quasi-religious tone through which burned threats of intimidation and terror. He and his well-chosen lieutenants used drama in a way that was riveting, converting die-hard skeptics from all levels of society. Many industrialists were very supportive because of the promised improvements in the economy, accompanied by the suppression of trade unions. Also supportive were many academics, one of the most notorious being the philosopher Martin Heidegger at the University of Freiburg who harangued his students that Nazism was exactly what Germany needed, insisting they give up their belief in free speech and join the party, as he himself did in 1933.

The physicist Carl Friedrich von Weizsäcker, a close colleague of Werner Heisenberg and an acquaintance of Karl-Friedrich, recounted to Fritz Stern many years after the war about the attraction that Nazism initially held for him. "It is true that I could not develop any interest in the Nazis' ideology either before or after their assumption of power. Nonetheless, I was very much tempted after 1933 to join the movement in some way or another." He attributed his feelings to an "elemental reaction" to what he called a "pseudo-outpouring of the Holy Spirit in 1933." For the rest of his life, Weizsäcker struggled to comprehend his compulsion and the dark paths it led him to take.[35]

Hitler knew exactly how to maximize the compulsion that Weizsäcker and many others felt. For example, from his early days in Munich, Hitler adored the spectacle of the torchlight parade. There was something atavistic about fire in the darkness and thousands of people marching through the streets with their blood inflamed. On the night of January 30, 1933, following the naming of Hitler as chancellor by President Paul von Hindenburg, a massive parade was held in Berlin with tens of thousands of storm troopers holding high their torches, the light flickering on their unsmiling faces as glowing sparks drifted in the wind (Fig. 6.2).

Fire also marked the next step in Hitler's progression to absolute power. On February 27, the Reichstag burned to the ground. To this day, there is controversy as to who was behind it, although Marinus van der Lubbe, a young, mentally ill communist from the Netherlands, was found guilty of the crime and beheaded. Whether or not he was a dupe of the Nazis, there is no doubt that Hitler seized on the conflagration as the reason to convince Hindenburg, then 85 years old and feeble, to sign emergency decrees that established the

Fig. 6.2 Marinus van der Lubbe on trial for Reichstag fire. United States Holocaust Memorial Museum, courtesy of National Archives and Records Administration, College Park

constitutional basis for dictatorship. On March 21, the Malicious Practices Act went into effect, making it illegal to say anything against the Reich. Two days later the Enabling Act, euphemistically titled The Law to Remove the Crisis of the People and the State, suspended the constitution. Free press, free speech, and the rights of opposition parties—all were swept away in an instant. Hitler became the only law. Declaring the Reichstag fire to have been caused by communists, Hitler had many of them imprisoned in hastily built concentration camps from which most never emerged. He then let it be widely known that anyone who spoke out against him and the glory of the Third Reich would join them there. He made good on his threat. From 1933 to 1945, approximately 800,000 Germans (non-Jewish) were imprisoned for showing some kind of resistance to Hitler.

The list of punishable offenses was very long. A person could be beaten up or jailed (often listed as taken into "protective custody") for accepting an anti-Nazi leaflet, for slipping a piece of bread to a Jew, for failing to report suspicious activity of a family member or friend, or for listening to a BBC radio broadcast. If there was no proof, it could be fabricated.[36]

On April 7, 1933, barely a month after Hitler's rise to power, the Aryan clause further transmogrified Germany. As part of the mildly titled The Law for the Restoration of the German Civil Service, it required that only non-Jews (Aryans) be employed in civil service, which encompassed all jobs in government, academia, and the state-supported church. The upheaval was made worse by its quick adoption by many private companies throughout the Reich. Hitler was not yet ready to enforce the Final Solution, murdering Jews by the millions in the death camps. His goal was simpler: to ruin them economically, terrorize them physically, deny them all human rights, and drive them out of Germany. So-called half-Jews were able to hold on longer than full-Jews, losing what little protection they had in 1935 with the enforcement of the Nuremberg laws.

The Aryan clause forced non-Jews to make an immediate ethical choice: They could support it; openly oppose it; or remain silent and do nothing, even though they disagreed. Too often, even when non-Jews *did* express their opposition to the expulsions, they were more concerned with the deleterious effects on their institutions and their fields of study than with the chaos in the lives of their expelled colleagues and friends. From then on, having compromised their consciences once, it became easier to do it twice, three times, as more draconian rules were imposed, until their compromises were no different than outright support.

The historian Klaus Hentschel in his book *Physics and National Socialism: An Anthology of Primary Sources* makes the strong point, based on scrupulous research, that among physicists only around twenty-four were flagrantly pro-Nazi. "It is nevertheless an interesting facet of the social dynamics within the National Socialist system that de facto collaboration with the Nazi regime was not only tolerated but in fact widely practiced among German physicists, while explicit ideological commitment to the Nazi phraseology was repudiated and invariably led to isolation within the physics community." From a pure science standpoint, they were uneasy with the Nazi ideology of race-based science, such as the claim that Aryan physics was true and Jewish physics was false.

Unable to give their outright support, they nevertheless conformed with the "good intention of attempting to prevent worse from happening, of avoiding having less-qualified persons assume important positions in science and administration." Hentschel concludes that the upshot was the transformation of "the vast majority of scientists and engineers into easy targets and convenient servants of a criminal state."[37]

In the Protestant churches, pastors with Jewish ancestors or married to Jewish spouses were to be banned from the pulpit. Their faith did not matter; their culture did not matter; only their "blood" mattered. Dietrich's close friend, Franz Hildebrandt, a Lutheran pastor whose mother came from a Jewish family, reluctantly went into exile in England following his release from prison. Deeply troubled, Dietrich's choice was to speak out boldly in opposition, initially focusing his efforts on blocking the adoption of the clause by the churches. Just a week after it went into effect, he wrote to Erwin Sutz, his good friend from his year at Union Theological Seminary, about the Aryan clause: "The Jewish question is also giving the church a great deal of trouble, and here the most intelligent people have totally lost their heads and their Bible."[38] Over the next few years, he often quoted Proverbs 31:8, "Speak out for those who cannot speak," asserting that this was the very least that the church was asked to do.[39] To Dietrich's dismay, the majority of non-Jews in academia and the churches chose compliance, among them his father and his brother, Karl-Friedrich. Both men hated what was going on and attempted in various ways to soften its depredations. They voiced their disapproval in private letters and behind closed doors, but they rarely spoke out. The main reason is undoubtedly linked to their idealistic belief that scientists should remain detached from politics. Another reason may have been that as civil servants, their jobs and pensions were at stake, to which Hitler added the threat of imprisonment. Scientists, along with everyone else in the Reich, were to "march in step" (*Gleichschaltung*), if not willingly, then by force.

During the 1930s, some scientists resisted, including the physicist and Quaker Herbert Jehle, a friend of both Karl-Friedrich and Dietrich, who turned down an academic position in Berlin, an action inspired by Dietrich whom he had met in England while doing postdoctoral research. In 1940, Jehle refused military service and was sent to a concentration camp in France where he almost died of starvation before making it to the USA with the intervention of Quakers, in particular, the British astrophysicist Sir Arthur Eddington, who had himself been a conscientious objector during the First World War.[40] Other non-Jewish scientists resigned and sought jobs outside Germany, including the physicist Paul Ewald, who gave up his position as rector at the Stuttgart Polytechnic, eventually going to Cambridge. Unfortunately, they were the exception rather than the rule. Most non-Jewish scientists were willing to "march in step," even if some sullenly dragged their feet.

By and large, Aryan physicists and chemists worried more about the damage done to German science, writ large, than the damage done to individual Jewish scientists deprived of their jobs and their rights. This was the argument used by Max Planck in his meeting with Hitler on May 16, 1933, following the enforcement of the Aryan clause. Planck was 75 years old. Respected worldwide, he was regarded as a kind of ethical grandfather of German science. Yet in the meeting, Planck tried to convince Hitler to stop the expulsion of Jewish scientists because it would do great damage to the institution of German science, while benefiting foreign science. He went so far as to claim "that forcing worthy Jews to emigrate would be equivalent to mutilating ourselves outright." Planck never mentioned the damage it would do to individual humans. That Planck had the audacity to challenge him made Hitler fly into a towering rage. Never again did Planck confront him. On the contrary, he counseled other scientists that it would do no good to protest; they should stay in their positions and attempt to save the citadel of pure science from within by being compliant—even as their Jewish colleagues were thrown from the parapets. At the meeting of the Kaiser Wilhelm Society on May 23, 1933, just a week after his meeting with Hitler, Planck read the following message sent by the society to Hitler: "The Kaiser Wilhelm Society for the Advancement of the Sciences begs leave to tender reverential greetings to the Chancellor, and its solemn pledge that German science is also ready to cooperate joyously in the reconstruction of the new national state."[41]

Even so, Planck's outward show of support for Hitler was tepid. Paul Ewald recalled seeing him perform the Hitler salute: "Planck stood on the rostrum and lifted his hand half high and let it sink again. He did it a second time. Then finally the hand came up, and he said 'Heil Hitler.'"[42]

Einstein wrote to his friend, Paul Ehrenfest, that "the lack of courage on the part of the educated class in Germany had been catastrophic." In April 1933, Einstein's bank account was seized, his house ransacked, and his portrait widely circulated on which appeared the words "not yet hanged."[43] Aspersions were even cast on the theory of relativity, with Philipp Lenard, himself a Nobel laureate, charging that it was "mathematically botched-up" because it was the work of a Jew who was not a good German. Then, on May 10 in yet another example of Hitler's propagandistic use of fire, 40,000 people in Berlin cheered as Einstein's writings were burned in a great conflagration along with books by many other authors who were anathema to Nazism, among them Thomas Mann, Sigmund Freud, and Erich Maria Remarque, the author of the anti-war novel *All Quiet on the Western Front*. A jubilant Joseph Goebbels, minister of propaganda, proclaimed: "These flames not only illuminate the final end of an old era; they also light up the new."[44] As a copy of a play by Heinrich Heine caught fire, so too did the words within: "Where they burn books, they will ultimately also burn people."[45]

Despite all that had happened, many non-Jewish German scientists sharply criticized Einstein for fomenting discord. An oft-quoted example is Max von Laue's letter to Einstein on May 14, 1933, in which he chastised him for speaking out politically. The ashes were not yet cold from the book-burning, yet here was one of the leading physicists in Germany insisting that the two worlds of politics and scientific research were separate and should remain so. Von Laue believed scientists should be "reserved," by which he meant silent.[46] To his credit, he soon realized that silence in the face of oppression was not the answer. Just four months after his letter to Einstein, on September 18, 1933, he gave a speech at a major physics conference, obliquely comparing the attitude of the Reich toward physics and Einstein to the Inquisition's attitude toward heliocentrism and Galileo. Von Laue stated that despite religious and political forces, Galileo believed that facts would ultimately prevail. He ended his speech with Galileo's apocryphal words: "And yet, it moves!" Von Laue did not mention Einstein by name, but everyone in the audience knew to whom he was referring.[47]

In November 1932, two months before Hitler became chancellor, Michael Polanyi had invited several prominent scientists and mathematicians at various institutes and universities to a dinner party in his Berlin home, among them Karl-Friedrich, Haber, Paul Schrödinger, Fritz Herzog, and Gustav Hertz. He made the serious suggestion to them that they should all resign together if Hitler came to power.[48] When none of the non-Aryan scientists resigned, including Karl-Friedrich who remained at the University of Frankfurt, Polanyi was astonished.[49] In his *Reminiscences*, the physicist Leo Szilard, himself a Jew, wrote of Polanyi's reaction:

> They all thought that civilized Germans would not stand for anything really rough happening. The reason that I took the opposite position was based on observations of rather small and insignificant things. What I noticed was that the Germans always took a utilitarian point of view. They asked, 'Well suppose I would oppose this, what good would I do? I wouldn't do very much good, I would just lose my influence. Then why should I oppose it?' You see, the moral point of view was completely absent, or very weak. . . . And on this basis did I reach the conclusion in 1931 that Hitler would get into power not because the forces of Nazi revolution were so strong, but rather because I thought there would be no resistance whatsoever.[50]

Polanyi was not the only person to suggest that everyone resign together. The chemist Otto Hahn had also thought about doing so but was restrained by Planck who was concerned that if everyone resigned, second-rate Nazi scientists would take their places. A similar call for mass resignation had gone out at Göttingen University, but according to Paul Rosbaud, who was there at the time, as soon as the Jewish professors were fired, all the other faculty members "remained sitting and had forgotten their intention to rise and protest."[51]

Polanyi urged Haber to take a stand against the Aryan clause, holding up as an example their former colleague and friend, James Franck, who had immediately resigned in public protest as chair of physics at the University of Göttingen even though he was not in danger of losing his job at that time because of an exemption in the clause for war veterans. When the local pro-Nazi newspaper in Göttingen found out what Franck had done, it published a denunciation letter from forty-three university lecturers who charged that Franck's resignation was "tantamount to an act of sabotage."[52] Apparently, the sabotage lay in his taking the initiative, denying the Nazis the chance to torment him into resigning. Despite Franck's example, Haber did not act immediately because he was conflicted as to what was the right course of action, not only for himself but also for his institute and staff. He was furious because he considered himself a war hero to whom Germany owed much. Furthermore, he was a convert to Christianity. Even his failed efforts to extract gold from seawater during the 1920s were on behalf of Germany to enable it to pay off its war debts. To his friend, Richard Willstätter, Haber wrote, "I am bitter as never before. I was German to the extent that I feel fully only now and I find it odious in the extreme that I can no longer work enough to begin confidently a new post in a different country."[53]

Haber tendered his resignation on April 30, 1933, to go into effect in six months in October. The reason for the delay was that Haber realized the far-reaching implications of what was happening. Approximately forty percent of his institute staff were Jewish. To fire them on the basis of race and religion was unconscionable. In his retirement request, Haber stated he had always sought scientists solely on the basis of character and qualification. He refused to do otherwise. He wanted to stay until October so that he could use his international reputation to find some of them positions outside of Germany. Furthermore, he hoped to be able to exert influence on the selection of his successor. Under the circumstances, Haber recommended that his institute be combined with the Kaiser Wilhelm Institute of Physics with the two directors being Karl-Friedrich Bonhoeffer and Max von Laue, both of whom he trusted implicitly. At the Kaiser Wilhelm Society, Otto Hahn was in favor of Haber's plan, but he was overruled by Gerhart Jander, who had been appointed as acting director by the Nazis.[54] It was this battle over Haber's successor to which Paula Bonhoeffer alluded in a letter to Dietrich: "Karl-Friedrich's chances of moving here [Berlin] have come to naught, since the position for which he was being considered will not be in his field at all, but in another altogether."[55] Subsequently, Jander redirected the focus of the institute to armaments research, which is what Haber had feared would happen and which he was trying to prevent.

Not long after Haber's resignation, Karl-Friedrich wrote to him expressing his dismay and gratitude: "but at least I want to assure you – what I hope you would know of me in any case – that I will always proudly and gratefully acknowledge that you have been my teacher and that I shall tell the future generation of your achievement on behalf of German science and technology."[56] Karl-Friedrich appreciated Haber's attempt to make him head of the institute, having decided already that he had to leave the University of Frankfurt.

So also, Haber's efforts on behalf of his fired staff were not very successful. Of his directors, Herbert Freundlich went to the Netherlands, then to England, and finally to the USA. Rudolf Ladenburg had left in 1932 for Princeton. Polanyi accepted the position at the University of Manchester that he had initially turned down. However, the rest of Haber's staff struggled to find positions. On April 19, 1933, Adalbert Farkas wrote to Karl-Friedrich from Berlin. "You see how difficult the situation here is. I fear that it will become extremely difficult for us to find something and there seems to be no option for me other than to seize some kind of practical profession. Of course I will not decide on anything until we talk with the Geheimrat [Haber] about my brother's future." Ladislaus and Adalbert first went to Cambridge on temporary assignments, publishing some papers together. Then, without prospects for permanent positions in Britain or the USA, they went to Palestine at the recommendation of Haber where Ladislaus established a physical chemistry institute at the Hebrew University and Adalbert taught courses.[57]

At a dinner at the home of Polanyi, before Polanyi resigned his position and went to England, Erwin Schrödinger had pointed to the tablecloth and declared, "This white tablecloth is beautiful but were I to throw a bottle of ink over it, what would it be like? To my mind that is what Hitler has done." Though not Jewish, Schrödinger chose to get out of Germany. In 1935, in response to a quantum mechanical paradox put forth by Einstein, Boris Podolsky, and Nathan Rosen (known as EPR), Schrödinger proposed a thought experiment. It involved a cat sealed in a box with a radioactive source, a Geiger counter, and hydrogen cyanide that would be released into the chamber if the counter indicated the decay of a single atom. Was the cat alive or dead? Or was the cat alive *and* dead? It seemed to depend on an observer. Although the thought experiment was a response to the EPR paradox, there is something chilling about it in terms of what was happening in the Third Reich. To describe the strange state that was quantum mechanically analogous to being alive and/or dead, Schrödinger used the word Verschränkung—entanglement.[58]

Notes

1. James, et al., *One Hundred Years*, 59.
2. Mary Jo Nye, "Historical Sources of Science-as-Social Practice: Michael Polanyi's Berlin," *Historical Studies in the Physical and Biological Sciences* (University of California Press, 2007), 423.
3. Karl-Friedrich was at the University of Frankfurt at the time. Scott and Moleski, *Polanyi*, 126.
4. Max Born, *The Born-Einstein Letters*, trans. Irene Born (New York: Walker and Company, 1971), 111.
5. Letter to Wentzel, Max Planck Gesellschaft Archives, 14.12.29. Also listed in Einstein Archives Online, number 87–205. http://alberteinstein.info/vufind1/Record/EAR000069124.

6. Karl-Friedrich's letter is dated June 25, 1931. Dietrich spent two days in Frankfurt in July 1931 after visiting Karl Barth and attending lectures. DBW, Vol. 11, 30–31, 33.
7. Nye, *Polanyi*, 70. See endnote 13.
8. Letter from Franck to Bonhoeffer, February 29, 1932, Abtielung Rep 23, Personal Correspondence, Nachlass Karl Friedrich Bonhoeffer, Archiv der Max-Planck-Gesellschaft.
9. Rabinowitch came to the USA in 1938, teaching at M.I.T. He then joined James Franck at the University of Chicago. en.wikipedia.org/wiki/Eugene_Rabinowitch.
10. DBW, Vol. 10, 243.
11. Ibid., 293, 420.
12. At the Nuremberg trials, Hjalmar Schacht was tried, but not convicted. See en.wikipedia.org/wiki/Hjalmar_Schacht.
13. DBW, Vol. 10, 269–270.
14. Schlingensiepen, *Dietrich Bonhoeffer*, 70.
15. Ibid.
16. William D. Apel, "Bonhoeffer and Merton on Obedience," in *Bonhoeffer for a New Day*, ed. John W. De Gruchy (Grand Rapids: Eerdmans, 1997), 160.
17. DBW, Vol. 10, 269. Bethge, *Bonhoeffer*, 427.
18. DBW, Vol. 8, 52.
19. Lee Cato was dean of the chapel at Prairie View A & M University in Prairie View, Texas, for more than 40 years. Locke's *The New Negro* was published in 1925. DBW, Vols. 10, 258; 17, 69.
20. Ibid., 269.
21. Terms such as *half-Jew* were defined under the Nuremberg Laws.
22. Bonhoeffer mentioned this case and the ethical failings of the unnamed official in *Ethics*. DBW, Vol. 6, 295.
23. Bethge, *Bonhoeffer*, 165. See footnote 66.
24. Karl-Friedrich's mention of insoluble questions in mathematics is a reference to the work of Kurt Gödel. Ibid., 276.
25. In the mid-1930s, Columbia University turned down Richard Feynman because he was culturally Jewish. Instead, he went to M.I.T. but later had trouble gaining entry to Princeton for his doctorate for the same reason. He won the Nobel Prize in Physics in 1965. See en.wikipedia.org/wiki/Richard_Feynman.
26. Ibid., 289–290.
27. DBW, Vol. 11, 4.
28. Arvid was the nephew of Adolf von Harnack. He was executed in December 1942, and Mildred was executed in February 1943. The relationship of communists to the resistance is very complex because

of ideologically charged perceptions both before and after the war. For information on Arvid and Mildred Harnack, see Shareen Blair Brysac, *Resisting Hitler: Mildred Harnack and the Red Orchestra* (New York: Oxford University Press, 2000). See also Anne Nelson, *Red Orchestra* (New York: Random House, 2009).

29. Bethge, *Bonhoeffer*, 126.
30. Ibid., 139.
31. Ibid., 126.
32. Delbrück interview, 3.
33. Meinecke was anti-Semitic and anti-Slavic, supporting Hitler after he came to power although he became disillusioned. Heuss became president of the Federal Republic of Germany (West Germany) in 1949. Brysac, *Resisting Hitler*, 79; see endnote 15, 406.
34. Leibholz-Bonhoeffer, *Bonhoeffers*, 93. See Sifton and Stern, *No Ordinary Men*, 17. See also Bethge, *Bonhoeffer*, 258.
35. Stern, *Dreams and Delusions*, 174.
36. Brysac states that the figure undoubtedly includes some Jews and disaffected Nazis. She quotes former German Chancellor Willy Brandt who defined a resister as a person who "takes serious risks to act illegally in a world where right has become wrong." Brysac, *Resisting Hitler*, 223; see endnote 2, 424.
37. Klaus Hentschel, ed., *Physics and National Socialism: An Anthology of Primary Sources* (Basel: Birkhäuser Verlag, 1996), lxxvi–lxxvii.
38. DBW, Vol. 11, 101.
39. DBW, Vol. 13, 217.
40. DBW, Vol. 11, 585. Bert Schroer, "Physicists in Times of War" (Berlin: Institute for Theoretical Physics, December 2005), 2.
41. Planck showed how deeply entrenched anti-Semitism was by trying to make the case with Hitler that there were "worthy" and "worthless" Jews, to which Hitler replied that "a Jew is a Jew." Hentschel, *Physics and National Socialism*, 359–360. Clark, *Einstein*, 470.
42. J. L. Heilbron, *The Dilemmas of an Upright Man: Max Planck as Spokesman for German Science* (Berkeley: University of California Press, 1986), 164.
43. Ibid., 471.
44. William L. Shirer, *The Rise and Fall of the Third Reich: A History of Nazi Germany* (New York: Simon and Schuster, 1960), 241.
45. DBW, Vol. 11, 4.
46. Clark, *Einstein*, 417.
47. Hentschel, *Physics and National Socialism*, 67–71.
48. Scott and Moleski, *Polanyi*, 134–135.

49. Friedrich, the second son of Karl-Friedrich and Grete, was baptized in October 1932, with Dietrich preaching. The fact that their family was growing may have influenced Karl-Friedrich's decision not to resign. DBW, Vol. 17, 74.
50. Nye, *Polanyi*, 75–76. See reference to Leo Szilard, *Reminiscences* (Boston: Harvard University Press, 1969), 95–96.
51. Heilbron, *Dilemmas*, 150. Paul Lawrence Rose, *Heisenberg and the Nazi Atomic Bomb Project* (Berkeley: University of California Press, 1998), 250.
52. Frank von Hippel, "James Franck: Science and Conscience," *Physics Today* American Institute of Physics (June 2010): 44.
53. Stern, *Dreams and Delusions*, 73.
54. James, *One Hundred Years*, 94.
55. DBW, Vol. 13, 61.
56. Stern, *Dreams and Delusions*, 73.
57. Michael Chayut, *From Berlin to Jerusalem*, 246.
58. Schrödinger's response to Nazism was not consistent, which he later regretted. In 1940, he took a position in Dublin, Ireland. Wigner, "Michael Polanyi," 416.

CHAPTER 7

Seizing the Wheel

At the University of Frankfurt, the terror of anti-Semitism began earlier than in Berlin, even before the enforcement of the Aryan clause. It had a particularly brutal aspect because the Nazis believed that to exert control in Frankfurt—with its high number of Jewish students and professors—would force all the other universities to comply. Of Jewish descent, Peter F. Drucker, the Austrian economist, was a lecturer in international and public law at Frankfurt. His description of the obligatory faculty meeting led by the new Nazi leader is chillingly illuminating, particularly in regard to a question asked by a professor of biochemistry. First, the Nazi leader declared that Jews were dismissed without salary and would not be permitted on the university premises:

> Then he launched into a tirade of abuse, filth, and four-letter words such as had rarely been heard even in the barracks and never before in academia . . . Next the new boss pointed his finger at one department chairman after another and said, "You either do what I tell you or we'll put you into a concentration camp." There was dead silence when he finished; everybody waited for the distinguished biochemist. The great liberal got up, cleared his throat, and said, "Very interesting, Mr. Commissar, and in some respects very illuminating. But one point I didn't get too clearly. Will there be more money for research in physiology?"[1]

K. L. Housley, *The Scientific World of Karl-Friedrich Bonhoeffer*, Palgrave Studies in the History of Science and Technology, https://doi.org/10.1007/978-3-319-95801-9_7

The Nazi leader assured the professor that there would be plenty of money for racially pure science, meaning that which upheld the superiority of Aryans and the inferiority of Jews, Slavs, and others. Then, with brown-shirted storm troopers outside, the meeting ended. Drucker was appalled to see that only a few professors had the courage to walk out together with their Jewish colleagues. Within forty-eight hours, Drucker left Germany.

There can be no doubt that Karl-Friedrich was at that meeting. However, he was not the biochemist to whom Drucker referred. In an undated letter written about that time to Harteck who was still at the Cavendish laboratory in England, Karl-Friedrich mentioned his concerns about politics and his inability to forget what was happening by burying himself in his work. He wrote of an article in the publication *German Economist* that stated that connections with the outside world would be forbidden after April 10. He also mentioned there was to be a university meeting the next day to which he planned to go and listen. He concluded, "I would not have thought that the anti-Semitism here would take on such vulgar forms [hurtful shape], even though I myself have often gotten annoyed about the Jews. These methods are somehow shameful for us. What will Polanyi do?"[2] Karl-Friedrich also expressed his anxiety that Adalbert Farkas might not be able to habilitate in Germany; furthermore, the prospects of Farkas finding a position in America were poor.

Despite the fact that Karl-Friedrich expressed his dismay in the private letter to Harteck, there is no evidence that he protested publicly. Immediately following the meeting, twenty-three professors at Frankfurt lost their jobs, among them the Jewish theologian, Martin Buber, who eventually settled in Palestine, and the two philosopher-sociologists, Theodor Adorno and Max Horkheimer, who ended up in the USA.[3] The mathematician Max Dehn hung on to his job until 1935 and then stayed in Germany until 1939 at which point he fled to the USA by way of Denmark, Norway, and the USSR via the trans-Siberian Railroad to Japan, and then across the Pacific, finally settling down in Idaho. Paul Tillich also lost his position because he spoke out against Hitler and spoke up for his Jewish colleagues. Tellingly, it was Dietrich, not Karl-Friedrich, who wrote to Erich Seeberg, a theology professor at University of Berlin under whom Dietrich had studied, to implore Seeberg and other professors to provide written testimonial about Tillich's scholarly significance so that he would not be driven out.[4]

Nothing came of Dietrich's efforts and Tillich was forced from Frankfurt, going to Union Theological Seminary.

The terror at the University of Frankfurt continued unabated. On May 10, in synchrony with the huge book-burning in Berlin, "un-German" books were loaded in manure carts pulled by oxen, dumped in Frankfurt's public square and ignited. Then, on May 23, Ernst Krieck, a staunch Nazi who had been educated only to the level of an elementary school teacher, became the new rector. At his investiture, he spoke in glowing terms of the renascence that was happening in German universities in which the humanistic ideal was being replaced by the national and political. "Nowadays the task of the universities is not to cultivate objective science but soldier-like militant science, and their foremost task is to form the will and character of their students."[5]

Not wanting to stay at Frankfurt, Karl-Friedrich immediately began looking for another position because to him "soldier-like militant science" was a travesty. He sought the chair of physical chemistry at the University of Leipzig that had been left vacant with the retirement of Max Le Blanc. The Physical Chemistry Institute had been founded by Wilhelm Ostwald who had won the Nobel Prize in 1909 for work on catalysis, chemical equilibria, and reaction velocities. These were the same areas in which Karl-Friedrich excelled, making him well qualified for the position. However, when he refused to swear the loyalty oath to Hitler, the process of his appointment became drawn out and difficult.[6] Finally, on October 12, 1934, Karl-Friedrich prevailed. One of his new colleagues at the University of Leipzig was Werner Heisenberg, head of the department of physics, whose Nobel Prize the year before had been awarded partially on the basis of Karl-Friedrich's corroborating work on hydrogen.

As the decade progressed, there was a marked change in Karl-Friedrich's involvement in anti-Nazis activities, including the harboring of half-Jews in his institute. The change was brought about in part by the influence of family members including his brother-in-law, Hans Dohnanyi, who, as a lawyer in a position of authority with the Reich court, began to keep a detailed record of Nazi crimes called "the chronicle of shame." Hans shared his horrific findings with his family and like-minded friends, opening their eyes and eventually enlisting many of them into the resistance.[7] Karl-Friedrich was also influenced by Dietrich's increasingly strong stand against Nazism. Then, there was the example set by Julie Bonhoeffer, Karl-Friedrich's grandmother, who had a brave and independent personality. "There is such an unusual vitality in her," Dietrich remarked.[8] Living most of her life in Tübingen, she had come to spend her last years with Karl and Paula, her son and daughter-in-law, in Berlin. On April 1, 1933, Hitler had called for a boycott of all Jewish businesses. Julie, at the age of 91, chose to pointedly ignore the boycott by walking straight through a belligerent cordon of storm troopers blocking entry to a store owned by a Jewish family. When she died two years later on January 16, 1936, Dietrich gave the homily at her funeral, recalling that incident and adding that his grandmother could not bear to see another person's rights violated and that her final years were clouded by the great suffering of the Jews.[9]

On April 3, 1933, a few days before the Aryan clause went into effect, Paula wrote to her daughter and son-in-law, Christel and Hans Dohnanyi, expressing her concern about the regression from intellectualism to primitivism that was occurring in Germany and her apprehension about Dietrich with whom she was in agreement as to the role of the church in opposition to the Reich: "But we need to think about holding onto humanism, as Christ's most urgent command, in response to the question 'Who is my neighbor?' I do worry a lot about Dietrich. It's time for the church to show that it is still a force in our lives, and fills a need in our lives! It's really making it hard for Dietrich."[10]

Around the same time, Paul Lehmann and his wife were in Germany on a visit and were invited to the Bonhoeffers for dinner. Because he was fluent in German, Lehmann had become close friends with Dietrich at Union Theological Seminary. That evening, Lehmann was surprised at the tension in the Bonhoeffer household because of the recent events. He recalled that Klaus would occasionally get up "in the course of conversations that, admittedly were not wholly innocuous," to make sure that none of the servants were listening behind the doors. Already there was anxiety that someone would report them to the Nazi authorities.[11]

According to Emmi Delbrück Bonhoeffer, it was during this period that Klaus found an anonymous note on his desk at Lufthansa, warning him to break off relations with two Jewish associates. "My husband rolled up the note and lit his cigarette with it."[12]

During the spring of 1933, even Dietrich was in turmoil as to what to do. On February 1, 1933, he delivered a radio speech titled, "The Younger Generation's Altered View of the Concept of Führer," in which he spoke at length on narcissism and the transmogrification of the idea of leadership. At one point, he asked the prescient question:

"To what extent is leading and being led healthy and genuine, and when does it become pathological and excessive?"[13] However, when Gerhard Leibholz's father died on April 11, 1933, Dietrich declined to conduct the funeral on the advice of his church general superintendent because Leibholz was a Jew, a decision Dietrich came to regret. Later that year, he wrote to Sabine and Gerhard from England where he had taken on the pastorate of two small churches, St. Paul's and Sydenham, whose members were Germans working in the London area:

> I am tormented even now by the thought that I didn't do as you asked me as a matter of course. To be frank, I can't think what made me behave as I did. How could I have been so horribly afraid at the time? It must have seemed equally incomprehensible to you both, and yet you said nothing. But it preys on my mind, because it's the kind of thing one can never make up for. So all I can do is ask you to forgive my weakness then. I know now for certain that I ought to have behaved differently.[14]

Also indicating Dietrich's confusion was his essay published in June 1933, titled "The Church and the Jewish Question," which made clear that he had two chief concerns during the early days of Hitler's rule. The first was the effect of the Aryan clause on Christians who had converted from Judaism, not the effect on Jews. The second was the forceful intrusion of the state into the affairs of the church. Attempting to mediate a position, he quoted Martin Luther's statement that "The Jews are the most miserable people on earth. They are plagued everywhere, and scattered about all countries, having no certain resting place." The prevailing belief among Christians was that Jews would have a "homecoming" with God when they accepted Jesus. It must be pointed out that the paragraph on the curse on the Jews is not in the three extant drafts of Dietrich's essay; it appears only in the published version. There is the possibility that either he or an editor added it under pressure from church leadership. However, in the essay, Dietrich pushed far beyond the theology of the Jewish curse. In discussing the church's role in the face of the threat to law and order, Dietrich stated there were three possibilities: The first was to question the state as to the legitimacy of its actions, making it responsible for what it did. The second was service to the victims of the state's actions. Quoting the Apostle Paul in Galatians 6:10, "Let us work for the good of all," Dietrich wrote: "The church has an unconditional obligation toward victims of any societal order, even if they do not belong to the Christian community." It was the last role that was the most radical, revealing the direction in which Dietrich was headed: "The third possibility is not just to bind up the wounds of the victims beneath the wheel but to seize the wheel itself. Such an action would be direct political action on the part of the church."[15]

Dietrich had gone to England in October 1933 in the desperate hope that physical distance from Germany, as well as "unobtrusively working as a pastor," would help him sort out his ideas that were putting him at odds not only with church leadership but also with some of his friends. In a letter to the theologian Karl Barth written after Dietrich had arrived in London (so as not to be influenced by him), he explained his reasons: "I felt that, in some way I don't understand, I found myself in radical opposition to all my friends; I was becoming increasingly isolated with my views of the matter, even though I was and remain personally close to these people. All this frightened me and shook my confidence so that I began to fear that dogmatism might be leading me astray . . ."[16] Barth's response was double-barreled, stating in the strongest terms that Dietrich should not have retreated to England: "And now that you have come to me with this after the fact, I truly cannot do otherwise than call to you—Get back to your post in Berlin straightaway!" Toward the end of the letter, Barth wrote in fury that "your church's house is on fire," and because Dietrich had the capability to help, he should return to his post by the next ship.[17]

Dietrich sent Barth's letter to his father, who replied that it showed a fiery spirit, but that it was his opinion that, "It is also valuable to be able to look at things from the outside whence, perhaps some influence can be brought to bear, and particularly to save oneself until the right moment."[18] Balancing Barth's forceful directive against his father's more measured advice, Dietrich chose to stay in England for a year and a half. It was not to be a time of quiet retreat; instead, it was seminal, one reason being that Jewish refugees were streaming into London, many finding their way to the doors of his two churches. Some of the exiled scientists, fearful and without money, had been directed there by his brother, Karl-Friedrich, in hopes that Dietrich could assist them in finding jobs, or at least food and shelter.[19] Dietrich did his best to help. In the process, he began to question his narrow, church-focused response to the Aryan clause. Barth had accused him of fleeing when the church's house was on fire, but Dietrich was now aware that Barth had it wrong—the house on fire was all of Germany, and in that house were Jews.

Widespread spying and the planting of fear went hand in hand with Hitler's assumption of power. The universities were swarming with informants ready to report anyone who said anything against Hitler or who defended a Jew. At the Charité Clinic, Karl discovered that Nazis, often young medical trainees, reported what the doctors said and did. "Professional relationships between the doctors belonging to the hospital, hitherto undisturbed, were to a large extent destroyed by mutual mistrust," he wrote. Although the entire staff was under pressure to join the party, Karl believed that most of the assistants withstood the pressure and that the hospital was a "thorn in the flesh" to the leadership of the university teachers. Karl even refused to hang a portrait of Hitler in the clinic. He stopped attending official functions after hearing a speech by Bernhard Rust, Hitler's minister of science education and culture. "Considering Rust's offensive attitude toward the professoriat, it was unfortunate that neither I nor other professors had the courage to leave the auditorium."[20]

Karl made a vain attempt to keep at least five of his highly respected Jewish associates on the staff at the Charité Clinic: Franz Kramer, Hans Pollnow, Paul Jossmann, Herta Seidemann, and Erwin Straus. Both Kramer and Pollnow were specialists in child psychopathology. In 1932, they had published a groundbreaking study on what they called a hyperkinetic disease in children that included inattention and poor impulse control. Called the Kramer–Pollnow syndrome, it is now identified as attention deficit hyperactivity disorder (ADHD). Kramer's association with Karl went back to their years in Breslau. In 1902, Kramer had received his doctorate under Carl Wernicke at the University of Breslau, and then in 1907, he had received his habilitation under Karl Bonhoeffer, who had replaced Wernicke. Totally trusted by Karl, Kramer went with him to Berlin in 1912, remaining one of his closest colleagues for over thirty years. Somehow Karl managed to keep Kramer on staff for two years after the enforcement of the Aryan clause. But in 1935, Kramer had only his struggling private practice to keep him going. He tried desperately to get a position in Turkey or the USA, petitioning Adolf Meyer, head of psychiatry at Johns Hopkins, for help, but nothing came of it. Unfortunately, it was more difficult for a Jewish research psychiatrist or neurologist to find an academic position outside Germany than it was for a physicist or chemist. In 1938, at the age of sixty with his license to practice medicine stripped from him by the Nazis, Kramer fled to the Netherlands with Karl's help.[21]

Pollnow was one of the youngest of Karl's colleagues, being only thirty-one-years old when he lost his position. Prior to earning his medical degree, he had received his doctorate in philosophy, making him one of the best educated men at the clinic. Because Pollnow's prospects were high for having a brilliant career in pediatric psychology, Karl had made a last-ditch effort to place him in what was dubbed a "vacant supplemental assistant's position," but when that fell through, all Karl could do was to write glowing letters of recommendation for him. In May 1933, Pollnow went to France with the letters in his pocket. To the best of his ability, he continued his research, although he was not allowed to practice medicine because he did not have a French license. Following Kristallnacht in November 1938, Pollnow turned his attention to psychological trauma suffered by Jewish children who had fled across the border into France without their parents. During the war, he was arrested by the Gestapo, sent to a concentration camp, and executed.[22]

Of those Karl tried to save, Erwin Straus (who was already well known for his work on phenomenology), Paul Jossmann, and Herta Seidemann managed to escape from Germany to the USA. Seidemann was working on her habilitation on brain pathology and neurological syndromes (dyslexia) under Karl, whom she admired greatly. Then in 1933, she was denied permission to finish her habilitation by the Nazis. She went to Switzerland carrying recommendations from Karl, but then returned to Berlin when an extension on her Swiss work permit was denied. With further assistance from Karl, who may have hidden her briefly, she made it out of Germany in 1938, sailing to the USA on the *Queen Mary*. Karl's recommendations helped her get a position at the Montefiore Hospital in New York City. For a while, she collaborated with Karen Horney, who had received her doctorate under Karl in 1914. Seidermann corresponded with Karl up until his death.[23]

Karl also interceded for his assistant Fred Quadfasel. According to Norman Geschwind (a medical colleague of Quadfasel's in the USA), Quadfasel was a Protestant from an East Prussian family who spoke out against Hitler, was arrested, and had his passport confiscated. Subsequently, Karl visited Quadfasel in prison and told him about the arrest for political reasons of Bonhoeffer ancestors in 1848, the year of revolution in Germany, reassuring him that it was no disgrace to be in prison. Karl then arranged for a relative who was in the diplomatic corps to intercede on Quadfasel's behalf to get his passport returned and to provide diplomatic immunity so he could get out of the country. It was just one more example of how Karl handled the problems at the Charité Clinic with "Apollonian equanimity" and "sovereign calm." When an assistant told him that he hated the Nazis, Karl replied, "A psychiatrist does not hate, he understands."[24]

In regard to his patients, Karl chose a course of action dictated by pure science combined with a sense of medical ethics that mingled responsibility to the state with responsibility to the individual. Within three months of the enforcement of the Aryan clause, Karl's cautious support of eugenics going back to the First World War led to his involvement in sterilization of the mentally ill. This practice increased enormously beginning in the summer of 1933 with the issuance of the Law for the Prevention of Children with Hereditary Diseases. Manipulating the language to transmogrify an evil into a good, the Nazis substituted the word *freedom* for the word *prevention*, renaming it the Law for the Freedom from Hereditary Diseases (Fig. 7.1). Karl was not fooled. He attempted to

Fig. 7.1 Nazi propaganda poster on forced sterilization. United States Holocaust Memorial Museum, courtesy of G. Howard Tellier

work within the system to prevent the law's misuse. Via his connection with the Kaiser Wilhelm Institute of Anthropology, Human Heredity, and Eugenics, Karl called for strict medical control over diagnostic decisions, insisting that there had to be an incontrovertible genetic basis. This was in line with his research going back more than twenty years on exogenous causes of mental illness. Also consistent across his entire career was his demand that diagnostic decisions be made by trained psychiatrists. "The danger of mistaken judgments by inadequately trained doctors was great," he wrote, "the more so because the only qualification taken into consideration in choosing a specialist was being a Nazi."[25]

Moreover, Karl was appalled that the new law made sterilization compulsory, removing the requirement set in place during the Weimar era that patients give voluntary consent. "Any withdrawal of the compulsion clause was inconceivable, given the National Socialist mentality," Karl wrote. "And thus the only possibility remaining open was to restrict its application, and, through publications and instructional procedures in psychiatry, to draw attention to the special difficulties of diagnosis in dealing with this area of 'freedom from hereditary disease.'" Karl even taught a course at the University of Berlin on sterilization law, a course that was eventually banned.[26]

However, Karl's involvement with the genetic courts as a medical assessor and scientific expert, which he himself took on of his own volition, did not appreciably affect the rate of sterilization. On the contrary, it provided a gloss of medical acceptability, in the long run making it easier for Hitler to proceed, especially after Karl's full retirement in 1938 and the appointment of the notorious Nazi doctor, Maximilian de Crinis.[27] From then on, sterilization of mentally ill people was replaced by outright murder—for which the term *euthanasia* is obscenely inadequate. In 1939, Karl came to realize what was happening and was appalled, choosing to work with Dietrich to prevent mass murder at the Bethel and the Lobetal institutes by providing non-genetic diagnoses that could be written on patient questionnaires. However, he never spoke out openly against sterilization and euthanasia. Had he done so when the laws first went into effect, it might have made a difference given his prestige.[28] Approximately 400,000 sterilizations were carried out on patients who were categorized as "useless eaters," including those with retardation, schizophrenia, epilepsy, dementia, and several other conditions. Beginning in 1939, approximately 275,000 were murdered by gassing, lethal injection, and starvation.[29]

In his article *Neuroscience in Nazi Europe: Resistance Against the Third Reich*, Lawrence A. Zeidman states that the overwhelming majority of neuroscientists either cooperated or collaborated, some in heinous ways. The neuroscientists who resisted fell into two categories: those who openly resisted and those who were ambivalent. Among those who resisted were: Oskar and Cécile Vogt who criticized the Nazi regime, lost their positions at the Kaiser Wilhelm Institute for Brain Research in Berlin, and hid Jewish friends; Alexander Mitscherlich who was jailed for anti-Nazi statements and who, following the war, was an observer at the Nuremberg trials of Nazi doctors; Haakon Saethre, a Norwegian neuropsychiatrist who hid Jews and was murdered by the Nazis at the end of the war; Walther Spielmeyer who helped Jewish neuroscientists find jobs outside of Germany; Jules Tinel, a Frenchman who was involved with the resistance by saving downed Allied pilots and whose son died in a concentration camp; and Johannes Pompe, a Dutch pathologist who was executed in April 1945. In the ambivalent category, Zeidman lists Karl Bonhoeffer, Hans Creutzfeldt (Karl's long-standing colleague at the Charité Clinic), Max Nonne who had been involved during the First World War on the treatment of shell-shocked soldiers, and Oswald Bumke, who was at the University of Munich. All of them were opposed to sterilization on the basis of genetic inheritance that could not be proven. They worked to save many patients by providing different diagnoses, for example, listing schizophrenic patients as neurotic, thereby keeping them from being sent to the "death asylums." Zeidman concludes that these men "were able to help some but not all of their colleagues and their patients, while affording some level of protection to themselves both personally and professionally." Karl Bonhoeffer's own words affirm this: "One of the points envisaged in the law was that the doctor was bound to notify the existence of any hereditary disease. With regard to this I was unable to consent to any setting aside of the doctor's duty to preserve professional secrecy, and in fact never notified the existence of any illness discovered in my consulting room to the medical authorities." Finding that Karl Bonhoeffer created an anti-Nazi atmosphere at the Charité Clinic that was "decent, professional, balanced," Zeidman concluded:

> For Bumke, Bonhoeffer, and Creutzfeldt, making false diagnoses and preventing patient transfer were limited tools for resistance and protest

> and could only be undertaken as exceptions. Otherwise, suspicion would be aroused and limited ward capacities of the university clinics would become overcrowded. They probably knew that if they resisted only some of the time it would more likely go unnoticed by the Nazi regime. They were pragmatic and did what was feasible under the circumstances to provide for and protect their patients or colleagues, while at the same time not exposing themselves to undue personal or professional risk.[30]

Just as Dietrich was clear in his opposition to the Aryan clause, although his initial focus was on its application to people who had converted to Christianity, so also was he clear in his opposition to sterilization. In August 1933, a month after the sterilization law had gone into effect, he spent time in Bethel in Westphalia meeting with Bishop Friedrich von Bodelschwingh to formulate a confession of faith embodying the church's position on the Aryan clause. Established by Bodelschwingh's father, Bethel was an institution that treated and housed epileptics, one of the groups of disabled people Hitler targeted for sterilization. On August 20, 1933, Dietrich wrote to his grandmother, Julie Bonhoeffer, to wish her a happy birthday. Having fulfilled that nicety in the letter's opening, he immediately launched into a description of the environment at Bethel and the deep impression it had made on him. "Here we have a part of the church that still knows what the church can be about and what it cannot be about. I have just come back from the worship service. It is an extraordinary sight, the whole church filled with crowds of epileptics and other ill persons, interspersed with the deaconesses and deacons who are there to help in case one of them falls." He told his grandmother that the patients were keen listeners and participants and that they were truly defenseless, which endowed them with clearer insight into the realities of human existence. Then, he wrote with prescient compassion: "What utter madness when some people today think that the sick can or ought to be legally eliminated. . . . What we see as 'sick' is actually healthier, in essential aspects of life and of insight, than health is."[31]

Karl's pragmatic caution was also evident when he and his colleague, Dr. Jürg Zutt, were called upon to conduct the psychological examination of Marinus van der Lubbe, the young man accused of burning down the Reichstag. In his memoir, Karl wrote that he was troubled at the trial by "the Nazi circus" in the face of which he staunchly maintained his medical equanimity. During the trial, Karl met several party members: "The types of faces one saw on this occasion were for the most part very unpleasant." He said that Hermann Goering was in a "state of uncontrollable fury" at the co-defendant Georgi Dimitroff, who was a communist. "Lubbe was a case of a young man who, though psychopathic, was humanly speaking not altogether unlikeable, a muddle-headed adventurer who, in the course of the proceedings, fell into a defiant stupor from which he only emerged shortly before his execution."[32]

Many people both inside and outside Germany believed that Lubbe was an unwitting tool and that Goering and his henchmen had been responsible for burning down the Reichstag. According to Bethge, there were high expectations that Karl Bonhoeffer would provide proof and high disappointment when he did not:

> With his reputation for common sense and incorruptibility, the eminent psychiatrist was expected to give the conclusive evidence that had not been forthcoming elsewhere. Karl Bonhoeffer's expert opinion was awaited with eager anticipation, and the disappointment was extreme when his medical and psychiatric findings omitted any mention of guilt or innocence, or any possibility that van der Lubbe had been incited by others or that drugs had been administered to the prisoner during the period of observation. Indignant letters, and even a scurrilous leaflet, were sent to the Bonhoeffers' home.[33]

For Karl's oldest son, Karl-Friedrich, pragmatism was also the order of the day as evidenced in his assistance to Paul Harteck. Still in England at the Cavendish laboratory, Harteck was in a quandary as to what to do about his future. Once his Rockefeller Foundation grant ran out, he was supposed to return to the Haber Institute, but Jander, the acting director, had refused him permission. Harteck wrote to Karl-Friedrich asking to whom he should appeal, or whether he should call his "idiot cousin" in Austria, who was a Nazi, to seek help.[34] On November 10, 1933, Karl-Friedrich wrote to Planck on Harteck's behalf to see if he would intervene: "As a scientific member of the Kaiser Wilhelm Institute for Physical Chemistry, I have an interest in ensuring that the name of the institution will not be completely discredited by such incomprehensible actions. Mr. Harteck is now certainly the best and most deserving young experimental physical chemist we have in Germany and meets all the requirements which are placed on ethnicity and nationalist sentiment today."[35] The last sentence is an oblique reference to Harteck's being non-Jewish and sufficiently in support of Nazism, although he was never a member of the party. An indication of Harteck's alignment with the "nationalist sentiment" occurs in a letter he wrote to Karl-Friedrich in May 1933 in which he was callous about what was happening to the Jews: "In London the Jews, the one-half and one-quarter Jews of Germany, are gathering. If these people have ever had a liking for Germany, it can only have been a very superficial one, because now you really don't notice anything of it."[36] Apparently, Harteck returned to Berlin to meet with Planck, Otto Hahn, and Lisa Meitner, but nothing came of their efforts on his behalf.

Several months before the controversy with Jander, Karl-Friedrich had written to Harteck on July 7, 1933, that the climate of slander, rumors, and suspicion had gotten so bad that even his wife, Grete, was suspected of being Jewish.[37] He told Harteck that both Otto Stern (a relative of Fritz Stern) and Franz Simon had decided to leave Germany, which brought the total number of vacancies in physical chemistry to six. This opened up employment possibilities for Harteck. "You must see that you get one," Karl-Friedrich wrote, recommending that Harteck should talk with his Austrian relative to ask if he could write in support.[38] Subsequently, in 1934, Harteck (who did not become a Nazi) took the position left empty by Stern as professor of physical chemistry at the University of Hamburg. For Karl-Friedrich, it was a win-lose outcome: Harteck was back in Germany with a good job, but Stern—whom Karl-Friedrich considered to be "one of our best men" as well as a relative of the Breslau family that he considered lifelong friends—was gone. Stern left for the USA where he became a professor at the Carnegie Institute of Technology in Pittsburgh, winning the Nobel Prize in Physics in 1943.

The historian Ute Deichmann makes the point in the essay *The Expulsion of German-Jewish Chemists and Biochemists and their correspondence with Colleagues in Germany after 1945: The Impossibility of Normalization?* that Harteck's experience was typical of many young non-Jewish scientists who suddenly had a chance at gaining better positions at the expense of Jewish academicians. "Against the background of unemployment among academics, the fact that many non-Jewish German scientists benefited from the expulsion of their Jewish colleagues was one of the most important reasons for the ready acceptance of the Nazi policy of expulsion at universities."[39]

Another explanation is presented by the historian David C. Cassidy in his article, "Heisenberg, German Science and the Third Reich." Instead of decrying the expulsion of Jews, Heisenberg and his Aryan associates tried to save their science, toward which they took an idealistic pure science stance, by seeking Aryan physicists who were working in other countries to return to Germany to fill the vacated positions. "Despite the obvious ethical problems with such a strategy, to say little of the support it ultimately lent to the regime, Heisenberg and colleagues pursued it with grim determination," Cassidy wrote. "The survival of even a portion of German physics, they seemed to believe, was more important than either outright opposition to the regime or outright support of their victimized friends and colleagues."[40] This was the same position Planck had taken. In fact, at one point he himself advised Heisenberg not to resist.

There is yet a third interpretation. Max Delbrück did not see the situation as one of careerism and capitulation. In 1933, he was still struggling to set his scientific direction in a Nazi world that required young scientists to attend indoctrination camps and swear an oath of loyalty to Hitler to be considered for academic positions. For the rest of his life, Delbrück felt guilty about coming to the USA on a Rockefeller Foundation grant in 1937, never living in Germany again. Instead, he became a professor at Caltech and at Vanderbilt University in Nashville, Tennessee, winning the Nobel Prize in Physiology or Medicine in 1969. In an interview forty-one years after his arrival in the USA, Delbrück responded to a question about the guilt of German scientists who remained behind: "If anybody feels guilt, I feel guilty for not having stayed, because I had so many friends who I admire for having stayed and having tried to save what was to save—rescue it across this disaster." One of the friends he mentioned was Karl-Friedrich (Fig. 7.2).[41]

Opportunism, compliance, stealthy resistance, idealism, guilt, anger—they were all entangled in the rationale of non-Jewish scientists for staying, even taking better positions vacated by Jews. No amount of historical and psychological analysis is sufficient to make total sense of the mentality and brutality that prevailed in Germany. From the distance provided by time and space, it is easy to blame; it is hard to understand. It is Fritz Stern's opinion that some scientists who stayed experienced an "inner emigration," which was a state of retreat into an endangered inner sphere, filled with ambiguity and contradiction. "One was neither

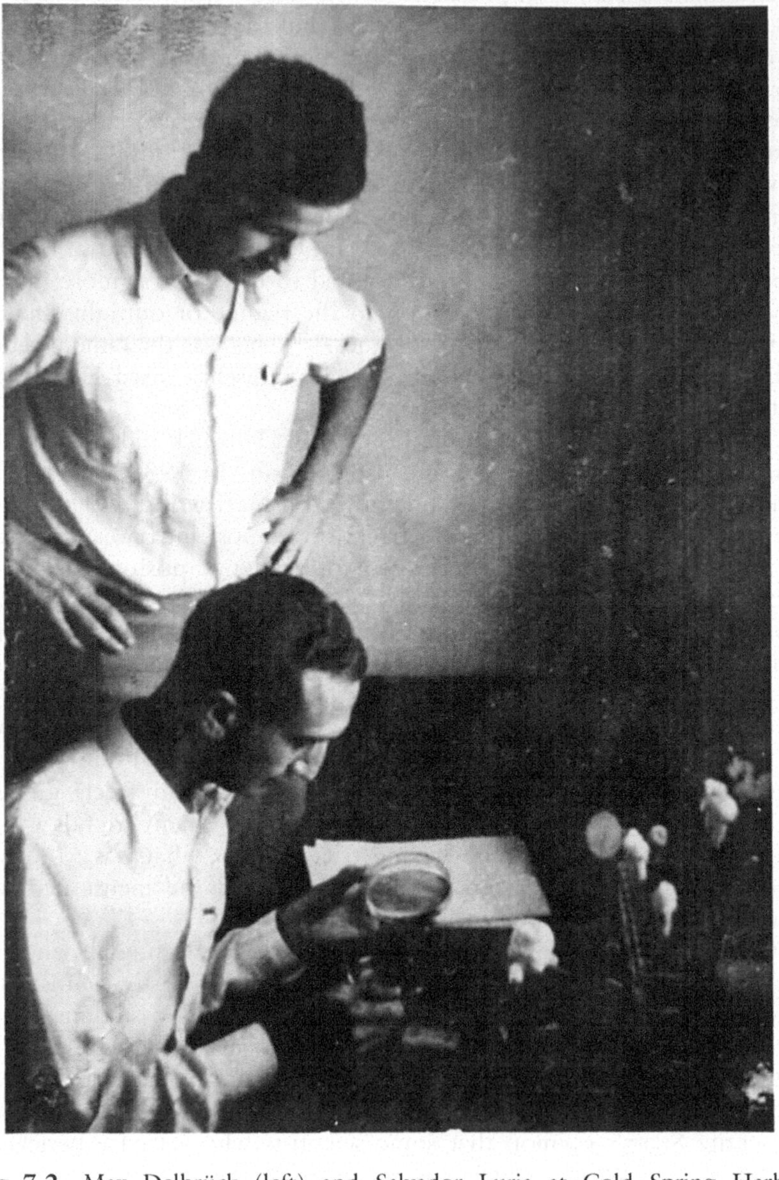

Fig. 7.2 Max Delbrück (left) and Salvador Luria at Cold Spring Harbor Laboratory in 1941. World History Archive/Alamy Stock Photo

an opponent of the regime—or one was so only in secret—nor was one a supporter." Stern, whose family was forced to leave Germany in 1938, then wrote, "Yet here too we should muster a certain empathy: A refusal to say yes was no small thing."[42]

Notes

1. Peter F. Drucker, *Adventures of a Bystander* (New Brunswick: Transaction Publishers, 1997), 161–162.
2. Bonhoeffer to Harteck, undated. PHP, Box 1, Folder 4, RPI. See also Nye, *Polanyi*, 75, endnote.
3. Adorno first went to England and Horkheimer to Switzerland before coming to USA. In 1949, both returned to the University of Frankfurt.
4. DBW, Vol. 11, 104.
5. Clark, *Einstein*, 470.
6. Ulf Messow, "Karl Friedrich Bonhoeffer," *Saxon Biography* (Institute for Saxon History and Folklore, April 8, 2014), 2. http://www.isgv.de/saebi.
7. Bethge, *Bonhoeffer*, 935–936.
8. DBW, Vol. 11, 78.
9. DBW, Vol. 15, 910.
10. DBW, Vol. 11, 101. See footnote.
11. Bethge, *Bonhoeffer*, 264.
12. Victoria Barnett, *For the Soul of the People: Protestant Protest Against Hitler* (Oxford: Oxford University Press, 1998), 180.
13. Dietrich's speech was cut off at the end although the reason may have been that he had run out of allotted radio time. A printed version was then released. DBW, Vol. 12, 267.
14. DBW, Vol. 13, 42.
15. The essay "The Church and the Jewish Question," is very important. However, there are many questions about it. In a footnote to the essay, the editors of *Dietrich Bonhoeffer Works* wrote: "At present it is impossible to determine the editorial history of the published essay, particularly whether changes were suggested by Bonhoeffer or someone else; given the extensive attention this essay receives, however, this would be a worthwhile research project." There is also some confusion about the phrase "dem Rad selbst in die Speichen zu fallen," frequently translated as "to fall within the spokes of the wheel." No matter how it is translated, the point is to totally stop an unjust state. DBW, Vol. 12, 361–370.
16. DBW, Vol. 13, 21.
17. Ibid., 39, 41.

18. Bethge, *Bonhoeffer*, 326.
19. On January 13, 1934, Dietrich wrote to Karl-Friedrich that "a Dr. Steiner was here, whom you directed to me." DBW, Vol. 13, 81. Steiner was a physicist. A letter from Steiner to Karl-Friedrich, dated December 12, 1935, is in the Max Planck Gesellschaft Archives.
20. Leibholz-Bonhoeffer, *Bonhoeffers*, 93–95.
21. In a postcard, dated December 23, 1938, sent to Dietrich from Haarlem, Paula Bonhoeffer wrote that Karl was in Ultrecht because of Kramer. DBW, Vol. 15. Kramer survived the war in the Netherlands; after the war, he continued to practice as a psychiatrist and neurologist in the Netherlands. In Germany, the Kramer-Pollnow Prize is awarded for biological research in pediatric psychology. See Klaus-Jürgen Neumärker, "The Kramer-Pollnow Syndrome: A Contribution on the Life and Work of Franz Kramer and Hans Pollnow," *History of Psychiatry* (London: Sage, 2005), 435–451.
22. See the Charité Portal that lists Jewish physicians and scientists who lost their positions, gedenkort.charite.de/en/people/hans_pollnow.
23. After coming to the USA, Straus continued his work on phenomenology at Black Mountain College and other places, struggling to get his ideas accepted by American psychologists. Jossmann worked at the Veterans Administration Clinic in Boston. He edited a special edition of a German psychiatry and neurology publication to commemorate Karl Bonhoeffer's eightieth birthday. He also nominated Bonhoeffer as an honorary member of the American Psychiatric Association. Arthur Kronfeld was another colleague whom Karl Bonhoeffer may have helped. He had habilitated under Bonhoeffer. He and his wife committed suicide in Moscow on October 15, 1941, as German troops were advancing on the city. For Kronfeld and Seidemann, see Herrn, gedenkort.charite.de/en/people.
24. For his essay "The Work and Influence of Wernicke," Geschwind had conversations with Quadfasel and Paul Jossmann about their mentor Karl Bonhoeffer, one of Wernicke's most renowned students. Robert S. Cohen and Marx W. Wartofsky, eds., *Proceedings of the Boston Colloquium for the Philosophy of Science 1966–1968* (Dordrecht, Holland: D. Reidel Publishing, 1969), 11.
25. Ibid., 94.
26. Leibholz-Bonhoeffer, *Bonhoeffers*, 95.
27. Karl reached mandatory retirement age in 1936, but he stayed on, celebrating his twenty-fifth anniversary in 1937 and giving his valedictory speech in 1938. His successor, Maximilian De Crinis, wrote the euthanasia decree that Hitler signed in 1939. Bethge, *Bonhoeffer*, 563.
28. Uwe Gerrens analyzed the records of the Reich Ministry of Justice and came to the conclusion that the moderating influence claimed by

Karl Bonhoeffer could not be proven. Uwe Gerrens, *Medizinisches Ethos und theologische Ethik: Karl und Dietrich Bonhoeffer in der Auseinandersetzung um Zwangssterilisation und 'Euthanasie' in Nationalsozialismus* (Munich: R. Oldenbourg, 1996.) See also Hans-Walter Schmuhl, *The Kaiser Wilhelm Institute for Anthropology, Human Heredity and Eugenics* (Berlin: Springer, 2008), 221–222. There is also the claim made by some that Karl Bonhoeffer was a war criminal, the evidence being a letter on the sterilization of Gottfried Hirschberg written by Bonhoeffer in December 1941; but even there, he made the point that schizophrenia was not genetic, which may have been an attempt to save Hirschberg from being murdered.

29. Zeidman, *Neuroscience*, 826.
30. Ibid., 826–838. Also see Appendix II for further information on neurologists discussed by Zeidman.
31. DBW, Vol. 11, 157–159.
32. Leibholz-Bonhoeffer, *Bonhoeffers*, 96.
33. Bethge, *Bonhoeffer*, 264–265.
34. Heilbron, *Dilemmas*, 165.
35. PHP, Box 1, Folder 4, RPI.
36. Nye, *Polanyi*, 75. See footnote 136.
37. Grete's father was Hungarian. The charge of being one-fourth Jew was also leveled at Hans. See "The Righteous Among the Nations, Dohnanyi, Hans von." db.yadvashem.org/righteous/family.
38. Bonhoeffer to Harteck, July 7, 1933. PHP, Box 1, Folder 4, RPI.
39. Ute Deichmann, "The Expulsion of German-Jewish Chemists and Biochemists and their correspondence with Colleagues in Germany after 1945: The Impossibility of Normalization?" in *Science in the Third Reich*, ed. Margit Szöllösi-Janze (Oxford and New York: Berg, 2001), 251.
40. David C. Cassidy, "Heisenberg, German Science and the Third Reich," *Social Research* 59, no. 3 (Fall 1992): 649–650.
41. Delbrück interview, 103.
42. Stern, *Dreams and Delusions*, 178–179.

CHAPTER 8

The Beginning of Resistance

After Haber resigned, he briefly went to England where he pondered what to do with his remaining years, trying to summon the strength to start over. Although there were friends and admirers in England, there were also people who had never forgiven him for his leading role in the use of poison gas in the First World War. Fifteen years had gone by since 1918 when the guns of war went silent, but no Englishman had forgotten what happened on the battlefield of Ypres where Haber himself had directed the first gas attack. In Germany, he had been considered a war hero. In England, he had been considered a war criminal.

Ironically, the man who had considered himself one hundred percent German found that the land that reached out to him, offering him shelter and the chance to resume his research, was Palestine. On learning of the invitation to Haber to settle there, Einstein chided him pointedly in a letter: "pleased . . . that your former love for the blond beast has cooled off a bit. Who would have thought that my dear Haber would appear before me as a defender of the Jewish, yes, even the Palestine cause." Einstein, who had a fraught relationship with Judaism and with Palestine, then ended his letter with the poignant hope that Haber would be able to find "a truly humane atmosphere" in which to regain his optimistic spirit. "For me the most beautiful thing is to be in contact with a few fine Jews—a few millennia of civilized past do mean something after all."[1]

K. L. Housley, *The Scientific World of Karl-Friedrich Bonhoeffer*, Palgrave Studies in the History of Science and Technology,
https://doi.org/10.1007/978-3-319-95801-9_8

It was not to be. On January 29, 1934, at the age of sixty-five, Haber died in a hotel in Basel, Switzerland. With him at the time of his death were his son, Hermann, his daughter-in-law, Marga, and his close friends, the physician, Rudolf Stern, and his wife, Käthe. After talking for a while, Haber retired to his bedroom, and then just minutes later he called out desperately to Rudolf for help. Rudolf rushed to his side, only to find there was nothing he could do. Haber had suffered a massive heart attack. Of Haber's death, Rudolf's son, Fritz Stern, wrote: "German in every fiber of his being, in his restless, thorough striving, in his devotion to friends and students, in his very soul and spirit, Haber died in involuntary exile, a broken man, alone, a fate tragically prophetic for Germany Jewry, for German history."[2]

It is here for the first time that Karl-Friedrich found the courage to break his silence. His resolve to do so was, as always, strengthened by correspondence with Dietrich, who was still in London, serving as pastor to the two German congregations, St. Paul's and Sydenham. In a letter dated January 13, 1934, two weeks before Haber's death, Dietrich had pondered the right course of action for himself, considering the ominous developments in Germany. Should he stay in England? He was enjoying it there, finding it to be "a sort of intermezzo." Yet under the circumstances, England was not a good place to be because "one is close enough [to Germany] to want to take part in everything and too far away for active participation." Should he travel to India and the Far East to study with Mahatma Gandhi? That was an enticing possibility because he was becoming convinced that in the western world Christianity was "approaching its end." Or should he return to Germany where Karl Barth had just been dismissed from the University of Bonn. "I can hardly believe it is true," Dietrich wrote to Karl-Friedrich, having read about the dismissal in *The Times*. "But if it is, perhaps I ought to come home, so that there's at least someone at the universities who's prepared to say these things (Fig. 8.1)."[3]

Those last words were not lost on Karl-Friedrich. When Haber died two weeks later, he wrote to Haber's children:

> I have just learned from the newspaper of the death of your respected father. I am so dismayed by it that I can hardly find words. [I am so

Fig. 8.1 Dietrich Bonhoeffer. Sueddeutsche Zeitung Photo/Alamy Stock Photo

> sorry] that he had to die at this moment! Everything good and beautiful which I owe to him comes back to me, and I cannot suppress the feeling of bitterness at our powerlessness to support him in these last difficult years. I shall always remember his wisdom and goodness, and as long as I live I shall do everything in my power to nurture and keep his memory alive in our field. I am grateful to the fate which for many years brought me close to such an extraordinary man as your father, and I hope that I will have an opportunity to express publicly how much Germany owed to him and also what I feel for him. Since I do not know where his grave is, I intend to have a wreath laid at the Haber-Linde [linden tree] as a sign of my gratitude.[4]

As the weeks went by and nothing was published to honor Haber, Karl-Friedrich decided to make for himself the "opportunity to express publicly" his gratitude. He sat down and wrote an obituary, breaking the silence. One of the few published obituaries for Haber, it appeared in *Chemiker Zeitung* on March 10, 1934. Karl-Friedrich wrote, "Free of all academic narrowness, he cherished in his work the close reciprocal relationship of technology and pure science. In this way he developed into a scientific personality whose intellectual concern was always devoted to preserving the ties between scientific progress and practical life."[5]

Another obituary was written by Haber's old friend, Max von Laue, in which he compared Haber to the tragic Greek hero, Themistocles, who at the end of his life was unjustly ostracized and driven into exile.[6] Laue's reference to a historic figure was similar to his reference to the inquisition of Galileo that he had included in his address to the physics conference in 1933. It was Laue's way of speaking truth obliquely. The first reference to Galileo had not gone unnoticed by the Reich. Neither did the second to Themistocles. Johannes Stark, head of the Imperial Physical Technical Institute, read the obituary and demanded that Laue be expelled immediately from the German Physical Society, which the society refused to do. Stark had won the Nobel Prize in Chemistry in 1919. He had also been an early supporter of Einstein's theory of relativity. Then Stark had changed, becoming an outspoken Nazi ideologue, aligning himself with Philipp Lenard who had won the Nobel Prize in Physics in 1905. Both men were so anti-Semitic that they would not stop at expulsion of Jewish scientists, but demanded that all Jewish scientific ideas be expunged from German science. However, not even Stark had the temerity to claim that the enormously important Haber-Bosch process was false. Not able to get rid of Haber's accomplishments, Stark attempted to wipe out his memory instead. Just as Stark criticized Laue for writing an obituary, it is likely that he criticized Karl-Friedrich for the same thing.

One year after Haber's death, Max Planck, president of the Kaiser Wilhelm Society, decided that the society should hold a memorial service for Haber on January 29, 1935, at the elegant Harnack Haus. Memorial services for illustrious people were commonplace in Germany, but Haber had died an enemy of the Reich, his name only to be spoken with spit and curses. Resolutely, Planck went on with his plans for the service, apparently not anticipating any major problems due to Haber's worldwide fame as a Nobel laureate. Because of the laudatory obituary, Karl-Friedrich had written a year before, Planck invited him to deliver a eulogy. He also invited Otto Hahn and Joseph Koeth, a retired colonel who had worked with Haber during the First World War on poison gas. Finally, Planck invited Haber's son, Hermann, to participate, but he formally declined. Later on, Hermann wrote confidentially to a professor in England explaining why he refused to take part: "We find—something we could not put in writing—that one has no right to celebrate a person *dead*, whom one would not tolerate *alive* today."[7]

Not long after the invitations had been sent out, Bernhard Rust, the Reich Minister for Science, Education, and Culture, issued a directive that civil servants were not allowed to attend the memorial unless they sought permission, which in all likelihood, he was not about to grant. Strongly worded notices were tacked to bulletin boards at the University of Berlin, the Kaiser Wilhelm institutes, and elsewhere. Obsequiously, the Union of German Chemists followed Rust's prohibition and forbade their members to attend. To his credit, Planck would not back down because Haber had been his friend for decades. He wrote to Rust defending the memorial: "The Kaiser Wilhelm Society has attested to its positive attitude toward the present State and its profession of loyalty to the Führer often enough by word and deed to be able to lay claim to receive the confidence that during its planned event any tendency giving rise to misconceptions be most strenuously avoided."[8] In other words, Planck planned to go ahead with the memorial, albeit judiciously.

This put Karl-Friedrich in a terrible bind. Having prepared his speech, he traveled two hours from Leipzig to Berlin. On the day of the memorial, he met Delbrück, who was then working for the physicist Lise Meitner at the Kaiser Wilhelm Institute of Chemistry. With the speech folded in his pocket, Karl-Friedrich and Delbrück headed to the Harnack Haus. Rust had forbidden the German press to report on the service, and Harnack Haus was under Nazi surveillance. However, the foreign press was very interested in what was occurring because of Haber's reputation; their presence may have helped keep the situation from getting out of control.

Separately from Delbrück and Karl-Friedrich, Planck and Hahn were walking together to the service, with Planck excited that it was going to take place. However, he was also very worried that Nazi thugs might prevent attendees from entering Harnack Haus, breaking up the service. At that time, Hahn was not teaching at the University of Berlin but was acting director of the private Kaiser Wilhelm Institute, which meant that he was not a civil servant; therefore, Rust's prohibition did not pertain to him. Neither did it pertain to Delbrück. However, that had not prevented Hahn from receiving a stern warning not to participate from the rector of the University of Berlin.

Despite the pervading threat of violence, Planck and Hahn entered Harnack Haus without incident, while Delbrück and Karl-Friedrich stayed outside, pacing up and down that cold winter's day. Karl-Friedrich was torn by agonizing uncertainty. He had not simply respected Haber, he had worshipped him. But how could he pay him honor without risking his career, his life, even the lives of his family? Delbrück remembered the moment vividly: "So here he was, he had come to Berlin and had received this strict order from the Minister not to attend the meeting. So what was finally arranged was Otto Hahn, who was a co-director with Lise Meitner of the Kaiser Wilhelm Institute for Chemistry, said he would read Bonhoeffer's speech. And Bonhoeffer and I walked around and around the place and tried to decide whether he should go in or not, and finally he decided not to go in, but I went in and sat in the back row and Hahn read the memorial."[9]

Inside Harnack Haus, the audience was a peculiar mix. Because professors could not attend, many of their wives were there in their stead, which was a means to pay respect to Haber without incurring punishment. Nor would the less identifiable wives come as readily to the attention of Nazi informers. Some executives from industry were there, including Carl Bosch with whom Haber had worked closely on the nitrogen fixing process. So also was the chemist Richard Willstätter, winner of the Nobel Prize in 1915 for describing the structure of chlorophyll. Willstätter had resigned his post at the University of Munich in 1924 in protest against the rising tide of anti-Semitism. He had withdrawn into semi-exile, carrying out research via telephone with his assistants, refusing on principle to set foot in his laboratory. For many Jewish scientists, Willstätter was the rare example of a man who openly confronted anti-Semitism, refusing to submit.[10] Therefore, his presence at the funeral went beyond his long-standing friendship with Haber, taking on symbolic importance. Also in attendance were scientists of Jewish lineage who were attempting to ignore anti-Semitism, among them Lise Meitner who mistakenly considered herself safe from Nazism because she was Austrian, a delusion that would be shattered when Hitler took over Austria in 1938. Others in attendance were Meitner's colleague, Fritz Strassmann, who had been blacklisted by the Union of German Chemists for making anti-Nazi statements, and Johannes Jaenicke who had assisted Haber on the ill-fated project to extract gold from seawater.[11]

Planck started the service with the obligatory Heil Hitler, probably in his standard half-hearted way, and then he delivered his introduction, closing with the words: "Haber remained loyal to us; we will remain loyal to him."[12] Hahn and Koeth spoke, the latter no doubt holding forth on Haber's research on Germany's behalf during the First World War. Then Hahn stood up again, unfolded the papers that Karl-Friedrich had handed to him and read the speech word-for-word. However, everyone in the audience knew who had written it and why its author was at that moment standing outside.[13]

In the book *One Hundred Years at the Intersection of Chemistry and Physics: The Fritz Haber Institute of the Max Planck Society*, Jeremiah James et al. write: "The Haber memorial service has often been praised, especially in the post-war era, as an act of political resistance, in particular on the part of Max Planck, against the arbitrary actions of the National Socialist state. Be that as it may, it is difficult to accept the series of events briefly outlined above as real resistance, and the majority of those did not show exceptional civil courage."[14]

Perhaps the courage it took was not exceptional, but it was courage nonetheless. When he had the opportunity, Planck spoke out again, lauding Haber and defending pure science against political intrusion. For example, one year later on January 11, 1936, the Kaiser Wilhelm Society celebrated its twenty-fifth anniversary at the Harnack Haus with significant pomp and circumstance. Scientists presented papers on their latest research. Two cabinet members attended: Baron von Neurath, foreign minister, and Dr. Hjalmar Schacht, economics minister, whom Dietrich had heard speak in New York City on his book *The End of Reparations*. A laudatory telegram was received from Hitler who promised to attend the celebration banquet. However, despite superficial signs of mutual admiration, Planck used the occasion to make some strong points about the independence of science in opposition to the collective tendencies of the Nazis: "New scientific ideas never spring from a communal body, however well organized, but only from the mind of the individual, divinely inspired research worker, who wrestles with his problems in lonely thought and concentrates all his powers on the single point on which for the moment he focuses his entire world." Then, setting caution aside, Planck spoke of the deaths of two Jews who had played important roles in the Kaiser Wilhelm Society: Fritz Haber and Franz von Mendelssohn, the society's treasurer.[15]

Otto D. Tolischus, a reporter for *The New York Times* in its Berlin Bureau, was on hand for the festivities and immediately wired a long article to the editors. It was published with the provocative headline: "Reich Scientists Uphold Freedom: Nazis are Cool to Gathering of Kaiser Wilhelm Society Which Bars Coordination." The subhead was "Late Dr. Haber Praised: Famous Jew Driven to Suicide by 'Racial' Decrees lauded by Planck, Presiding." Tolischus may have been wrong about Haber committing suicide, although the extreme stress he was under more than likely contributed to his heart attack. The rest of the article was correct, exposing the schism between Planck's and Hitler's approach to science, and the threat to the existence of the Kaiser Wilhelm Society. At the end of the article, Tolischus quoted from a Nazi newspaper, which charged that the Kaiser Wilhelm Society had become a playground for Catholics, socialists, and Jews. "We must ask ourselves," the Nazi newspaper warned, "how far there is still room in the National Socialist State for the Kaiser Wilhelm Society in its present state."[16]

Planck's remarks at the twenty-fifth anniversary make clear that long after the Haber memorial he continued to speak out judiciously from within the system to defend not only the ideal of pure science but also, and more importantly, his friend. Planck was an old man of the old school, trying to uphold old ways in perilous times. The problem was that Planck underestimated how vituperative Hitler could be, as made clear by the following small incident: On December 2, 1928, a large linden tree had been planted in front of the Haber Institute in honor of Haber's sixtieth birthday. It was that tree that Karl-Friedrich had mentioned in his condolence letter to Haber's children, writing that he would place a wreath under it. So thorough was Hitler's intent of wiping out all memory of the man, the tree was cut down.[17]

As to Karl-Friedrich's part in the Haber memorial and what motivated him both then and immediately afterward, it is important to consider yet another powerful letter from Dietrich, still in England, written a few weeks prior to the memorial. Dated January 14, 1935, it was in response to Karl-Friedrich's expression of concern for Dietrich's safety because of his outspokenness. "Perhaps I seem to you rather fanatical and mad about a number of things," Dietrich replied. "I myself am sometimes afraid of that. But I know that the day I become 'more reasonable,' to be honest I should have to chuck my entire theology." He wrote of taking the Sermon on the Mount seriously because it had "the force that can blow all this hocus-pocus sky-high – like fireworks, leaving only a few burnt-out shells behind." By hocus-pocus, he meant Nazism. "Things do exist that are worth standing up for without compromise. To me it seems that peace and social justice are such things, as is Christ himself." He closed the letter with a reference to the fairy tale "The Emperor's New Clothes" by Hans-Christian Andersen, in which no one in the empire, except a child, has the courage to tell the emperor that he has been lied to—he is not dressed in the finest raiment but is totally naked. "All we are lacking today is the child who speaks up at the end."[18] No reply from Karl-Friedrich is extant, but Dietrich's letter is yet one more indication of his unwavering efforts to encourage his brother, indeed everyone, to stand firm.

That Karl-Friedrich began to do so is made clear by the recollections of some of his friends. Years later, Walther Jaenicke, who as a boy had met Karl-Friedrich through Walther's father, Johannes, at the Haber Institute, and who received his doctorate under Karl-Friedrich at the University of Leipzig, described him as a gentle man with a cheerful disposition who did not have a fighting spirit but who had an aversion to anything false, not caring for intrigue. Nevertheless, when Karl-Friedrich realized what was happening under Hitler, he "began unswervingly and without concessions to help the disenfranchised and to work for the purification of science."[19] The physical chemist Heinz Gerischer also received his doctorate under Karl-Friedrich under difficult circumstances. Beginning in 1937, two years after the Haber memorial, Gerischer, only eighteen years old, found shelter as a student in Karl-Friedrich's laboratory, staying with him as an assistant through the war and into the postwar period, doing research on electrochemistry.

During the war, Gerischer's mother, who was Jewish, committed suicide; his sister escaped from a Gestapo prison only to be killed in an air raid.[20] Perhaps because he harbored half-Jews, Karl-Friedrich was "badly denounced" during the war by a Nazi scientist at the University of Leipzig. The incident filled Karl-Friedrich's friend, Paul Rosbaud, with disgust, particularly when following the war the informant was cleared of any wrongdoing.[21]

However, Karl-Friedrich's harboring of half-Jews was still a few years away. Nearer at hand (only three months after the Haber memorial service) was the crisis caused by another purge of the civil service. This time there was no exemption for scientists of Jewish lineage who had served in the First World War. In his book *Beyond Uncertainty*, Cassidy writes that at the University of Leipzig, Werner Heisenberg, who was professor of theoretical physics and chair of the department of physics, reacted with "shock and dismay" because he had convinced himself that the enforcement of the Aryan clause in 1933 had been the worst thing that could happen, and that there would be no more dismissals of professors. Furthermore, he was a close colleague and friend of some of the men who would be driven out. According to Cassidy, "Heisenberg reviewed the situation, which would prove to be a turning point in his relationship to the state, with his closest colleagues, Weizsäcker, Hund, B.L. van der Waerden, and Karl-Friedrich Bonhoeffer. He then met with another member of his professors' club, Dean Helmut Berve, to formulate a response." The professors voiced their objections to the dismissals at the next faculty meeting on May 8, 1935, but were unable to stop the Jewish evictions, which was to be expected because Berve was a Nazi. Some of the professors received formal reprimands.[22]

As with the memorial service for Haber, the protests of Heisenberg, Karl-Friedrich, and others can be seen as too-little-too-late. It led, however, to even worse problems for Heisenberg, who was viewed at this time with great mistrust by the Nazis. Some of them, including Johannes Stark, considered him a "white Jew" meaning a person who was sympathetic to Jews to the point of thinking like them—the "thinking" in question being the totality of theoretical physics that had been promulgated by Jews such as Einstein. From the Nazi standpoint, such thinking was trash because science should be based on correct ideology, not truth. If Heisenberg had agreed, he would have been forced to throw out all his work on quantum mechanics, along with the Nobel that had been conferred on the basis of that work.

After they were forced out, the Leipzig professors scattered worldwide, literally to the four winds. For example, Joachim Wach, a Lutheran theologian who had served as a cavalry officer in the First World War and who had Jewish ancestors, made his way to Brown University in the USA. The mathematician Friedrich Levi, who had received the Iron Cross for his service in the First World War and who was famous for his work on abstract algebra, went to the University of Calcutta. Benno Landsberger, a highly regarded Assyriologist, went first to the Turkish University of Ankara and then to the University of Chicago.

It was the loss of the physical chemist Fritz Weigert that troubled Karl-Friedrich the most because they shared an interest in both photochemistry and the chemistry of living organisms. Weigert had come to Leipzig in 1914 to work with Wilhelm Ostwald, one of the founders of the field of physical chemistry and a Nobel Prize laureate. On December 19, 1935, Weigert wrote to Karl-Friedrich from the safety of Bohr's laboratory in Copenhagen, sending him greetings from colleagues and best wishes for the coming year. Weigert also inquired about apparatus and laboratory equipment he had been forced to leave behind in the basement of the institute, asking Karl-Friedrich if it could remain there a little longer because he had not yet found a way to retrieve it. There is no extant reply from Karl-Friedrich, but chances are that Weigert never got his equipment out of Leipzig because the Nazis would not have allowed it. As did so many other scientists who found temporary shelter with Bohr, Weigert ended up going to England where he became research director of the physiochemical department at Mount Vernon Hospital in Northwood.[23]

In August of 1935, Arnold Berliner, who had founded the prestigious journal *Die Naturwissenschaften*, was forced to quit because he was Jewish. He had been under intense pressure not to publish research papers and books by Jews. Furthermore, the publisher, Springer-Verlag, was supposed to fire all its Jewish employees, which was a major issue given that the Springers were themselves of Jewish descent. (Julius Springer would be forced to quit the firm just two months after Berliner.) When Karl-Friedrich learned of Berliner's firing, probably from Paul Rosbaud who worked with Berliner, he wrote to express his dismay. The groundbreaking paper by Karl-Friedrich and Paul Harteck on ortho- and para-hydrogen had been published in *Die Naturwissenschaften* in 1929. So also had Karl-Friedrich's papers on heavy hydrogen in 1934. As was Karl-Friedrich, Berliner was from Breslau, receiving his doctorate in physics from the University of Breslau in 1886. In 1911, he had approached Ferdinand Springer about founding a scientific journal similar to the English-language journal *Nature*. He wanted to bring knowledge together from all branches of science, a goal he achieved through *Die Naturwissenschaften*. Berliner was a friend of Albert Einstein, Max Born, Max von Laue, and Fritz Haber, in fact, he knew everyone in the world of German science, a world to which he was dedicated. Einstein said about Berliner on his seventieth birthday in 1932 that the scientific life of that period in Germany was inconceivable without *Die Naturwissenschaften*. "It is just as important to make knowledge live and to keep it alive as to solve specific problems," Einstein said. "We are all conscious of what we owe to Arnold Berliner."[24] However, by 1935, Hitler's answer to what was owed Berliner was—nothing. In his reply to Karl-Friedrich, Berliner wrote that he could not talk about his resignation except to say he had no choice, that in the face of "brutal pressure simply everything had to be given up. How difficult it is for me, you can easily imagine. I have often thought of Fritz Haber in these days."[25] In particular, he recalled the obituary and the memorial service at the Harnack Haus the preceding winter. Berliner closed his letter to Karl-Friedrich by stating that he had remained steadfast to natural scientists right up until the last minute. Six years later on March 22, 1942, under house arrest, starving to death, and waiting to be sent to a death camp, Berliner killed himself.

It was not only Jewish publishers and professors who were forced to leave Germany but also students, some in the last stages of their doctoral work. Bernard Katz, who would be awarded the Nobel Prize in Physiology or Medicine in 1970, was attempting to finish his medical degree at the University of Leipzig when the Aryan clause went into effect. Anti-Semitism had hindered his studies from his boyhood, including a threat made in the early 1920s when he was only eleven years old. He learned that his non-Jewish classmates had hatched a plan to assemble all the Jews in Leipzig in an underground hall that would be filled with poison gas. In an autobiographical essay, Katz wrote, "This episode has never been erased from my mind, and it gives an indication of ideas that some people were harbouring in their heads for 20 years before they were able to put them in practice." Doggedly, Katz kept going with his medical studies. At Leipzig, his academic work was so exemplary he was awarded a major prize in 1933, only to be denied it because he was Jewish. Taking his side was his mentor, the physiologist Martin

Gildemeister, head of the Physiological Institute, who secretly made sure that Katz received the prize money. Gildemeister's laboratory was one of the few places that Katz felt safe, completing his medical degree in 1934. The next year, Katz packed his bags, carrying in his pocket a temporary British visa, his League of Nations stateless-persons pass (his father had been Russian), a letter of recommendation from Gildemeister, and £4 in his pocket, making his way to London where he was given a position in the physiology laboratory of Archibald Hill, winner of the Nobel Prize.[26] All this may seem unrelated to Karl-Friedrich except that not long after he arrived at Leipzig, Gildemeister became a close friend and colleague. On Gildemeister's death in November 1943, Karl-Friedrich wrote to Dietrich disconsolately that he was a man "with whom I had a great deal to do precisely in my new area of responsibility and whom I miss terribly." Karl-Friedrich's new area of responsibility was building a model that would bridge physiology and physical chemistry, just as Gildemeister had built a bridge between physiology and mathematics. Both men were particularly interested in the local excitation of myelinated nerve fibers. One of the principal reasons for Karl-Friedrich's shift in research was his serious concern about being pulled, against his will and conscience, into military research. The German high command was more interested in heavy water than in myelinated nerve fibers.

Notes

1. Stern, *Dreams and Delusions*, 49. See endnote 28. "Letter in the Einstein Archives, Boston."
2. Ibid., 54. See also Stern, *Five Germanys*, 103.
3. The *Times* article was dated January 13, 1934, but Barth did not lose his position until the end of 1934, at which point he went to Basel, Switzerland. DBW, Vol. 13, 81.
4. Eberhard Bethge, "The Nonreligious Scientist and the Confessing Theologian: The Influence of Karl-Friedrich Bonhoeffer on His Younger Brother Dietrich," in *Bonhoeffer for a New Day: Theology in a Time of Transition*, ed. John W. De Gruchy (Grand Rapids: Eerdmans, 1997), 43.
5. Ibid., 57.
6. Ute Deichmann, "To the Duce, the Tenno and Our Führer," in *The German Physical Society in the Third Reich*, ed. Dieter Hoffmann and Mark Walker (Cambridge: Cambridge University Press, 2012), 312.
7. Ibid., 306.
8. Ibid.

9. Delbrück interview, 49.
10. Jean Medawar and David Pyke, *Hitler's Gift* (New York: Arcade, 2000), 130.
11. Deichmann, "To the Duce," in *The German Physical Society*, 313.
12. Stern, *Dreams and Delusions*, see footnote 38, 170.
13. There is no known copy of Karl-Friedrich's speech.
14. James, *One Hundred Years*, 94.
15. Otto D. Tolischus, "Reich Scientists Uphold Freedom," *New York Times* (January 11, 1936), n.p.
16. Ibid., n.p.
17. The fate of the tree is uncertain. Stoltzenberg states that in January 1946, a new plaque was placed by Harmut Kallmann on the "so-called Haber linden tree." Perhaps a new tree had been planted. Stoltzenberg, *Haber*, 210.
18. DBW, Vol. 13, 284–285.
19. Walther Jaenicke, "Karl-Friedrich Bonhoeffer," *Personliches* (1957), 369.
20. Adam Heller, Dieter Kolb, and Krishman Rajeshwar, "The Life and Work of Heinz Gerischer," *ECS Classics* (The Electrochemical Society, 2010), 1.
21. The Nazi scientist was Martin Kröger. Letter from Rosbaud to Samuel Goudsmit, June 22, 1947. Samuel A. Goudsmit Papers, 1921–1979. Series IV. Alsos Mission. Subseries B. Paul Rosbaud. Rosbaud correspondence and manuscripts 1946–1949 (Box 28, Folder 43). Niels Bohr Library and Archives, American Institute of Physics.
22. Cassidy, *Beyond Uncertainty*, 229.
23. Letter from Weigert to Bonhoeffer, December 19, 1935, Abtielung Rep 23, Personal Correspondence, Nachlass Karl-Friedrich Bonhoeffer, Archiv der Max-Planck-Gesellschaft.
24. Sven Thatje, "Dr. Arnold Berliner (1862–1942), Physicist and Founding Editor of Naturwissenschaften," *Naturwissenschaften* 100 (2013): 1105–1107. https://doi.org/10.1007/s00114-013-1124-4.
25. Berliner to Bonhoeffer, August 28, 1935, Abtielung Rep 23, Personal Correspondence, Nachlass Karl-Friedrich Bonhoeffer, Archiv der Max-Planck-Gesellschaft.
26. David Colquhoun, "Obituary Bernard Katz, 1911–2003," *Psychology News* (2003). www-ucl-sib.ucl.ac.uk/pharmacology.

CHAPTER 9

Heavy Water and the Atomic Bomb

It is remarkable how many things scientists in the early 1930s correctly understood about the atom considering all the particles they did not know existed. Heisenberg may have presented his paper on quantum mechanics in 1925, and Bohr may have pondered complementarity shortly thereafter, but they did so without knowledge of neutrons or positrons, hypothetical denizens of the atomic world. Science was advancing by leaps and bounds, but it was leaping over cavernous gaps in classical physics and chemistry.

One of those gaps was the atomic weight of hydrogen that wasn't quite right, making it impossible to determine a fixed value for the density of pure water. A maddening conundrum, it had been one of the driving forces behind Karl-Friedrich's work on para- and ortho-hydrogen. It had also led to his research on heavy water undertaken first at the University of Frankfurt and then at the University of Leipzig.

The term heavy water is a slight misnomer because it is only eleven percent heavier than common water. Heavy water (deuterium oxide) contains more of the hydrogen isotope deuterium than the isotope protium. Deuterium's nucleus is composed of a neutron and a proton, making it heavier than the nucleus of a protium that contains only a proton. If this seems to be confusing now, it was far more confusing to the scientists working on hydrogen then, none of whom had any idea that heavy water would play a major role in nuclear fission and the development of the atomic bomb during World War II. Theoretically, the potential of

K. L. Housley, *The Scientific World of Karl-Friedrich Bonhoeffer*, Palgrave Studies in the History of Science and Technology, https://doi.org/10.1007/978-3-319-95801-9_9

releasing an enormous amount of energy by means of fission had been discussed, but it was considered a pipedream until 1938 when it would be achieved. And fission was the easy part. Until then, to Karl-Friedrich and all the rest of the scientists in this highly competitive area of research, it was pure science, yet again, that spurred them on in hopes of discovering a piece of the puzzle as to how the physical universe works.

Here is a partial and oversimplified list of events related to heavy water in one brief period from late 1931 to 1934, showing the illogical progression of discovery, the complex interaction of scientists on both sides of the Atlantic, and the *impurity* of pure science: In November 1931, Harold Urey, at Columbia University in New York, discovered deuterium. Although he was a physical chemist, Urey had studied physics with Bohr in Copenhagen about which he liked to quip that Bohr would never have let him into his laboratory had he known he was not a physicist. In the not too distant future, it would be understood that isotopes of an element have the same atomic number but a different number of neutrons. However, when Urey began his research, neutrons had not yet been discovered. Urey had been spurred to undertake his research by the discrepancy in hydrogen's atomic weight. In the summer of 1931, Urey had read a thought-provoking letter written to the editor of *The Physical Review* by two University of California scientists, the physicist Raymond T. Birge and the astrophysicist Donald H. Menzel, suggesting that isotopes of hydrogen might provide an explanation. That letter launched Urey on his search, which was more like a highly competitive race.[1]

Simultaneously in Paris, the husband-and-wife team of Frédéric Joliot and Irène Curie were hard at work on beryllium radiation, making some inexplicable high-energy findings. These findings attracted the perplexed attention of many scientists around the world, among them James Chadwick at the Cavendish laboratory in England. In February 1932, Chadwick discovered the neutron, having worked night and day, thereby providing the key to understanding deuterium and nearly everything else.[2] Also that year in a photograph taken of a cloud-chamber at Caltech, Carl Anderson noticed the telltale track through the vapor of a new particle, the positron. Its existence had been predicted by Paul Dirac, just as Wolfgang Pauli had predicted the neutrino, meaning in Italian (thanks to a suggestion by Enrico Fermi) *little neutron*, although it would be another quarter century before the neutrino's existence would be proven. Meanwhile, Fermi was beginning to study nuclear reactions caused by collisions of slow neutrons. It is no surprise that

Dirac shared the Nobel Prize with Erwin Schrödinger in 1933 "for the discovery of new productive forms of atomic theory." Urey won the Nobel in 1934, beating out Karl-Friedrich who had been nominated. Chadwick won in 1935, Anderson in 1936, and Fermi in 1938. Pauli would have to wait until 1945. And the cascade of discoveries and Nobel prizes continued.[3]

These discoveries would not have been possible without corresponding improvements in instrumentation and the ability to make precise measurements. In 1928, Hans Geiger made major improvements to his counter that was sensitive to beta and gamma rays. Walther Bothe, who was then studying cosmic rays, made additional improvements to the counter. But if a scientist needed a Geiger counter, he was going to have to build it himself. When Urey needed a spectrograph for his experiments, he and his colleague, George Murphy, designed and built one, although they were worried (unnecessarily) that the spectral lines they were seeing were due to irregularities in the spectrograph's 21-foot-long grating.[4] Scientists often used spare parts knocking around the laboratory or retrofitted another piece of equipment, always with great ingenuity, and often with a fair amount of personal risk. At the University of California, Ernest Lawrence and his graduate student, Stanley Livingston, built the first cyclotron in 1931, accelerating charged particles in a spiral between the poles of an electromagnet. No sooner had the cyclotron been switched off than work was begun to build a more powerful model. The first electromagnet weighed only two tons. The second was an 80-ton behemoth that Lawrence found in a junkyard in Palo Alto, having been built in the First World War to power transatlantic radio communications, and subsequently discarded. The cyclotron revolutionized physics, blowing open doors to the study of atomic particles.[5]

Into this homemade equipment the scientists poured, packed, or dribbled elements that were often toxic, as was the beryllium the Curies were using; caustic, eating its way through seals on bottles; or reactive, making unfortunate things happen with metal shipping containers. For example, heavy water could only be shipped in containers welded without cadmium or boron; otherwise, it would arrive at its destination utterly useless as a neutron moderator. There were also the substances that lived out their half-lives in dark closets, their radiation undetected by inaccurate Geiger counters. Even though the dangers of radiation were well known, the researchers were cavalier about their own safety. Pure science must be served even at the cost of their shortened lives. A single story is sufficient to make the point: when the team of Joliot-Curie discovered artificial radiation on January 15, 1934, one person who raced into their basement laboratory in Paris to see them repeat the experiment was Irène's mother, Marie Curie, who was dying of aplastic anemia caused by radiation exposure. Without regard to the danger, she snatched the vial of phosphorus and held it up to the Geiger counter that crackled wildly. Marie died six months later at the age of 57. The Joliot-Curies received the Nobel Prize in chemistry in 1935, the third Nobel in the family.[6] Sadly, Irène died of leukemia in 1956 at the age of 58. Frédéric died two years later in 1958, also at the age of 58.

To the list of events related to heavy water must be added the following: In early 1934 at the Cavendish laboratory, Ernest Rutherford, the Australian physicist Marcus Oliphant, and Harteck bombarded deuterium with ions of heavy hydrogen (deuterons) and discovered tritium. This meant that hydrogen had three isotopes: protium (hydrogen 1), deuterium (hydrogen 2), and tritium (hydrogen 3). The first two are stable, but the third is unstable because it has two neutrons in the nucleus and a single proton. A weak beta emitter, it is radioactive. Significantly, it was produced by the thermonuclear *fusion* of the bombarded deuterium and the projectile deuterons, which was the first time fusion had been experimentally demonstrated. In assessing his lifework, Harteck rated the deuterium experiment as first in importance, and the para- and ortho-hydrogen experiment, with Karl-Friedrich, as second.[7]

Less obvious than these discoveries, but equally powerful in terms of the shaping of ideas, was the shift in the meaning of certain scientific words, sometimes without their users being fully cognizant of it. This shift brought about changes in what was considered to be true. Michael Polanyi wrote of a Royal Society meeting in 1934 at which Frederic Soddy, the discoverer of isotopy, objected to the statement that deuterium was a new isotope of hydrogen. Soddy had defined the isotopes of an element as chemically inseparable from each other. However, heavy hydrogen was chemically separable from light hydrogen, so, according to Soddy, it could not be an isotope. "No attention was paid to this protest, and a new meaning of the term 'isotope' was tacitly accepted instead," wrote Polanyi. The new meaning, which was grounded solely on nuclear charge, made Soddy's meaning untenable. "For to retain the original conception of isotopy, by which the chemical differences between light and heavy hydrogen would be classed with the chemical difference between two elements filling different places of the periodic system, would have been misleading to the point of absurdity," wrote Polanyi.[8]

Unlike Polanyi, whose mind could be fully engaged by the shifting definition of a single word, Harteck focused on hard results. When his employment at the Haber Institute was threatened because of the worsening depression, Harteck had been relieved to receive the Rockefeller Foundation grant, enabling him to spend a year in Rutherford's laboratory, even though the grant was a paltry $150 a month.[9] Considered the father of nuclear physics, Rutherford was a bit of a character in the laboratory, quick to lose his temper and quick to apologize for doing so. In 1934, he was 62 years old and had developed a slight tremble in his hands that resulted in the ash from his pipe falling onto the scientific apparatus. However, what he lacked in physical steadiness, he made up for with technical brilliance, even though a major part of his day was taken up with administration and not experimentation. Harteck found him to be "a man of tremendous vitality and strength," despite his age.[10] "I wanted to work with Rutherford because I had the feeling that the nuclear physics would become important," Harteck said. "When you are in Berlin in [Haber's and Nernst's] colloquia, you got the fever. It was obvious to me that a new era in science had begun."[11] It was Rutherford and Soddy who, thirty-one years before, had roughly estimated how much energy the atomic nuclei contained: "at least twenty-thousand times and maybe a million times as great as the energy of any molecular change." Could that inconceivable amount of energy be unlocked in some way, either benignly for a civilian energy source, or malignly for a military bomb? "Moonshine," Rutherford brusquely declared at a meeting of the British Association for the Advancement of Science in 1933. But younger researchers such as Harteck had begun to reply "maybe."[12]

As he had proven in his experiments with Karl-Friedrich, Harteck was a gifted physical chemist who could figure out how to get things done in the laboratory and was ingenious enough to design needed equipment and procedures. Due to his penchant for the experimental, he was never at ease with pure theoreticians, such as Heisenberg and Weizsäcker, who he thought had their heads in the clouds and couldn't find their way around laboratory equipment. It also explains why he was at ease with Rutherford, considered one of the greatest experimentalists of the age. It was this practical skill that would make Harteck one of the key figures in atomic bomb research, in some aspects more important than Heisenberg himself.

Over and over, he put his knowledge of physical chemistry to work to increase output and reaction times in ways that never occurred to the theoretical physicists. But there is a downside when experimental practicality goes too far. In his book *Physics and National Socialism: An Anthology of Primary Sources,* the physicist and historian of science Klaus Hentschel listed Harteck among the "cold pragmatists" who took over from the hotheaded Nazi demagogues in 1940–41. Harteck was "much more interested in an efficiently running research and development sector in science and industry than in any politically correct programmatics."[13]

However, Harteck's cold pragmatism was in the future. The immediate problem for which he had to find a practical solution was that heavy water was in very short supply. Deuterium occurs naturally in water at an extremely low rate. The concentration can be increased by electrolysis, but it takes an enormous amount of energy to do so. Harteck had ingeniously designed a series of very small electrolytic cells. According to Per F. Dahl in *Heavy Water and the Wartime Race for Nuclear Energy,* "An electric current was passed through them for weeks on end, reducing many gallons of water spiked with a small amount of NaOH, to about 30 grams of very pure (up to 98%) heavy water. In this he was considerably assisted by Gilbert Lewis of Berkeley, who prior to then had shipped several precious batches of heavy water to Cambridge, and who supplied him with a detailed description of his own electrolytic procedure."[14]

Unfortunately, Harteck hardly had enough heavy water to conduct his own experiments, let alone enough to fill the numerous requests for samples he was receiving from other scientists. In California, Lewis had the same problem. Ralph Fowler, chair of theoretical physics at Cavendish, wrote to Lewis that, "Harteck, who is running our small generating plant here very successfully on your lines, suffers in just the way you describe from the importunate demands from everybody who has a nice little experiment to perform."[15]

There was only one place in the world that had the requisite power capacity to make heavy water in quantity: the Norsk Hydro plant in Vemork, near the town of Rjukan, Norway. Vemork was located in a rugged defile so deep it only received direct sunlight for a brief period midday. Completed in 1911, the plant was a gloomy place perched on a rock ledge beside the thundering Rjukanfossen, a 104-meter waterfall. Using the Haber-Bosch process to which it had converted in 1929, the plant

Fig. 9.1 Norsk Hydro, Vemork, Norway. Pictorial Press Ltd./Alamy Stock Photo

produced hydrogen and ammonia mainly for use in fertilizer. This process produced as residue large amounts of electrolyzed water in which deuterium was concentrated in the older electrolytic cells. It was technically possible to feed this residue back into the system to concentrate it further. It was an inefficient process, but in the early 1930s it was the only way (Fig. 9.1).

Realizing the potential of the Vemork plant, Karl-Friedrich was the first scientist to request deuterium in the form of potash-lye samples (concentrated aqueous solution of potassium hydroxide) from Norsk Hydro in May 1933. The samples were shipped to him in June, and he immediately set to work, by coincidence beating out Odd Hassel, a brilliant but peculiar Norwegian colleague who was a physical chemist at Oslo University. Hassel knew Karl-Friedrich from his days at the University of Berlin where he had received his doctorate under Fritz Haber in 1924 at the time that Karl-Friedrich was Haber's assistant.

Like Karl-Friedrich, Hassel was first interested in physics but then switched to physical chemistry. His thesis was on X-ray crystallography, one of Polanyi's specialties. Back in Norway, Hassel became interested in deuterium following Urey's discovery. Hassel was socially withdrawn perhaps because he was an albino who had to stay out of light. Very thin with a distant look in his light blue eyes, he had white hair and very pale skin, giving him an otherworldly appearance. His solitary ways may have been heightened by his having been orphaned early, forcing him to cobble together his education. By choice, he liked to work alone. Sir Derek Barton, who shared the Nobel Prize in Chemistry with Hassel in 1969, found him to be strangely negative, almost "as if he were not there."[16]

Recognizing, as did Karl-Friedrich, that heavy water was being concentrated slowly in the electrolytic cells at the Norsk Hydro plant, Hassel was interested in increasing that rate, but when he made a proposal to Norsk Hydro he was turned down. Not to be put off, he began concentrating heavy water on his own in space provided by the Geologisk Museum in Oslo, using a laborious process similar to Harteck's. By the time that Norsk Hydro reconsidered his proposal and provided Hassel with an electrolytic cell, Karl-Friedrich was well along with his research at the University of Frankfurt using the potash-lye samples he had received. This led to the publication that year of Karl-Friedrich's paper titled "Heavy Water"(1933) in *Angew. Chem.*, summarizing the method of separating the isotopes of hydrogen and the properties of heavy water. This was followed by a second paper the same year, "The Exchange of Hydrogen Between Water and Hydrogen Compounds Dissolved In It," co-authored with G.W. Brown and published in *Z. Physick. Chem.* The compounds tested were ammonia (NH_3), ammonium chloride (NH_4Cl), and cane sugar ($C_{12}H_{22}O_{11}$).[17]

Jomar Brun, head of the electrolysis plant at Vemork, did not think anything of Hassel's and Bonhoeffer's requests until he received a third request, this time from I.G. Farben, the huge German industrial firm that was part owner of Norsk Hydro.[18] Realizing that something was brewing and that there might be a marketing opportunity for heavy water, Brun raised the prospect with Leif Tronstad who was working as a scientific consultant for Norsk Hydro. In the small world of physical chemistry, Tronstad and Karl-Friedrich knew each other because Tronstad had (like Hassel) studied at the Haber Institute under the guidance of Herbert Freundlich in 1928–29, the period when Karl-Friedrich and Harteck made their breakthrough on para- and ortho-hydrogen.

Subsequently, Tronstad spent ten months studying under Rutherford at the Cavendish laboratory, although not at the same time as Harteck. The interconnections between all these men would prove to be of great importance. According to Dahl:

> The decisive factor in convincing Tronstad and his Hydro colleague that Vemork was just the place for the production of heavy water was probably the results of Bonhoeffer's analysis of the potash-lye samples dispatched to him back in June, which indicated a heavy-to-normal hydrogen ratio of 1:2300, compared to 1:6000 for normal water, according to the best available data at the time. They decided that Tronstad, with his scientific reputation, should approach Hydro with a specific proposal for the production of heavy water by the repeated concentration of water from the electrolyzers at Vemork.[19]

The upshot was that in 1934 Norsk Hydro built a plant that produced heavy water concentrated to approximately 99% purity. Further expansion took the output from 40 kg of concentrated liquid in 1936 to 80 kg in 1938. Heavy water was expensive, but sales, though never high, were fairly steady during this period considering its use was solely for research.

From 1934 to 1939, Karl-Friedrich co-authored at least seven more papers on heavy water, of which two were in Paul Rosbaud's publication *Naturwissenschaften,* published by Springer.[20] However, Karl-Friedrich's interest was already turning toward living organisms, applying what he knew about kinetics to organic chemistry. One of the reasons he was interested in heavy water was to explore its potential as a radioactive tracer or biomarker in organisms, although it turned out to be the tritium in heavy water that was radioactive, not the deuterium. Eventually, Karl-Friedrich used tracers to label organic molecules, to study the synthesis of polysaccharides in yeast cells, and enzyme hydrolysis of glucosides.[21]

The other focus of Karl-Friedrich's heavy water papers was measuring the velocity of exchange. What was becoming clear about heavy water was its ability to slow down neutrons without absorbing or capturing them, which would put them out of action. In late 1938, this became enormously important. The seismic shift occurred when Otto Hahn and Fritz Strassmann fissioned the atom at the Kaiser Wilhelm Institute of Chemistry, building on research done by the nuclear physicist, Lise Meitner. However, they did not realize what they had done.

On December 22, 1938, Hahn telephoned Rosbaud about the perplexing results and to get his thinking about when, and if, to publish in *Naturwissenschaften*.

Five months earlier, Rosbaud had played a crucial role in spiriting Meitner out of Germany, getting her across the border at night into Holland with the help of Dirk Coster, a Dutch physicist.[22] Meitner was forced to flee after Hitler took over Austria because at that point her Jewish heritage put her in the Nazi crosshairs. Up until then, she had considered herself to be immune from Nazi persecution, protected by her Austrian citizenship. Tucked in her pocket at the time of her escape was a diamond ring that had belonged to Otto Hahn's mother, which he had given to Meitner to bribe a border guard if all else failed. She made it across without need of the ring, thanks to some fast-talking by Coster and Rosbaud who convinced the border guards that she had permission. From Holland, she traveled to the safety of Sweden where she accepted an invitation to work at the Nobel Institute in Stockholm. According to her nephew, the nuclear physicist Otto Frisch, Meitner had not left Germany willingly. Instead, she had to be "perhaps stampeded" by Rosbaud, Costner, and Hahn on very short notice. Only after her escape did Meitner realize the danger she had been in, excoriating herself for her naïveté.[23]

Rosbaud was Austrian as well, but he was never naive. His eyes were wide open. With the assistance of Francis Edward (Frank) Foley, whom Rosbaud first met when Foley was in Berlin as passport control officer while covertly running the British Secret Service (MI6) station, Rosbaud had sent his Jewish wife and daughter to England. Their safety relieved him of the stress of trying to protect them, freeing him to concentrate on espionage. His spy name *The Griffin* was apropos because Rosbaud possessed eagle eyes and a lion's courage. Arvid Brodersen, a member of the Norwegian resistance, never met Rosbaud personally, but he knew him well through several vital intelligence agents who smuggled out Rosbaud's reports. Among those agents was Sverre Bergh, a Norwegian engineering student who attended Dresden Technische Hochschule in Germany during the war. It was Brodersen's opinion that "Rosbaud was no spy in the sense of a controlled and paid agent of a foreign power, but an ethical person acting on his own deeply felt need: to do his best to prevent Hitler from winning the war."[24] With his doctorate in metallurgy, Rosbaud was up-to-date on what was happening in science all across Europe. He was urbane, socially adept, multi-lingual, and possessed a cool head in dangerous situations, such as Meitner's escape.

With Meitner gone, Hahn and Strassmann were in a quandary. As a nuclear physicist, Meitner had brought a unique and highly valued perspective to their chemical research. In his phone call to Rosbaud, Hahn confided that he had just written a paper on experiments in which a totally different element was created when a slow neutron collided with a uranium atom. He and Strassmann knew something profound had occurred, but they were utterly perplexed. How was it possible that radium isotopes could behave as if they were another element entirely—barium? It made no sense.

A few days before phoning Rosbaud, Hahn had mailed the paper to Meitner who was spending Christmas vacation in Sweden with Otto Frisch. "Perhaps you can suggest some fantastic explanation," Hahn asked her hopefully. Going for a walk in the snow, with Meitner on foot and Frisch on skies, they pondered Hahn's letter. They sat down on a stump of a tree and began to do complex calculations on scraps of paper they had in their pockets. Suddenly, they saw what Hahn and Strassmann had missed: The atom had been split.[25] That was the end of their walk. Immediately, Meitner and Frisch headed back to let Hahn know.

Even before that moment of insight, Rosbaud had recognized the enormity of what Hahn and Strassmann had done. To make the deadline for the January 1939 issue of *Naturwissenschaften*, he pulled another paper that had already been typeset, substituting the fission paper. Meitner and Frisch followed up in subsequent issues with papers of their own that provided the theoretical interpretation. In fact, it was Frisch who first used the term fission, having asked a biologist what was the term used when a cell divided.

Arnold Kramish in his biography of Rosbaud, *The Griffin: The Greatest Untold Espionage Story of World War II*, states that Rosbaud was playing a strategic game. "He wanted the rest of the world to know of the significance of the work at least as soon as Nazi planners did."[26] That may be true, but it is just as likely that Rosbaud valued the achievement from the standpoint of pure science. There were other scientists, foremost among them being Leo Szilard, who upon hearing the news immediately set to work to suppress it, hoping thereby to prevent the devastation they foresaw. In fact, Szilard was so worried he immediately wrote to Fermi (who had by then fled from Italy to the USA) and the team of Joliot-Curie, asking them not to publish their results should they discover that fission could be maintained by a chain reaction. Fermi agreed not to publish, but the Joliot-Curies (having made the discovery) did publish for reasons that remain obscure. One thing was certain: the idea of destructive atomic force caused by a chain reaction had been in the wind for a long time. Publication or no publication, research into the building of a bomb was about to begin.

Shortly after Meitner and Frisch solved the puzzle of what had happened in Hahn's laboratory, Frisch raced from Sweden to Copenhagen to tell Bohr, who was preparing to go to the Institute for Advanced Study in Princeton for several months. "He had only a few minutes with me," Frisch recalled, "but I had hardly begun to tell him when he smote his forehead with his hand and exclaimed: 'Oh, what idiots we all have been! Oh but this is wonderful! This is just as it must be!'"[27] Not long after arriving in the USA, Bohr gave a presentation at a conference attended by many famous scientists already involved in nuclear research, including Fermi. When Bohr stood up and announced that the atom had been fissioned, the effect on the audience was like the chain reaction Bohr thought was highly unlikely to be achieved. They raced back to their laboratories and began to attempt to duplicate the discovery. Bohr also cautioned that it would take "the entire efforts" of a country to build a bomb. He was wrong about a chain reaction being unlikely. He was right about it taking the entire efforts of a country—the USA.

In Germany, it was Harteck who first proposed to German Army Ordnance (Heereswaffenamt, known by the initials HWA) that atomic power might be used as a weapon. He had read a scientific paper on secondary neutrons and had instantly realized that an atomic bomb could be built because in the act of being fissioned, the uranium nucleus emitted several neutrons, setting in motion a chain reaction.[28] Harteck's letter was dated April 24, 1939 and was written with his assistant, Wilhelm Groth: "We take the liberty of calling your attention to the newest developments on nuclear physics, which, in our opinion, will probably make it possible to produce an explosive of many orders of magnitude more powerful than conventional ones. . . . That country which first makes use of it has an unsurpassable advantage over the others."[29] The grim irony is that to Harteck and Groth the proposal was nothing more than a means of coming up with a research project that would prevent them and their scientific colleagues at the University of Hamburg from being drafted into the German army. It would keep them busy and literally out of the line of fire. A few days before the letter to HWA, Harteck had written to Karl-Friedrich that, "I will be a soldier in fourteen days learning how operate a cannon. Please don't laugh." He then shared his thoughts that some uranium isotopes might be better than others for fission.[30] The immense ramifications of what he was proposing to HWA escaped him. In an interview conducted in 1967 by Joseph J. Ermenc, Harteck was asked why he and Groth wrote the letter. "I tell you that in those days in Germany we got no support for pure science," Harteck said. "We were very, very poor. So we had to go to an agency where money was to be obtained. I was always realistic about such things. The war office had the money and so we went to them. If we had gone somewhere else, we would have got nothing."[31]

Although Harteck didn't receive an answer for a long time, the letter had a major impact. In combination with meetings of the Reich Research Council, under the leadership of the physicist, Abraham Esau, and the HWA, whose main scientific advisor was the physicist, Kurt Diebner, the letter played a part in the establishment of the Uranium Club under the direction of Heisenberg. At the end of April 1939, an initial meeting was held on the prospects of atomic energy. One of the results was that Esau, a man with administrative experience as president of the national bureau of standards (Physikalisch-Technische Reichsanstalt) and who also had a better sense of the big picture than most of the other participants, immediately worked to ban the export of uranium from Germany and to gain control over the flow of uranium from Czechoslovakia.[32]

Josef Mattauch, Meitner's successor at the Kaiser Wilhelm Institute, attended the initial meeting with Esau as did Walther Bothe, Hans Geiger, and Robert Döpel. The next day he mentioned what had transpired at the meeting to Rosbaud who passed the information to Robert Hutton of Cambridge who passed it to John Cockcroft, head of Great Britain's newly formed committee on nuclear research, which became known by the odd name of MAUD (arising from a misinterpretation of a message Meitner had sent from Sweden).[33] The clouds of war were gathering thickly over Europe, and Rosbaud was already doing everything he could to keep communication on all types of scientific research (not just nuclear but also rocketry) flowing in one direction only: from Germany to England.[34]

The fact that Heisenberg's name was mentioned in connection with the formation of the Uranium Club (although he did not attend the April meeting) worried British intelligence and everyone involved with nuclear research outside Germany, as well as more than a few people inside.

His brilliance was legendary and, although certainly not a Nazi, he was at heart an ardent patriot who supported all things German and who, once war had begun, could not countenance Germany losing. Fritz Haber had declared during the First World War that the purpose of his scientific institute was, "For humanity in time of peace; for the fatherland in time of war." Heisenberg would have agreed, putting the emphasis on "fatherland," an ideal that predated Hitler. Besides fatherland, there was Heisenberg's unquestioning loyalty to pure science. According to David Cassidy in *Beyond Uncertainty: Heisenberg, Quantum Physics, and the Bomb,* Heisenberg turned around the official slogan of the government "we must make use of physics for warfare" to "we must make use of warfare for physics."[35] This was dangerously naïve. Yet he was strongly supported in his delusional thinking by his young associate, Carl Friedrich von Weizsäcker, who wrote after the war about his own strange idealism:

> I followed the path of the Third Reich in quiet conformity on the outside and with a great tension inside that never relaxed. I did not share a single opinion of the Nazi creed, came easily into emotional contact and conversation with regime opponents and expected something inexpressibly extraordinary out of the events. This was the background of my hopes of gaining access to power through the uranium researches: the dream of making a better future understandable to a man like Hitler.[36]

What did Weizsäcker, who knew without a doubt that he was researching a bomb, mean by "a better future"? Strongly influenced by Nazi ideology (his father was a career diplomat under Hitler) even though he himself never joined the party, Weizsäcker spent the rest of his life trying to accept guilt while subtly shifting the blame to the point of becoming blameless. Using obfuscating language was one method of achieving that goal, a method he and Heisenberg practiced from the start. For example, they spoke often of researching a nuclear *machine,* as if the bomb was a mechanical thing humming and whirling innocently—much like the use of the term *gadget* by the Allies. They claimed that this machine (actually a breeder reactor) could potentially be used to power submarines. All of this was true. However, the machine could also produce plutonium that could be used in a bomb. This language helped them gain some protective psychological distance, allowing them to dream "of making a better future."

Rosbaud emphatically trusted Karl-Friedrich as well as Hahn, Mattauch, Strassmann, and von Laue (although as Kramish phrases it, Laue was "the confidant of his *lesser* secrets").[37] But he never trusted Heisenberg, finding him egotistical in the extreme. He did not like being in the same room with him, a feeling that was mutual. Nor did he trust Weizsäcker, whom he considered far too close to the Nazis, and who was always whispering in Heisenberg's ear. This mistrust had taken root in the early 1930s and was solidified by several incidents during the war, including the highly controversial visit of Heisenberg and Weizsäcker with Bohr in occupied Copenhagen in September 1941. In January 1944, Rosbaud wrote to his brother, Hans, about the faculty at the Nazi-dominated University of Strasbourg, of which Weizsäcker was a member: "The physicist is by far the most cunning and diplomatic. His father is the well-known Herr v. W. Give him my best regards. He is a pure theoretician with a strong philosophical inclination, a good friend of Heisenberg. But here too, I keep my distance. . . . I recommend to you to restrict yourself to very objective things, and even there a certain reticence is desirable."[38]

However, at this point before the outbreak of war, Karl-Friedrich and Heisenberg were personal friends as well as colleagues at the University of Leipzig. Karl-Friedrich had known Heisenberg since his time at the Haber Institute when Heisenberg presented his groundbreaking work on quantum mechanics at Nernst's physics colloquium. Indeed, Karl-Friedrich's research on molecular hydrogen was built on Heisenberg's quantum mechanics. At Leipzig, they were heads of different institutes, but they saw each other frequently and were members of the Wednesday club that met on that particular day once a month in Berlin to discuss their research. Karl-Friedrich's focus had remained on heavy water, even though as has been pointed out already, he was changing course. One of the last papers Karl-Friedrich published before the war was written with J. Hochberg and was titled "Hydrogen Exchange of DiMethyl Sulfone and Sodium-Methane-Sulfonate in Alkaline Solution," which showed a gradual exchange of hydrogen atoms with those of the deuterium solvent. Meanwhile Heisenberg was researching cosmic rays and what were called explosion showers, seeking a revolutionary new theory, but it wasn't quite working out, and he was experiencing much resistance from other theoretical physicists. Not that Karl-Friedrich, a physical chemist, would have had much to add to the discussion on cosmic rays, yet he was aware of Heisenberg's research as Heisenberg was of his.

Heisenberg and his wife were invited to Karl-Friedrich's fortieth birthday party that was attended as well by Karl and Paula Bonhoeffer, who had traveled from Berlin for the special occasion. On January 23, 1939, Paula wrote to Dietrich, "We have been with Karl-Friedrich in Leipzig on his fortieth birthday. In the evening we had a very happy time with Heisenberg and his wife."[39] However, in the same letter Paula hinted at darker things. Without going into any detail, she expressed her concern about Sabine and Gert Leibholz, then in exile in England. She spoke of a pleasant weekend visit the Leibholz family had spent with Uncle George, without mentioning that Uncle George was Bishop George Bell of the Church of England and Dietrich's primary contact for the resistance. She warned Dietrich that the display kiosks in Berlin listed military conscription for those born in 1906 (the year Dietrich was born) and 1907. She mentioned that Dohnanyi was coming for a brief visit, that he was depressed and very sorry about the transfer of Fritz Wiedemann to San Francisco. She did not add that Wiedemann was a Nazi adjutant who had fallen out of favor with Hitler. She wrote, "What will happen with Uncle Rudi is truly very worrisome." Uncle Rudi was an elderly uncle whose name was used in family correspondence as the code name for Hitler. Then Paula mentioned a Jewish friend who had lost his job as an attorney. By the time Paula signed "Your Mother" at the end of the letter, there was much information conveyed to Dietrich slantwise, almost none of it good.

Notes

1. Dahl, *Heavy Water*, 24.
2. "James Chadwick," en.wikipedia.org/wiki/James_Chadwick.
3. For information on Nobel Prizes, see www.nobelprize.org/nobel-prizes/lists/all.
4. Harold Urey, en.wikpedia.org/wiki/Harold_Urey.
5. Ernest Lawrence, en.wikipedia.org/wiki/Ernest_Lawrence.
6. Dahl, *Heavy Water*, 55.
7. Paul Harteck interview with Joseph J. Ermenc, July 6, 1967, PHP, Box 10, Folder 2, RPI.
8. Michael Polanyi, *Personal Knowledge* (Chicago: University of Chicago Press, 1958), 111.
9. Dahl, *Heavy Water*, 327. See endnote 9-8. "W. E. Tisdale to F Haber, 20 February 1933: PHP, Acc. 85-22, Box 1, Folder 3."
10. Dahl, *Heavy Water*, 135. See endnote 9-21, Harteck cited by Ermenc.

11. PHP, Ermenc interview, RPI.
12. Jonothan Logan, "The Critical Mass," *American Scientist* 84, no. 3 (May–June 1996): 264.
13. Klaus Hentschel, ed., *Physics and National Socialism: An Anthology of Primary Sources* (Basel: Birkäuser Verlag, 1996), lxxvi.
14. Dahl, *Heavy Water*, 134–135. See endnotes 9-18, 9-19.
15. Ibid., 31. See endnote 4-37.
16. Ibid., 42. See endnote 4-57.
17. Others with whom Karl-Friedrich co-authored papers were R. Klar, K.W. Rummel, K. Wirtz, W.D. Waters, J.C. Jungers, and J. Hochberg. Jungers was head of the department of physical chemistry at the University of Louvein in Belgium. G.M. Begun, ed. *Isotope Separate and Isotope Exchange: A Bibliography with Abstracts* (ORNL-2852 October 14, 1959), n.p.
18. Dahl, *Heavy Water*, 43. See endnote 4-62.
19. Ibid., 43. See endnote 4-65. "Bonhoeffer to Hydro, 29 July and 20 November 1933; NHAO, K03. 50. n. 2.1: Tungt vann, teknisk, Hefte 1".
20. Begun, *Isotope Separation*, n.p.
21. Ulf Messow, "Bonhoeffer," 2.
22. Coster helped save many Jews during the German occupation of the Netherlands.
23. Frisch, *What Little I Remember*, 113–114.
24. Kramish, *The Griffin*, 228.
25. Frisch, *What Little I Remember*, 116.
26. Kramish, *The Griffin*, 51.
27. Frisch, *What Little I Remember*, 116.
28. A paper on chain reactions was published by Siegfried Flügge and Gottfried von Droste. Flügge became a member of the Uranium Club.
29. The original of this letter has proven hard to locate; Harteck's own copy was lost when the British authorities emptied his desk in his Hamburg Institute in 1945. Also influential was a letter from Georg Joos to the education minister, proposing that uranium fission be studied for energy production. Dahl, *Heavy Water*, 132, 327. See endnote 9-4: "Harteck P and Groth W to Army Ordnance, 24 April 1939; PHP, Acc. 85-22, Box 2.
30. Letter from Harteck to Karl-Friedrich Bonhoeffer, April 21, 1939. PHP, Collection MC17, Box 2, Folder 1. RPI.
31. In a letter to his son written in late 1946, Max von Laue corroborated Harteck's statement about the funding of research that was designated as "decisive for the war effort" in official files, explaining that even his books on X-ray and electron interference and Heisenberg's book on cosmic rays were so designated, otherwise they would not have been published.

"And if someone wanted to research persistently through the files of the final years of the war, he would notice that absolutely everything conducted in science was 'decisive for the war effort.' Otherwise, the state and party agencies would have granted neither the means nor the co-workers necessary." That designation saved many young scientists from going to the front. Alan D. Beyerchen, *Scientists Under Hitler: Politics and the Physics Community in the Third Reich* (New Haven: Yale University Press), 190–191. See also PHP, Ermenc interview, RPI.

32. Lawrence Badash, Elizabeth Hodes and Adolph Tiddens, "Nuclear Fission: Reaction to the Discovery in 1939," *Proceedings of the American Philosophical Society* 84, no. 3 (June 1986): 215.
33. Kramish, *The Griffin*, 54.
34. Dahl, *Heavy Water*, 130–131.
35. Transcript of interview of Heisenberg by J.J. Ermenc, August 29, 1967. See Cassidy, *Beyond Uncertainty*, 659.
36. Carl Friedrich von Weizsäcker, *Pioneer of Physics, Philosophy, Religion*, 11. Online *Springer Briefs on Pioneers in Science and Practice* (2015), www.springer.com/us/book/9783319134451.
37. Kramish, *The Griffin*, 212.
38. Ibid., 116.
39. DBW, Vol. 15, 113.

CHAPTER 10

The Summer of Decision

In the summer of 1939, Dietrich, Karl-Friedrich, and Werner Heisenberg visited the USA. The brothers sailed a few weeks apart on the *SS Bremen*, the same ship that Karl-Friedrich had taken ten years before in 1929 on his triumphal trip to the annual meeting of the American Chemical Society. Heisenberg sailed on June 17 on the maiden voyage of the *RMS Mauritania*, the largest ship ever built in England and the pride of the Cunard-White Star Line.[1] However, these voyages were not celebratory; they were freighted with unrelenting questions of great magnitude, particularly for the Bonhoeffers. Was war unavoidable, or was there a chance, even if slim, to maintain peace? Could the terror with which Nazism had inundated Germany be stopped before it flooded all Europe? Did the fissioning of the atom mean that a bomb of horrific destructiveness would inevitably be built? What was God's will? Or did God's will have no place in the world of willful men?

Although the Bonhoeffers and Heisenberg had different reasons for coming to the USA, they would all leave with one goal: to get back to Germany before war broke out. Only Heisenberg would consider his return as an opportunity for personal glory, giving him the chance to use his undisputed brilliance to help the fatherland be victorious. Dietrich and Karl-Friedrich would give up any hope for personal glory and would return to help the fatherland be defeated. For Dietrich, that meant the time had come "not just to bind up the wounds of the victims beneath the wheel, but to seize the wheel itself."[2]

K. L. Housley, *The Scientific World of Karl-Friedrich Bonhoeffer*, Palgrave Studies in the History of Science and Technology, https://doi.org/10.1007/978-3-319-95801-9_10

Dietrich had accepted a short-term lecture position at Union Theological Seminary, which had been offered to him with the express purpose of getting him out of harm's way. Sailing on June 2 from Southampton, England, on the five-day cruise to New York City, Dietrich jotted in his diary that the weather was "glorious and the sea utterly still." Perhaps this was his chance to start anew, or at least to catch his breath without the constant fear of arrest. He hoped to be able to return to his fledging seminary to continue training pastors for the dangerously beset Confessing Church, which had been struggling to maintain its autonomy in the face of the forced Nazification of Protestant churches.[3] Perhaps he would return as early as the fall, perhaps in a year. Originally located in an abandoned school building in the town of Finkenwalde near the Baltic Sea, the seminary had been shuttered by the Gestapo in September 1937. Undeterred, Dietrich had begun again, gathering a small group of men as inconspicuously as possible in rural Pomerania, living first in an empty parsonage and then in a remote farmhouse without electricity. To disguise their true scholastic purpose, they were not called seminarians but "apprentice vicars."[4] As for the seminarians trained by Dietrich at Finkenwalde, many had already been conscripted into the army or had been arrested. In fact, it was the imminent danger of his own conscription or arrest that had finally forced Dietrich to come to the USA. In Germany, conscientious objectors were executed.[5]

In March 1939, while visiting England, Dietrich had asked advice of George Bell, Bishop of Chichester, whom he had met in the early 1930s. Wise and kind, Bell was one of the few men Dietrich could turn to because he was outside Germany, had a broad ecumenical viewpoint, and had numerous contacts around the world. Dietrich had explained to him that he did not wish to escape from the situation in Germany, but neither did he want to stay if he was doing more harm than good:

> It seems to me conscientiously impossible to join in a war under the present circumstances. On the other hand, the Confessional Church as such has not taken any definite attitude in this respect and probably cannot take it as things are. So I should cause a tremendous damage to my brethren if I would make a stand on this point which would be regarded by the regime as typical of the hostility of our church towards the state. Perhaps the worst thing of all is the military oath [to Hitler] which I should have to swear. So I am rather puzzled in this situation, and perhaps even more,

> because I feel, it is really only on Christian grounds that I find it difficult to do military service under the present conditions, and yet there are only a very few friends who would approve of my attitude. In spite of much reading and thinking concerning this matter I have not yet made up my mind what I would do under different circumstances. But actually as things are I should have to do violence to my Christian conviction, if I would take up arms "now and here." I have been thinking of going to the Missionfield, not as an escape out of the situation, but because I wish to serve somewhere where service is really wanted.[6]

The news of Dietrich's troubles had also reached Reinhold Niebuhr who was in Edinburgh, Scotland, in the spring of 1939, delivering the prestigious Gifford Lectures, his topic being "The Nature and Destiny of Man." Meeting with Dietrich in Sussex in early April, Niebuhr realized the severity of the situation and immediately began the process of finding him employment in the USA. Niebuhr wrote to Henry Leiper, executive secretary of the Federal Council of Churches in New York City, that Dietrich embodied "the spirit of the struggle over there [Germany] so well in his person that I believe his coming would be valuable for our sake as well as for his."[7]

Despite the interventions on Dietrich's behalf, nothing worked out as planned. The distance—actual, psychological, spiritual—between the precarious seminary in Pomerania and Union Theological Seminary up on the rocky heights by the Hudson River proved to be too great. As will be seen, the storm that immediately engulfed Dietrich upon his arrival in New York City threatened to drown him in self-doubt and guilt. Not until he prayerfully had made up his mind and was back on the ship heading home did the storm subside and stillness enter his heart.

For his part, Karl-Friedrich was considering a professorship in physical chemistry he had been offered at the University of Chicago. His friend, James Franck, who had taken a position there in 1938, was probably instrumental in extending the job offer. An outstanding experimental physicist, Franck had resigned from his position at the University of Göttingen in 1933 instead of waiting to be hounded out by the Nazis. Franck's courage had served as an example to Haber and many others because he did not go silently, but publicly decried the outrageous treatment of Jews as "aliens and enemies of the fatherland."[8]

By the late 1930s, Franck was in the process of shifting his research focus to the physics of living tissue, just as Karl-Friedrich was doing.

Interested in photosynthesis, Franck's first experiments at Chicago had to do with the effect of light intensity on hydrangea leaves. "I'm possessed again with photosynthesis," he wrote to a friend, "which is proving to be a cross between a Moloch and a Medusa."[9] His coauthor was Edward Teller, with whom Franck had worked at the University of Göttingen and who had come to George Washington University in 1935. According to Frank von Hippel, Franck's grandson, Franck wrote to Lise Meitner about his new field of research:

> It's completely different from physics, where the simplest solution is almost always the right one. But in living tissue, that's absolutely not the case. Nature has had the time to take very complicated paths to the solution, because they mostly offer safeguards against derailment. I believe I understand the physico-chemical side of photosynthesis. The chemistry, which actually interests me less, I try to understand as far as I can.[10]

A sea change was happening in biology, driven not solely by biologists but by physicists, chemists, and physical chemists who were intrigued by the congruence of their fields. So also Max Delbrück, then at Caltech on a Rockefeller Foundation grant, was in the process of switching from theoretical physics to studying the genetics of the fruit fly and bacteriophages, about which he had kept Karl-Friedrich informed by letter. In fact, it had been Karl-Friedrich and the physicist Max Born who had initiated Delbrück's major change in direction. Eight years before in 1931, they had jointly recommended to the Rockefeller Foundation that Delbrück study with Niels Bohr in Copenhagen. Paradoxically, Delbrück's time with Bohr ignited his interest in biology. Listening to Bohr struggle to extend the quantum mechanical implications of complementarity to living organisms (as well as a host of other fields, including psychology and philosophy), Delbrück had found himself intrigued, disturbed, and annoyed all at the same time, his annoyance stemming from Bohr's vagueness. "It was significantly intriguing for me, though, to decide to look more deeply specifically into the relation of atomic physics and biology. And that means learn some biology."[11]

Even more important than the prospect of doing research on living organisms, Karl-Friedrich longed to get his family to a place of safety, as Franck had done for his family. Complicating the matter were Karl-Friedrich's parents who would probably not be willing to move to the USA and for whom he felt responsible as the eldest son.[12] Because

Karl-Friedrich had turned forty in January, 1939, he could not be conscripted into the military. Even so, the danger for his family was high.

Then, too, there was Karl-Friedrich's research on heavy water. Being a recognized expert in Germany meant that he was ipso facto a recognized expert in the world, and as such, his advice was sought by scientists in the USA, among them Enrico Fermi. In the summer of 1939, Fermi was still at Columbia University, pondering chain reaction experiments and the efficacy of heavy water as a moderator. Since the two men met at least once during Karl-Friedrich's visit, it is possible that Fermi consulted him. In a letter dated July 9, 1939, on letterhead of the University of Michigan where he was visiting, Fermi wrote to Leo Szilard about potentially using carbon in the form of graphite bricks to slow down neutrons instead of using heavy water. "I have discarded heavy water as too expensive; but if you can easily get several tons of it, it would work very nicely."[13] Over the next few years, the USA would increase the production of heavy water, relying initially on a Canadian source, an electrolytic-hydrogen plant on the Columbia River in British Columbia.[14] However, in 1939, the amount available was small. Just a few years later, the University of Chicago would become the center for atomic research under Fermi's guidance. There on December 2, 1942, Fermi and his team would achieve the first sustained chain reaction using graphite instead of heavy water in a uranium pile. The experiment was conducted on an unused squash court under the stands at the football field. Had Karl-Friedrich accepted the position at the University of Chicago, it is possible that he would have become involved in the Manhattan Project as both Fermi and Franck did. In what would have been an ironic twist of fate, in attempting to avoid working on a bomb for Germany, he might have ended up working on a bomb for the USA.

Secrecy had not yet descended on the topic of achieving an explosive nuclear chain reaction, despite Szilard's valiant attempt to convince Fermi and the Joliot-Curies not to publish their findings. In fact, the opposite was the case. The newspapers were full of speculation about the destructive power of the atom. Reporting on a regional meeting of the American Association for the Advancement of Science, William L. Laurence, a science writer for the *New York Times*, wrote an article titled "New Key is Found to Atomic Energy" dated May 5, 1939. He reported that a tiny amount of what was called actino-uranium (actually U-235) "would blow a hole in the earth 100 miles in diameter." Laurence likened it to a philosopher's stone that when separated in pure form could yield 235 billion volts per atom. "It would wipe out the entire City of New York, leaving a deep crater half way to Philadelphia and a third of the way to Albany and out to Long Island as far as Patchogue."[15]

There was little discussion at this point about the ethics of building an atomic bomb. An editorial with the prescient title "Incomparable promise or awful threat?" in the July 1939 issue of *Scientific American* warned that scientists could not control their own discoveries and that "if the human race won't leave its new playthings alone, and gets badly hurt, that's its own funeral. In a few years we may have the answer." Yet the editorial writer also recognized that if physicists decided not to pursue atomic research, there was the danger that "world conquerors"—meaning Germany—would pursue it.[16] This was the threat that was bedeviling the exiled physicists who knew, by their own terrifying experiences, what Hitler was capable of. They also knew of the daunting intellectual power of German science even in its diminished form. In his memoir *My Life & My Views*, Max Born, who had been Franck's colleague at Göttingen, summed up the magnitude of the threat: "I had to leave Germany like many others, and I have witnessed the terror that engulfed the rest of the world when Hitler's initial successes made it appear possible that he might subjugate all the peoples of the world. Had Germany been able to construct an atomic bomb before the other nations there would have been no salvation."[17]

On the issue of ethical responsibility, Lawrence Badash, Elizabeth Hodes, and Adolph Tiddens provide valuable insight in their article "Fission: Reaction to the Discovery in 1939" in the *Proceedings of the American Philosophical Society*. They point out that the problems of achieving a bomb were staggeringly large. Much basic research was required before any practical application could be considered. In fact, it was not at all clear that U-235 could be separated at all. The possibility was there, but it would be a very long time before it could be achieved. (No one could foresee the herculean effort of the Manhattan Project funded by the government.) Furthermore, scientists preferred to stick to their data from which they extrapolated carefully. They were not prone to ethical speculation:

> While scientists are creative people in their own realm, they tend elsewhere to be no more imaginative, prescient, or possessed of extraordinary foresight than other humans. To expect that in 1939 they should have focused

> quickly upon the significance of a new discovery, explored carefully its potential consequences, and weighed in the balance the question of their own involvement exaggerates the abilities and orientations of scientists.[18]

While in 1939 "it was simply too soon for morality and ethics to be of more than evanescent concern" to scientists, there were remarkable exceptions, mostly exiles, one of them being Szilard, and another being Einstein, who up until then had been a pacifist. On August 2, 1939, Szilard drafted a letter to President Franklin Roosevelt, with input from Einstein and Alexander Sachs, an economist and lawyer with connections to the White House. For maximum effect, the letter was signed by Einstein. It raised the possibility of a chain reaction and the development of a bomb. Szilard wrote, "A single bomb of this type, carried by boat and exploded in a port, might very well destroy the whole port together with some surrounding territory." The suggestion was made that a liaison be appointed between the Roosevelt administration and physicists. With all the crises confronting the president, Sachs did not deliver the letter immediately, waiting for the right time. He finally delivered it personally on October 11 when he spoke to the president and ordnance experts about its content. This led to the establishment of an advisory committee under Lyman Briggs, director of the National Bureau of Standards, which would lead inevitably to the Manhattan Project.[19]

At that time, no one knew where Heisenberg stood on atomic fission because he had not been deeply involved in nuclear research, nor had he shown any interest in practical applications, being to his core a pure theoretician (Fig. 10.1). However, because he was brilliant in one sphere, many people (including himself) thought erroneously he was brilliant in all spheres. Heisenberg had not come to the USA to look for a job, as had Dietrich and Karl-Friedrich. He had been offered several positions over the years and had turned them down. Instead, he was on a major lecture tour, one of his topics being his controversial new cosmic-ray meson theory. His itinerary took him across the country, including a physics symposium at the University of Michigan attended by Edward Teller, who had studied under Heisenberg, seminars at Purdue University in Indiana, a conference at the University of Chicago on cosmic rays, and lectures at Columbia, one of the universities that had offered him a position the previous year. More times than not, Heisenberg lectured to standing-room-only crowds. In Chicago, he argued vehemently with J. Robert Oppenheimer about his meson

Fig. 10.1 Werner Heisenberg. Alpha Historica/Alamy Stock Photo

theory.[20] The men had first met when Oppenheimer had studied under Max Born at the University of Göttingen in 1926. Heisenberg was personable, but when it came to defending one of his scientific ideas, he could be implacable.

In a letter to his wife, Elisabeth, whom he affectionately called Li, Heisenberg said nothing about scientific arguments, writing instead about the unbearably hot and humid weather and how nothing had changed since he had been in Chicago ten years before. "The day before yesterday, I was invited to the house in which I had lived for half a year. The piano stood in the same place, and I had to, of course, play it. —I also met Bonhoeffer today. He is soon going back home and will bring you greetings and catch you up."[21]

An accomplished pianist, Heisenberg never missed an opportunity to play no matter where he was or what the circumstance, nor did he fail to mention to Elisabeth what composers he had played (Handel, Brahms, Mozart, Debussy, Vivaldi, etc.) as well as the poor quality of American pianos, a flaw he attributed to the steamy weather. Much to his joy, his hosts at Purdue had managed to round up two violinists and a cellist to form a quartet. "We went over the Beethoven and the two Mozarts: that was such a great delight. In spite of the bad piano (on some keys the ivory was missing, in the higher ranges the tone was reminiscent of a stone axe), I believe I played quite decently."[22]

Heisenberg was the man of the hour, but the hour was quickly passing, and physicists wanted to see him one last time. Some suspected him of being a spy. Others suspected he was being spied on. Heisenberg's serious run-in with Nazis over his support of theoretical physics (considered to be of Jewish origin and therefore false) as opposed to experimental, results-oriented Aryan physics, had cost him dearly. He had been refused the coveted position of chair of theoretical physics at the University of Munich, which had become vacant on the retirement of the eminent physicist, Arnold Sommerfeld. By the time of his trip to the USA, Heisenberg had been exonerated of the Nazi charges (with the intervention of Heinrich Himmler whose mother was a friend of Heisenberg's mother), but there was still mistrust and resentment. Physicists in the USA, particularly those who had been forced to leave Europe, including Fermi who had come from Italy, could not comprehend how Heisenberg could continue to be so loyal to Germany.[23]

Most troubling were the confounding moral inconsistencies in the man, for example, while he had fought against the imposition of Aryan physics, he had acquiesced to Nazi demands that Einstein's name not be used in reference to relativity, as if that theory had sprung into existence on its own, instead of arising in the mind of a Jew.[24] By 1939, Aryan physics reached its peak in terms of influence on research, but it continued to exert heavy psychological pressure on scientists to silence and self-police themselves, one indication being either the use of an asterisk in a scientific publication to denote a Jewish name, or the deletion of the name completely.[25]

Also unsettling to émigré scientists was Heisenberg's strange intervention on behalf of Max Born, who had been his mentor and with whom Heisenberg had collaborated on quantum mechanics. Heisenberg had managed to get dispensation for Born from the Aryan clause so he could return to his position as head of the physics department at Göttingen from which he had been expelled in 1933. Subsequently, Born was astonished to discover that the offer to return did not include his family. It had not been made out of a sense of compassion but because Heisenberg believed German science would suffer a serious setback without Born. According to Paul Lawrence Rose in *Heisenberg and the Nazi Atomic Bomb Project*, "Heisenberg clearly realized the shamefulness of what he was doing, but nevertheless he felt justified in making the proposal by the knowledge that he was serving the great god of physics—and of Germany."[26]

Everywhere Heisenberg went in the USA he met with colleagues who had been driven out of Germany, including Lothar and Gertrud Nordheim, former physics colleagues from Göttingen who had gone first to Purdue and then to Duke, and the physicist Hans Bethe who was at Cornell. Some, including Bethe, tried to convince Heisenberg to stay in the USA. Others sought his opinion about the prospects of war. Depending on who remembered the conversation, Heisenberg came across as evasive, patriotic, egotistical, sincere, or detached. He crowed to Arthur Compton at the University of Chicago that Germany was ready for war, an opinion he also voiced to Rudolf Peierls. He gave Isidor Rabi, then at Columbia, the impression that he was worried about losing his high academic standing in Germany if he did not return. To some people, he seemed more concerned with the potential takeover of Europe by the Soviet Union than by Germany, as if Germany were the lesser of two evils, an argument he is said to have made to Niels Bohr on his highly controversial visit to occupied Copenhagen in September 1941.[27]

While in Ann Arbor, traveling at that point with the Nordheims, Heisenberg stayed in the home of Samuel Goudsmit, a Dutch physicist of Jewish descent, who had come to the University of Michigan in 1927. Toward the end of World War II, Goudsmit would become scientific head of the Allied intelligence mission in Europe named Alsos, meaning *grove* in Greek, an oblique reference to Leslie Groves, military leader of the Manhattan Project. In that position, Goudsmit would have many more dealings with Heisenberg, few of them salutary, but at this point they were still friends. On a hot July day, Goudsmit and Fermi asked Heisenberg directly why he did not come to the USA. Goudsmit recalled him saying "Germany needs me,"[28] adding that Heisenberg expected Hitler's excesses to blow over, as if the six years that had gone by since Hitler's rise to power in 1933—years filled with expulsions, concentration camps, the burning of synagogues, torture, and murder—only added up to a passing inconvenience. This belief struck Goudsmit as "stupidly optimistic."[29]

In his autobiography written years after the war, Heisenberg recalled his having responded as follows: "If I abandoned them [the German scientists] now, I would feel like a traitor. I don't think I have much choice in the matter. I firmly believe that one must be consistent . . . People must learn to prevent catastrophe not to run away from them. Perhaps we ought even to insist that everyone braves what storms there are in his own country."[30] As with so many of Heisenberg's statements, this one sounds uplifting until it is analyzed, whereas it falls apart. For example, it was disingenuous of him to look at Szilard, Bethe, Stern, Franck, the Nordheims, and all the others who had been mercilessly hounded from Germany and say they were running away. If he was referring to non-Jewish scientists, who did Heisenberg consider to have run away? Herbert Jehle, the Quaker physicist who was a friend of both Karl-Friedrich and Dietrich, who was imprisoned in a concentration camp in France? Max Delbrück who had taken the Rockefeller Foundation grant to Caltech, having been denied academic employment in Germany for refusing to sign the loyalty oath? Paul Ewald who spoke out against the banishment of Jewish scholars and then resigned his position, knowing he was about to be fired? None of these men could have worked from *inside* to save the world of pure science because the moment they took an ethical stand they were no longer *inside*, they were thrown *outside* without any power to effect change.

Furthermore, Heisenberg made other statements on his trip that indicated he did not believe he was headed back into a catastrophe that he personally had to prevent in solidarity with the German scientists who had gathered around him. Rather, he believed he was returning to the potential for success in which he would be a contributing factor. In *Beyond Uncertainty*, David C. Cassidy states that to Heisenberg, remaining in Germany was worth any price as long as he could continue to work and teach, and that he saw his survival in the Third Reich as the survival of decent physics. "After all, Planck had given him the task of preserving an 'island of existence' in the sea of anti-scientific chaos about him. And as long as he survived, that island would survive. But by seeing himself in such a grandiose rationalization for remaining in Germany, he more easily succumbed to further compromises and ingratiation with the regime."[31]

To his wife, Elisabeth, Heisenberg wrote rather bland letters from the USA, avoiding controversial subjects and revealing no angst on his part about returning home. For example, on July 13, he wrote from Purdue University that people had asked him under what conditions he would stay in the USA; but Heisenberg was sure that he and his family would not feel at home outside of Germany, the children would grow up speaking English, and the atmosphere would be foreign. "That would not be nice at all, and so we are just staying put." On July 14, he wrote from New York City that, "In Ann Arbor I met up with the Fermis right away at a party, and they told me much about their experiences during their recent months. I also met their children; they already spoke English rather well. To see a few old acquaintances again gave me great joy. Much has changed there but the friendly calm of this town, with gardens everywhere, has remained the same." He had further talks with Fermi and other scientists on a visit to the Brookhaven National Laboratory on Long Island. Then, on August 8, Heisenberg wrote to Elisabeth about preparing to sail home. He had taken the opportunity of free time in New York City after his lectures at Columbia to visit the Empire State Building, which he liked very much, comparing it and the other skyscrapers to mountains such as the Dolomites. When they were lit up at night, they reminded him of fairy castles. "From up above, I have also spotted the ship in the harbor that brings me to Germany. Tonight I go on board and after midnight, we will depart. I am looking forward to the sea voyage, and of course so much more to seeing you all again."[32]

The summer trip was an utterly different experience for Dietrich and Karl-Friedrich than it was for Heisenberg. From the moment he arrived at Union Theological Seminary, Dietrich was torn apart about his decision to come. In Germany, he had been officially prohibited (or unofficially warned away) from teaching, speaking publically, preaching, traveling to certain locations, and publishing his articles and books, of which he had thus far written five, the most recent being *Discipleship* (1937) and *Life Together* (1939).[33] The international ecumenical work in which he had been involved during the 1930s was faltering. So also was his pastoral work with the Confessing Church, which had been so hemmed in by destructive decrees, its very existence hung by a frayed thread. Hundreds of pastors had been jailed. Martin Niemöller, one of the founders of the Pastors' Emergency League, was in a concentration camp. Yet Dietrich's work with his outlawed seminary in Pomerania gave him great personal fulfillment, even hope. In New York City, he missed profoundly the daily interaction with the small group of dedicated students.

In a letter dated January 29, 1939, four months before the brothers sailed for the USA, Dietrich wrote to Karl-Friedrich from one of the small parishes in Pomerania. He started by offering Karl-Friedrich best wishes on his fortieth birthday, which put him beyond the age for compulsory military service, writing, "I envy you for having reached this number!" (Dietrich and Sabine turned thirty-three on February 4, 1939.) Then, Dietrich shared with him both his discouragement and resolve:

> It has been at times very depressing in the last few weeks, when one must see how many people are seeking quiet and security at all costs and using all kinds of pretexts and reasons. In such times, which come again and again, there is always a great deal to do with visits, lectures, etc. It is entirely certain to me that for the church it all depends upon whether we now hold on, even with great sacrifices. The great sacrifices are now small compared to what we would lose by wrongly giving in. I wouldn't know anything worth a full commitment today if not this. It matters not how many there are but rather only that there are still some who are steadfast. Of course, some things are more difficult for the married people, but I think sometimes much is also easier. I admire greatly the courage of the pastors' wives, who are willing to take everything on rather than to advise their husbands to give in. Often too, the parishioners are much clearer and more resolved than their pastors.[34]

In the struggle to keep the frightened pastors of the Confessing Church from capitulating to the Reich Church, Bonhoeffer had found an ally in Ruth von Kleist-Retzow, who had a home in Stettin where Finkenwalde Seminary was located. Aristocratic and strong-minded, she was opposed to Nazism and supported the Confessing Church. This support often included providing the seminary with food and supplies. She invited Dietrich to spend his vacations at her estate so that he could pursue his theological writing in solitude. It was there that Dietrich came to know several of Ruth's children and grandchildren, including Maria von Wedemeyer, to whom he would become engaged in January 1943, when she was only eighteen years old and he was about to turn thirty-seven. He would also come to know Ruth's son, Hans Jürgen, who would be arrested for his role in the attempt on Hitler's life in July 1944.[35]

Besides his connections through Ruth von Kleist-Retzow, Dietrich had become involved in the resistance through his brother-in-law, Hans Dohnanyi, who had played a role in planning the aborted 1938 coup attempt against Hitler. That attempt was to be triggered by Hitler's order to attack Czechoslovakia, which the conspirators presumed would force Great Britain and France to declare war against Germany. They also presumed that German military leaders would find the prospect of such a war so absurd that they would support Hitler's overthrow. Then, he would be arrested, at which point, Karl Bonhoeffer, in his role as a psychiatrist, would diagnose him as insane, which would be used as evidence in a subsequent civil trial. That was the plan until September 1938 when British Prime Minister Neville Chamberlain, desperately attempting to maintain peace, signed the Munich Agreement, indicating thereby that neither Great Britain nor France would declare war in the event Hitler seized Czechoslovakia, which is exactly what he did in March 1939.

Dohnanyi was serving as a supreme court judge in Leipzig while traveling weekly to Berlin to give lectures, which provided the cover to meet with resistance leaders, among them Hans Oster and Carl Goerdeler. When in Berlin, he stayed with his in-laws. During the summer of 1939, he joined the staff of the Abwehr (the military counterintelligence branch) under the leadership of Admiral Wilhelm Canaris. His appointment as "special leader" was actually a way to keep him from being called up for military service.[36] He continued to keep the detailed record of Nazi crimes, known as "the chronicle of shame," which was stored in a safe at the Zossen military base with the intent of using it in war crimes trials after the war.[37] Klaus Bonhoeffer and Rüdiger Schleicher were also involved in overthrow plans that now included the murder—not just the arrest—of Hitler. So also was Christel Dohnanyi.

The extent of Dietrich's involvement during this period is unknown because many of the pages of his diary for 1938–39 were torn out.[38] What is known is that he was perplexed about what part he, a theologian, could or *should* play. He knew he was not as legally canny as Dohnanyi. Would he be more a liability than an asset? Dohnanyi was linked to the leading figures of the resistance. However, Dietrich had a few important contacts of his own, chief among them being Bishop George Bell who had the ear of British political leaders, although some of them feared he might be too gullible and overly sympathetic to the German resistance.[39]

There were also the personal, day-to-day struggles. Dietrich was receiving financial support from his parents. Behind the scenes, his father had pulled many official strings to keep Dietrich out of trouble. For example, in 1938 Dietrich had been banned not just from living in Berlin but also from visiting his family there. Even so, Karl had gotten him permission to stay with them when he was in the city. So also Karl managed in 1939 to get Dietrich's military physical examination postponed for a year, the reason being Dietrich's upcoming trip to the USA.[40] Over and over, there can be glimpsed in the letters the hint of the father's preserving hand upon the shoulder of his youngest son.

For all these reasons, it seemed prudent for Dietrich to accept the offer from Union as well as an offer from the Federal Council of Churches. But no sooner had the ship docked in New York than Dietrich decided he had made a huge mistake. His longing to return immediately was acute, filling him with despair. He wrote in his diary on June 15, "The full force of self-reproaches about a wrong decision comes back up and is almost suffocating."[41] He read scripture, walked the sweltering streets of New York, and smoked pack after pack of cigarettes, all the while berating himself for being indecisive.

The next day, June 16, Dietrich visited the New York World's Fair, perhaps to shake his mind loose from the turmoil. However, instead of distracting him from the problems of the world, the fair only served to focus his attention on them. The fair had opened in late April in Flushing Meadows on the site of a former ash dump transformed into a fairyland, showcasing worldwide peace and prosperity. Every attendee was given an "I have seen the future" button, but it is doubtful Dietrich wore his because the future trumpeted at the fair bore no resemblance to the future he saw coming.

The New York World's Fair presented a future scrubbed clean, literally. One of its most popular attractions was a seven-foot robot named Elektro that did household chores. There was also General Motors' immaculate automobile-based metropolis. Most unbelievable of all, there was something called a television on display at the pavilions of RCA, General Electric, and Westinghouse, on which President Roosevelt's opening-day speech was telecast, to the amazement of viewers, some of whom doubted they were seeing it live. At the fair, there was no talk of persecution, concentration camps, and dictatorships. There was middle-class bliss.

At the center of the fair soared the futuristic tower, the Trylon, while next to it curved the dome of the Perisphere, within which visitors could view a "model city of tomorrow." Avenues radiated out like spokes along which were modernist pavilions for corporations to unveil their new technological wares and for nations to showcase their cultural wealth. Each avenue was lit at night with a different color that deepened as a visitor moved further from the center, for example, changing from light rose to deep magenta and from cerulean blue to ultramarine, creating a surreal aura. As the fair's official pamphlet declared: "The eyes of the Fair are on the future – not in the sense of peering toward the unknown nor attempting to foretell the events of tomorrow and the shape of things to come, but in the sense of presenting a new and clearer view of today in preparation for tomorrow; a view of the forces and ideas that prevail as well as the machines."[42]

Dietrich noted in his diary that he found the Russian pavilion too pompous and bourgeois. "The other pavilions were too commercialized." Political posturing and subtle messages about readiness for war were rife. For example, the Polish pavilion was among the most spectacular, overflowing with gleaming medieval armor, ancient weaponry, and artworks, all shipped over just before the fair opened. The overt intent was to make a statement about Poland's long illustrious history and its importance in modern Europe. The covert intent was to ward off attack. At the pavilion's entrance was a replica of a monument in Warsaw commemorating a victorious battle in the Middle Ages. When the fair opened, the Polish ambassador made the connection between the statue and his country's acute awareness of the rising threats from Germany and Russia, delivering a sharp warning that was reported in

The New York Times. "Over the miniature Poland with its story of peaceful pursuits and achievement, there stands on guard the symbol of armed might which once before saved Poland from an armed invasion." Sadly, most of the prized possessions inside the pavilion would never return to Poland, being sold off in the USA after Poland was overrun by Germany in September 1939. The replica at the entrance would find its way to Central Park, while the original monument in Warsaw would be melted down for bullets by the Nazis.[43]

Tellingly, there was no German pavilion, supposedly for financial reasons, although the primary reason was Hitler's anger at anti-German sentiment in the USA. None of these issues were lost on Dietrich, who was aware of the darkness beneath the surface glitter. In particular, he was nonplused by the Temple of Religion wherein Protestants, Catholics, and Jews took turns giving talks. The entrance was inscribed with the words, "For all who love God and prize religious freedom," but Dietrich found it to be more like a "dreadful movie theatre," crowded with a thousand folding chairs.[44]

After arriving in New York City, Heisenberg also had the opportunity to visit the World's Fair before heading to Chicago by train. On Sunday, June 25, he wrote to Elisabeth that he had been unimpressed by the emphasis on technology and propaganda, particularly at the Russian pavilion. He had also been repelled by the American focus on always having a good time. However, he found the crowds of people at the fair to be fascinating: "How many incredibly different people there are here! Whites, blacks, all kinds of mixes, ugly mugs, and also wonderfully strong, fresh faces—everything mixed together. That this mixture can amount to a nation seems like a miracle, when you think of what we denote with the word *Volk*."[45]

It was precisely the dangerous narrowness of the word *Volk* in Germany and the phrase "We the people" in the USA that was troubling Dietrich to the point of sleeplessness. Furthermore, he found the contrast between what was happening in Germany and the USA to be extreme. With the economy beginning to improve after ten years of breadlines and dust bowls, the USA had turned strongly isolationist. The majority view was that Germany had caused the First World War, at a steep price in American lives. That could not be allowed to happen again. Europe's problems were Europe's problems. The USA was not going to get involved. In the face of a severe refugee crisis, it had virtually shut down immigration for Jews. For example, on June 4, 1939 (as Dietrich was sailing across the Atlantic), the *MS St. Louis* carrying over 900 Jewish refugees was denied permission to land in Florida, having already been turned away from Cuba. The ship was forced back to Europe with the inevitable result being concentration camps and death for many of the passengers.[46]

After attending worship services at various churches, Dietrich complained to his parents that in the USA "everything is dreadfully sensationalized and full of hatred and dreadfully pharisaic," and he meant what he said. It was not a matter of being out of sorts and resorting to hyperbole. Coming from Germany where the churches were denied all freedoms and in which pastors had been murdered for speaking out, he found that American ministers threw around the word freedom without any appreciation of what it really meant. "The Americans in their sermons speak so much about freedom. Freedom, as a possession is for the church a dubious thing; freedom must be won under the constraint of necessity. The freedom of the church arrives from the necessity of God's word." It was pharisaic because it applauded freedom—but not for all. Here again, Dietrich's strong connection to African-Americans is clear. To find out what had occurred with racism since he was last in the USA, he spoke at length with David Roberts, a professor at Union Theological Seminary, and his wife and was troubled to hear that black churches were not included in the recent Methodist merger. Roberts also warned that there was the danger of revolution in the south.[47]

Perhaps for solace and to help regain his theological footing, Dietrich crossed the color line as he had done in 1930–31, visiting his old friends at Abyssinian Baptist Church in Harlem. Although he did not mention it in his diary or his letters, Dietrich apparently preached there one Sunday morning, recounting to a student at Union that it was thrilling and joyful to have "members of the black congregation respond to his message" by punctuating his sermon with Amens and Hallelujahs. It is odd that he did not write about this event, and it is possible that the student remembered it in error, but in comparison with the sermons he was hearing at white churches, some of which he found "simply unbearable," the sermons at Abyssinian Baptist Church would have been gospel-based and delivered from the heart.[48]

On June 19, in despair over not having received any news about Germany, Dietrich wrote in his diary, "I want to know what is happening to the work over there, whether all is well or whether I am needed."

The next day, June 20, 1939, Dietrich made up his mind, writing to Niebuhr who had been instrumental in getting him the position at Union:

> I have come to the conclusion that I have made a mistake in coming to America. I must live through this difficult period of our national history with the Christian people of Germany. I will have no right to participate in the reconstruction of Christian life in Germany after the war if I do not share the trials of this time with my people. The brothers in the Confessional Synod wanted me to go. They may have been right in urging me to do so; but I was wrong in going. Such a decision each man must make for himself. Christians in Germany will face the terrible alternative of either willing the defeat of their nation in order that Christian civilization may survive, or willing the victory of their nation and thereby destroying civilization. I know which of these alternatives I must choose; but I cannot make this choice in security.[49]

That same day, he delivered his refusal in person to Henry Leiper who had offered him a position at the Federal Council of Churches working with German refugees. Leiper, Niebuhr, and Paul Lehmann had worked hard to get him employment and to make his trip to the USA possible, thinking that he would refuse military service and would be sent to a concentration camp should they fail. They had assumed everything was set and that Dietrich had been snatched from the jaws of death so that he could continue his important theological writing. Yet here he was, abruptly turning them down. About his meeting with Leiper, Dietrich wrote in his diary on June 20, 1939:

> He was visibly disappointed and indeed somewhat put out. For me it may mean something more than I am able to foresee at the moment, God alone knows. It is strange that in all my decisions I am never completely clear about my motives. Is that a sign of lack of clarity, inner dishonesty, or is it a sign we are led beyond that which we can discern, or is it both?[50]

Paul Lehmann, with whom Dietrich had been friends since his first trip to America, and who had visited the Bonhoeffers in Berlin in 1933, tried desperately to change Dietrich's mind. In fact, Lehmann tore up his own vacation plans, rushing from Columbus, Ohio, to New York for that purpose, accepting travel money from his father. Lehmann arrived the day before Dietrich sailed for Germany, but it was to no avail. Dietrich's mind was made up.

On June 26, Dietrich mailed a postcard to Karl-Friedrich, who was still in Chicago, telling him he had turned down the offer. He proposed sailing back to England with Karl-Friedrich on July 7 because he was appalled by the news he had received from Europe. "In case of war I want to be not over here but back in Germany."[51] In a letter to his parents on July 1, 1939, he asked them the date of "Uncle Rudi's birthday, so that I can eventually still congratulate him personally?" using the family's code name for Hitler and asking, via the reference to birth date, when war was expected to break out.[52]

Karl-Friedrich had to be surprised to receive the postcard from Dietrich. He must have understood some of Dietrich's motivations, even if he did not share the religious ones, because he was about to do the same thing: turn down an offer from the University of Chicago. However, when he arrived in New York City on July 1, he had not yet decided, as indicated by an entry in Dietrich's diary on June 30: "At 11 a call from Karl-Friedrich, who is coming from Chicago. There is much to talk about. He received an excellent offer of a professorship: it means a permanent decision. Then my questions."[53]

What made up the minds of the two brothers? A remarkable letter written by Paul Lehmann to his father just ten days after their departure provides the answer that appears in no other document. The letter begins with Lehmann thanking his father for the money to go to New York:

> It was really very worthwhile my going even though I could not persuade Bonhoeffer to change his mind. I simply could not bring myself to allow him to have come and gone, perhaps without any possibility of meeting again, without seeing him while he was here. . . . As for Dietrich himself, I can only say that I understand his position completely. He had agreed with his friends that he could come back without fail within one year, unless a mutually agreed upon code formula came to him. This formula was to express their conviction that war was imminent. This formula came and there was no alternative. He fully expects war in September and will thus have one month to make the necessary connections so vital to the ongoing of the Confessing Church. Meanwhile, there had come to Bonhoeffer the clear perception that he who wishes to have any voice in the new Germany must live through this worst of agonies with her. It was unthinkable to him to desert his friends and his task simply in order to safeguard his own life.[54]

As to what the code was, Dietrich gave a compelling clue. On June 19, he was still pacing the floors in an agony of indecision. The next day, he wrote in his diary: "In the morning a letter from my parents in southern Germany. Booklets from Stettin. Visit with Leiper. With that the decision has been made. I turned it down."[55] So the code was probably the booklets from friends in the town of Stettin in Pomerania, although it could also have been in his parents' letter (not extant) informing him of Uncle Rudi's birthday. It was the arrival of the coded information that Dietrich shared with Karl-Friedrich, and as a result, Karl-Friedrich likewise chose to turn down the offer from the University of Chicago. The brothers were going home together.

With the decision made and with time on their hands until the ship sailed, the brothers saw the sights, visiting the Empire State Building, watching newsreels at the movie theaters, and going to Radio City Music Hall. "Horrible, obtrusive, showy, sumptuous colors, music, and flesh," Dietrich declared. "Only in such a metropolitan atmosphere can one put on such fantasies. Karl-Friedrich is of a different opinion." Dietrich's spiritual upheaval did not pass until he walked up the gangplank with Karl-Friedrich in the middle of the night. Paul Lehmann continued to try to dissuade him. Dietrich wrote in his diary: "Last day. Paul is still trying to keep me here. It is no longer possible. . . . Drive to the ship with Paul. Good-byes at 11:30, departure at 12:30 [July 8]. Manhattan at night, the moon stands above the skyscrapers. It is very hot. The journey is over. I am glad that I was there, and glad that I am on my way home again." Later he wrote, "Since I have been on the ship my inner ambiguity about the future has ceased. I can think of my shortened time in America without reproaches."[56]

Even on shipboard, Dietrich continued to concentrate on ethical issues, this time turning to the subject of Protestantism and the Reformation in America. Writing while sitting on a deck-chair in the sunlight as the ship steamed eastward, he turned his thoughts to the concept of refuge, exploring the idea of pilgrims leaving Europe for America because of religious persecution. He concluded that fleeing was not a weakness but required courage to face the unknown and to make a new start in an unknown wilderness. "The flight of a Christian in times of persecution is not apostasy and disgrace as such, for God does not call every person into martyrdom."[57] Undoubtedly, it was martyrdom that was on Dietrich's mind. However, he did not seek it; he simply accepted it, as had some of his friends. In fact, on arriving in England, Dietrich would learn of the murder by lethal injection in Buchenwald of Pastor Paul Schneider for his strong resistance to Nazism and his refusal to step down as pastor of his church.[58]

While Dietrich wrote and read and Karl-Friedrich walked the decks, German pocket battleships and submarines were patrolling the North Atlantic. Panzer divisions with over 2,400 powerful tanks and the greatest firepower in the world were heading eastward toward the Polish border, awaiting orders to attack. Dive bombers by the thousands were lined up on runways in Germany. As the ship headed out into the North Atlantic, a preternatural calm was on the waters. Disembarking in Southampton, Karl-Friedrich flew back to Germany where he began to prepare for the start of classes, while Dietrich stayed briefly in England to say last good-byes, the most poignant being to his twin sister Sabine, whom he would never see again.

After giving lectures at Columbia University in late July, Heisenberg headed back to Germany on the night of August 8–9 on the luxury liner *SS Europa*, the *SS Bremen's* sister ship. The year before it would have been full of passengers; now it was nearly empty. As for the *RMS Mauritania* that had brought Heisenberg to the USA seven weeks earlier on its maiden voyage, its days of glory were already over. It was soon to be stripped of all signs of luxury, becoming a British troop carrier.

The statement that Dietrich made to Niebuhr about "willing the defeat" of Germany would be one Dietrich would repeat secretly to others and which would resonate with many people in the resistance movement. In his biography of Dietrich Bonhoeffer, Eberhard Bethge mentions a private conversation Dietrich had one evening in Switzerland with Willem Visser't Hooft, general secretary of the World Council of Churches, in which Visser't Hooft asked him, "What do you really pray for in the present situation?" Bonhoeffer spoke directly: "If you want to know, I pray for the defeat of my country, for I think that is the only possibility of paying for all the suffering that my country has caused in the world."[59]

After his death in 1945, Lise Meitner quoted Dietrich's words in a commemorative booklet she wrote. Receiving a copy from Meitner, Paul Rosbaud was deeply moved, writing on August 9, 1946, to Samuel Goudsmit. "You will understand when I tell you that exactly the same was my daily prayer from the first day of the war to the last. I don't make this confession to other people in this country, they might not understand it."[60]

All the Bonhoeffers agreed with Dietrich, including Karl-Friedrich. By believing that Germany had to be defeated, they knew that in the eyes of most Germans they were traitors. When Karl-Friedrich died in 1958,

Walther Jaenicke, whom Karl-Friedrich had harbored in his Leipzig laboratory, wrote in a memoriam, "The same love of his country with which Karl-Friedrich went into the field as a volunteer in 1917 made him wish the defeat of Germany in the Second World War."[61]

Eberhard Bethge made an insightful observation of what being a traitor and a patriot meant in the context of Germany under Hitler:

> For men and nations who have never found themselves in that most unhappy state of divided loyalty, it is difficult to realize in retrospect that abnormal situation in which the most conscientious person has to accept disgrace. Anyone who hastily measures the borderline situation of that time by the yardstick of his own principles, or studiously overlooks its peculiar features, distorts its contours and does not see the realities of those months. Germany's name could no longer be saved by ordinary but blind responsibility. All the world over, "treason" is normally thought of as a mean attitude, speculation for one's personal advantage, and the intention to injure one's own country. The opposite holds good for Oster, Dohnanyi, and Bonhoeffer.[62]

It can be said with certainty that the one thing of immense importance that Heisenberg lacked and Karl-Friedrich possessed was—Dietrich. He was the voice of conscience, constantly analyzing his own motives and attempting to discern in all things the will of God. Nor did Dietrich struggle in silence. He shared the struggle with his family and friends.

Notes

1. Heisenberg sailed from Liverpool, writing to his wife that the departure was very festive. Anna Maria Hirsch-Heisenberg, ed., *My Dear Li: Werner Heisenberg, Elisabeth Heisenberg Correspondence 1937–1946*, trans. Irene Heisenberg (New Haven: Yale University Press, 2016), 51.
2. The quote is from the essay "The Church and the Jewish Question." DBW, Vol. 12, 365.
3. The history of the Confessing Church is complex. Never a unified movement, it arose in part as a reaction to Hitler's establishment of The Reich Church. The Reich Church took away financial and administrative independence from the churches, enforced the Aryan clause, and attempted to reshape theology into a Teutonic form. See Victoria Barnett, *For the Soul of the People: Protestant Protest Against Hitler* (Oxford: Oxford University Press, 1998).

4. Bethge, *Bonhoeffer*, 589, 591.
5. Hermann Stöhr, a member of the International Fellowship of Reconciliation in Stettin, applied for conscientious objection status. He was arrested and executed in 1940. Ibid., 635.
6. DBW, Vol. 15, 156–157.
7. Ibid., 162.
8. Hippel, "James Franck," 44.
9. Jost Lemmerich, *Science and Conscience: The Life of James Franck* (Stanford: Stanford University Press, 2011), 228.
10. Hippel, "James Franck," 45.
11. The grant also covered time in Zurich with Wolfgang Pauli. Delbrück interview, 41.
12. Sabine Leibholz-Bonhoeffer wrote that one of the reasons Karl-Friedrich did not take the position at the University of Chicago was "he could not make up his mind in these uncertain times to leave his parents in Germany." Leibholz-Bonhoeffer, *Bonhoeffers*, 125.
13. Fermi to Szilard, July 9, 1939, Enrico Fermi Collection, Archives University of Chicago.
14. Dahl, *Heavy Water*, 178.
15. W. L. Laurence, "New Key Is Found to Atomic Energy," *New York Times* (5 May 1939): 25.
16. Lawrence Badash, Elizabeth Hodes, and Adolph Tiddens, "Fission: Reaction to the Discovery in 1939." *Proceedings of the American Philosophical Society*, Vol. 130, No. 2 (June 1986): 217.
17. Max Born, *My Life & My Views* (New York: Charles Scribner's Sons, 1968), 101.
18. Ibid., 223–224.
19. https://en.wikipedia.org/wiki/Einstein-Szilardletter.
20. Cassidy, *Beyond Uncertainty*, 293.
21. Hirsch-Heisenberg, *My Dear Li*, 57.
22. Ibid., 63, 69.
23. Cassidy, *Beyond Uncertainty*, 295.
24. At a meeting in Seefeld in November 1942, Heisenberg agreed with other German scientists not to use Einstein's name in public. He and Weizsäcker believed that the discovery of relativity would have occurred in due time without Einstein. Cassidy, *Beyond Uncertainty*, 330.
25. Beyerchen, *Scientists Under Hitler*, 170.
26. Rose, *Heisenberg*, 243–244.
27. Heisenberg made a strange statement to his wife Elisabeth in a letter dated January 15, 1946, that confirms his belief that Russia was a far greater threat than Germany. "At any rate, no American could seriously claim that Germany had been a threat to America, or that a Hitler

Europe was more dangerous, in terms of power, than a Eurasian continent under the political leadership of Russia." Hirsch-Heisenberg, *My Dear Li*, 286. See also Cornwall, *Hitler's Scientists*, 230.

28. Cassidy, *Beyond Uncertainty*, 295.
29. Samuel A. Goudsmit, *Alsos* (Woodbury, NY: AIP Press, 1996), 114.
30. Cornwell, *Hitler's Scientists*, 230.
31. Cassidy, *Beyond Uncertainty*, 280–281.
32. Hirsch-Heisenberg, *My Dear Li*, 65, 71, 72.
33. The other books were *Sanctorum Communio* (doctoral thesis), *Act and Being* (habilitation thesis), and *Creation and Fall.* The ban on public speaking occurred in September 1940. The ban on publishing occurred in March 1941. DBW, 17, 96, 98.
34. DBW, Vol. 15, 115.
35. Hans Jürgen von Kleist-Retzow was not executed; in 1945, he became a prisoner of war in the USSR; subsequently, he returned to Germany. DBW, Vol. 16, 794.
36. This was the same reason that Dietrich Bonhoeffer would become part of the Abwehr. It enabled him to avoid conscription while carrying out assignments for the resistance.
37. Bethge, *Bonhoeffer*, 672, 810.
38. Ibid., 626.
39. Ibid., 768. There is a fine line between the words resistance and opposition. Generally, the word resistance is used for an organized effort, extensive in planning, that carried with it high personal risk for its participants. It was possible for a German to be opposed to Nazism and to resist individually (e.g., giving food to Jews) without being part of the resistance.
40. DBW, Vol. 7, see footnote 34, 43.
41. DBW, Vol. 16, 222.
42. Wikipedia, *New York World's Fair.*
43. Bogdan Horbal, "World's Fair: Enter the World of Tomorrow, A Monument to a King and a Nation." http://exhibition.nypl.org/biblion/worldsfair/moment-time/brink-war/essay-horbal-polish-pavillion.
44. DBW, Vol. 15, 223.
45. Hirsch-Heisenberg, *My Dear Li*, 54.
46. *MS St. Louis.* en.wikipedia.org/wiki/MS_St.Louis.
47. DBW, Vol. 15, 194, 230, 235.
48. The student was Rudolf Schade who was at Union in the summer of 1939. DBW, Vol. 10, 31–32. See footnote 146. DBW, Vol. 15, 224.
49. This is an excerpt of a letter to Niebuhr; the original letter has been lost. DBW, Vol. 15, 210. See footnote 1.
50. Ibid., 227.
51. Ibid., 200.

52. Ibid., 211.
53. DBW, Vol. 15, 234.
54. Ibid., 253.
55. Ibid., 226.
56. Ibid., 230, 234, 235, 237.
57. DBW, Vol. 15, 447.
58. Schlingensiepen, *Bonhoeffer*, 233.
59. DBW, Vol. 16, 178. Bethge, *Bonhoeffer*, 744.
60. Goudsmit papers, Series IV Alsos Mission. Box 28. Niels Bohr Library & Archives with the Center for History and Physics, American Institute of Physics.
61. Walter Jaenicke, "Personliches: Karl Friedrich Bonhoeffer," *Physikalisch* (1957), 369. Onlineelibary.wiley.com/doi/10.1002.
62. Bethge, *Bonhoeffer*, 580.

CHAPTER 11

The Uranium Club

The Uranium Club was reorganized and put on a firmer footing on September 1, 1939, the day Germany attacked Poland or, as Hitler angrily proclaimed, Poland attacked Germany. At the Nuremberg trials after the war, it was revealed that on August 31, S.S. men dressed in Polish uniforms seized the German radio station in Gleiwitz near the Polish border. The Germans used as evidence of Polish aggression a bloody victim found at the radio station. The victim was actually an inmate from a concentration camp who had been dressed in civilian clothes, transported to the radio station, and beaten unconscious. Claiming that Poland had ignored his peace proposals, Hitler thundered on the radio on September 1, "This night for the first time Polish regular soldiers fired on our own territory. Since 5:45 a.m. we have been returning the fire, and from now on bombs will be met with bombs."[1] As good as his word in a malign way, Hitler poured one and a half million soldiers into Poland.

Language can always be twisted, and the Nazis did so with great skill. An invasion could be made to look like a defensive action. An agreement between the USSR and Germany, signed on August 24, 1939, just nine days before the invasion of Poland, could be called a non-aggression pact as if it indicated nothing more than Soviet neutrality in the event Britain or France declared war on Germany. In fact, a secret additional protocol to the pact divided up Eastern Europe as well as Finland between the two nations, beginning with the partition of Poland.

K. L. Housley, *The Scientific World of Karl-Friedrich Bonhoeffer*,
Palgrave Studies in the History of Science and Technology,
https://doi.org/10.1007/978-3-319-95801-9_11

Similarly, the Uranium Club was not what it seemed. Its innocuous name made it appear to be no different than a congenial hiking club, whose members liked to climb together to the sparkling snowfields of the Bavarian Alps, or perhaps it was a casual academic discussion group, gathering after work in a beer hall, united by a shared interest in an unusual element. The club's members were equally innocuous, just loyal scientists willing to do their part for the Reich. The club would undergo several changes during the war as the German high command reordered military priorities, and competing factions pulled the club in different directions. Progress was also impeded by a lack of heavy water—the Norwegians called it *juice*—as well as major technical challenges with isotopes. However, one aspect would not change: From the beginning of the war until its last days—when the Uranium Club was trying desperately to achieve a chain reaction and very nearly succeeded—Heisenberg would be its leader, if not always in name, then in spirit.

From the start, there was much animosity about Heisenberg being the leader. Harteck charged that Heisenberg was the wrong man for the job because he was a theoretician who had never run an experiment in his life.[2] Harteck was not the only man who felt that way. Heisenberg also lacked the administrative and diplomatic skills that the role of leader required. At one of the earliest meetings of the Uranium Club, the leadership position was offered to Harteck who declined, making the case that the undertaking should be led by a committee, not an individual. He added that he was not qualified because he was a chemist, not a physicist. Had Harteck accepted—with his gift for coming up with ingenious solutions to intractable technical problems—the outcome of the German atomic bomb effort may well have been different and, therefore, the outcome of the war.[3]

On September 15, Heisenberg wrote from Leipzig to his wife who was at the family's summer home in Urfeld. He mentioned that Weizsäcker's brother, Heinrich, who was an officer in the military, had been killed at the beginning of September, one of the earliest casualties of the war. Then, he wrote nebulously, "Bonhoeffer was in Berlin yesterday. He said he had heard through Debye [director at the Kaiser Wilhelm Institute of Physics] that in the next few days I would be commissioned to Berlin to some military armament agency or something similar. Tomorrow in Berlin I will hear more. If that is right, I would perhaps be based in Berlin for a longer period. I am not entirely comfortable with this thought."[4] None of his letters to her from then to the end of the war carried any detail about the Uranium Club, although he mentioned various meetings with leaders of the Reich including Albert Speer.[5]

By October 1939, the Uranium Club included four core members: Heisenberg, Weizsäcker, Karl Wirtz, and Harteck. Karl-Friedrich's closeness to the core group is undeniable. Wirtz had been Karl-Friedrich's assistant at the University of Leipzig in 1935–37 (Fig. 11.1). They had co-authored two papers on heavy water: "The Reaction of Formaldehyde with Hydrogen Peroxide in Heavy Water: Proof of the Lack of Exchange of the Hydrogen Atoms in Formaldehyde" (1936) and "The Liberation of Protons from Molecular Hydrogen with the Aid of Hydroxyl Ions" (1936).[6] Wirtz was a member of the National Socialist Teachers League, which was supportive of Nazism, but he was not a member of the Nazi Party. Of the four core members, only Weizsäcker stood at a distance from Karl-Friedrich, having been in orbit around Heisenberg since the

Fig. 11.1 Carl Friedrich von Weizsäcker in 1971. Interfoto/Alamy Stock Photo

age of fourteen when Heisenberg had convinced him to study physics instead of astronomy.[7]

Beyond the four men at its core, the Uranium Club was a loose organization without clear membership. Scientists identified with the club were: Erich Bagge, Walther Bothe, Klaus Clusius, Kurt Diebner, Robert and Clara Döpel, Erich Fischer, Siegfried Flügge, Hans Geiger, Walther Gerlach, Otto Hahn, Georg Joos, Gerhard Hoffmann, Wolfgang Gentner, Horst Korsching, Max von Laue (a marginal member at best), Josef Mattauch, and Georg Stetter. However, there were probably around seventy German scientists involved in some aspect of nuclear research, most of whom had only passing contact with the Uranium Club, including Hans Kopfermann, who had been at Haber's institute, and Nikolaus Riehl, who worked for the Auer Company. Karl-Friedrich was in that larger group.[8] His name appears and reappears in official records; he is listed among the attendees at an important meeting on bomb research; he is asked for advice on building a heavy water plant; and he receives government funding for heavy water research.[9] However, in the history of the German atomic bomb project, Karl-Friedrich was never at the center. As will be shown, even in his interactions with Harteck, there was an unwillingness to cooperate. Only after the war when the espionage of Paul Rosbaud came to light did the reason for Karl-Friedrich's reluctance become clear.

The scientists involved with atomic bomb research continued performing their regular jobs at universities, companies, and institutes, carrying on their own research not connected to that of the Uranium Club. For example, Heisenberg published three important papers on scatter matrix during the war and also published a textbook. Weizsäcker continued his work on astrophysics at the Nazified University of Strasbourg, which had been seized from the French. He also wrote on philosophy, science, and religion in relation to relativity and quantum mechanics. They taught students, oversaw doctoral studies of students who had not been drafted, and fulfilled administrative responsibilities under worsening wartime conditions and severe shortages of supplies. Klaus Hentschel writes in *Physics and National Socialism: An Anthology of Primary Sources* that compared to the unity of the Manhattan Project, the Uranium Club was "only a loosely knit, decentralized network of researchers with quite different research agendas. Rather than teamwork as was the model on the American end, on the German side we find cut-throat competition, personal rivalries, and fighting over the limited resources."[10]

Many books have dealt with the subject of why the Germans failed to build an atomic bomb while the Allies succeeded—as if the words *fail* and *succeed* have any meaning in this context. Among the historians who have plumbed this subject are Mark Walker, David C. Cassidy, John Cornwall, and Paul Lawrence Rose. While they disagree on the level of culpability of Heisenberg and others, they tend to agree on the following points: First, the Germans had the intellectual firepower to unlock the secrets of the atom, despite the loss of the Jewish scientists. Second, they had the willpower even though they knew they were working on a bomb and not a benign reactor. Third, the majority were not Nazis, but neither were they resisters; instead, they were "fellow travelers," a term that came to be attached to people who went along in silence.[11] Many strongly disliked Hitler, but once the war started, they were even more strongly German and wanted their country to win. Fourth, after the war, they tried hard to present themselves as moral, having chosen to prevent the bomb from being built by slowing down their research, and not demanding the financial and manpower commitment from the German high command that was actually needed. However, wartime records show otherwise. The Uranium Club scientists wanted to succeed for Germany, for pure science, and for themselves, but they were not at all certain that the inordinately complex technological problems could be solved before the war ended. In short, they were convinced they could build an atomic bomb—but not in time. They failed for numerous reasons, chief among them being a lack of administrative unity, an inability to recognize and correct fundamental errors, and a strong belief that the research was so complex it would take years and, therefore, would not affect the outcome of the war. According to Mark Walker in his book *Nazi Science: Myth, Truth and the German Atomic Bomb*: "Throughout the Lightning War [September 1939–November 1941] the overwhelming majority of Germans (and most likely German scientists as well) believed that the war would end in victory. 'Wonder weapons' were not needed."[12]

There is one other compelling argument that is made by Nikolaus Riehl and Frederick Seitz in the book *Stalin's Captive: Nikolaus Riehl and the Soviet Race for the Bomb*. Riehl had received his doctorate under Lise Meitner and was a friend of Otto Hahn. Working for the Auer Company, which had been taken over by the Degussa Company after

Hitler came to power, Riehl's responsibility was uranium isotopes. The book is partially his memoir, supplemented by Seitz's commentary. Seitz was a distinguished American physicist who worked in technical intelligence at the end of the war, meeting many German scientists. Assigned to General Dwight Eisenhower's headquarters in Versailles, Seitz came to know all the members of the Alsos team, including Samuel Goudsmit.[13] Seitz wrote that following the military debacle at Stalingrad in 1942, the Reich severely downgraded its interest in the atomic bomb program. Industry became burdened with the production of more conventional equipment in the midst of material shortages. At that point, Hitler and his advisors turned toward rocketry, focusing their attention on the V1 (buzz bomb) and V2 (liquid-fueled rocket) developed by Wernher von Braun.[14]

Riehl's experiences buttress Seitz's statement. Because he had a unique industry perspective instead of an academic perspective, Riehl was frequently asked after the war why the German uranium project did not make more progress, and why the Nazi regime did not support it more energetically:

> I believe that the relative laxity in the pursuit of the program rests primarily on the primitive level of intellectual understanding of Hitler and his advisers. They had, perhaps, a good comprehension of rockets that roared with a great deal of noise and whose operations were apparent to them, but they had no real understanding of the unfamiliar, abstract concepts associated with the release of energy through nuclear fission. The lack of any requirement on the part of the government caused many of us not to expect any significant results from the uranium project before the collapse of Hitler's government. As a result, we never felt any compulsion to bring the matter to the attention of our consciences.[15]

Some historians have pointed to the vast amount of money and manpower it took the USA, through the Manhattan Project, to build the atomic bomb. They make the case that Germany did not have the financial resources and scientific manpower to do so. In fact, Germany poured enormous money and manpower into its rocketry program. Klaus Hentschel states that the Peenemünde rocket research facility on the Baltic Sea employed "1,960 scientists and engineers and some 3,852 workers by 1942, and at its peak had a total of nearly 6,000 operational staff, set against a total of 5,000 employees at Los Alamos." To these figures, Hentschel adds a third: The hazardous rocketry production process involved 60,000 slave laborers under horrible conditions, leading to the deaths of approximately 20,000. These figures support Riehl's statement that rocketry made more sense to the German high command than did atomic research, which they supported, albeit at a much lower level. Too meticulous a scholar to miss the irony embedded in the Reich's support of both rocketry and atomic weapons, Hentschel insightfully adds in the introduction to his book that the fusion of the two technologies "is still with us in the form of the intercontinental nuclear missile."[16]

It must be noted, however, that Hentschel was not the first to make that connection. In June 1944, just after D-Day, Lise Meitner wrote to the physicist, Eva von Bahr-Bergius, that she feared what Germany would do as their situation became hopeless. With her uncanny ability to see things that others missed, Meitner wrote: "I am very anxious to know if the robot bombs [rockets] have something to do with uranium bombs."[17]

Another important reason why the Germans failed to build a bomb while the Americans succeeded is almost too obvious to mention: The continental USA was not under direct attack. The bombings, the battles, and the severe deprivations were "over there" to borrow a line from a First World War song. The scientists at Los Alamos were not concerned about their laboratories being blown up, other than by their own carelessness. They were not concerned about being conscripted into the military, as were so many German scientists (even Heisenberg was called up at one point). Their families were safe; there was food on their tables; they could sleep through the night; and they could concentrate on their research with adequate supplies and manpower. In contrast, Germany came under direct attack from the night sky, as Allied long-range bombers, taking off from bases in Great Britain, crossed the channel. The heavy bombing of German cities, particularly Leipzig and Berlin (both of which had prototype atomic reactors), severely hampered the work of the Uranium Club.

As evidence of the difficulty of doing any kind of science under extreme wartime conditions, there is a poignant letter from Karl-Friedrich to Dietrich written on March 26, 1944, almost a year after Dietrich had been arrested and incarcerated in Tegel prison. Karl-Friedrich was in Berlin on business for Osram (a company that prior to the war made lighting) and was writing the letter from his parents' living room.

With great circumspection, considering that the letter would be read by prison censors with information passed on to the Gestapo, Karl-Friedrich made it seem as if it were only a matter of time until Dietrich would be found innocent and released. He mentioned his parents' courage and that they were away visiting Ruth von Wedemeyer, the mother of Dietrich's fiancée, Maria, but would be returning to Berlin early because they had received permission to visit Dietrich. Then, Karl-Friedrich wrote of his own work:

> You'll be amazed, when you get out, to see how the world has changed in that one year. It's becoming very difficult to do any scientific work. Libraries have evacuated their books, making access to them a major problem; institutes have been damaged, and people are distracted by all sorts of worries at home. It takes more energy than one can summon to force oneself to concentrate. I've left things that were 80 percent finished lying around for months and can't get them done. Yet in our field it's crucial, for technology as well, that the main purely scientific things are kept moving forward. In the long run we cannot go on living off our capital. I should be happy if I could find a quiet, suitable place to which to move, somewhere that would let me carry on my scientific and technical research, and perhaps take Grete and the children there as well. I would then concentrate my lectures at Leipzig on a few days and become a visiting lecturer there, so to speak. But I'm afraid that no such place can be found anymore.[18]

Having considered some of the reasons why the Uranium Club ultimately failed in its mission to build a bomb, it is necessary at this point to go back to 1939 when Heisenberg first laid out a plan to achieve a chain reaction. Following the September meetings in 1939, Heisenberg began to study nuclear fission to the exclusion of all his other quantum mechanical work. On December 6, 1939, he circulated a report titled *The Possibility of the Technical Acquisition of Energy from Uranium Fission*. That he did so in less than three months indicates how hard he had been concentrating. In the report, he explained what was already known and the potential for enrichment of U-235, including the production of a bomb "the explosive power of which exceeds that of the strongest available explosive by several powers of ten."[19]

In an article published in *Nature* in 1947, two years after the war, Heisenberg provided an account of atomic research in Germany.

Although biased and self-exculpatory, the article reveals the scope of Heisenberg's thinking:

> As a result of the first conferences in the autumn of 1939, it was clear that there were two lines of attack possible in the exploitation of nuclear energy. One could attempt the separation of the rare isotope U(235) from ordinary uranium. This isotope, following theoretical arguments due to Bohr must be immediately applicable either to the controlled production of energy, using primarily the slow-neutron reaction, or directly as an explosive in bombs, using the fast-neutron reaction; the separation of U(235) was, however, a problem which made the greatest possible demands on engineering technique. Secondly, one could mix ordinary uranium with a substance which would slow down the neutrons produced in the nuclear fission without absorbing them. These slow neutrons give rise preferentially to the fission of U(235) and thus maintain the chain reaction. The advantage of this arrangement is that the chain reaction can be controlled through the heat developed thereby, so that energy can be abstracted in amounts sufficient for technical applications.[20]

What Heisenberg was saying was that if a nuclear reaction happened too fast, only a small fraction of the potential energy would be released before it fizzled out. A moderator could slow down neutrons without absorbing them, enabling a chain reaction in what was called a *Uranbrenner* or pile. The Germans were interested in heavy water as a moderator for the rare isotope uranium-235, which could be enriched from the more common isotope uranium-238. It was also possible to use natural uranium in a pile with a moderator, thereby generating nuclear power. Then, too (and not mentioned at this point by Heisenberg), there were other "substances" that could potentially be brought into existence, including what would come to be known as plutonium. This research was soon to undertaken by Weizsäcker who, in July 1940, reported on the possibility of breeding a new type of explosive radioactive material inside a uranium reactor (also called an engine). Heisenberg wrote that "As soon as such an engine is in operation, the question of producing an explosive gets a new twist, according to an idea of Von Weizsäcker." This "new substance" would be an explosive of "unimaginable effectiveness."[21] Per Dahl in his book *Heavy Water* provides a cogent explanation of the process that Heisenberg, in a perverse understatement, dubbed "a new twist":

> By 1940, it was generally realized . . . that if uranium-238 forms uranium-239 by neutron resonance capture [the colliding of neutrons to form a heavier nucleus], the latter isotope could transmute via beta emission [decay] to element 93, eka-rhenium. This was soon shown to be the case at Berkeley, and in due course in Germany as well. Element 93, neptunium as it became known, and its daughter product, element 94, eka-osmium or plutonium, assuming it too could be produced, ought to be fully as fissionable as uranium-235, according to the all-important Bohr-Wheeler theory. There appeared to be no end of nuclear explosives; the more pressing question was, could they be produced in time for use in the war under way in a situation which, in the German view, favored only short-term technical developments?[22]

Obtaining common uranium was not difficult. The Germans had gained easy access to uranium in March 1939, before the outbreak of the war, with the ruthless occupation of Czechoslovakia, which gave them possession of the Joachimstahl mines, the world's largest uranium reserves. Many years earlier, it was from those mines that Marie Curie had obtained the pitchblende in which she discovered radium. Getting men to work the mines presented no difficulty either: Germany used slave labor and prisoners of war. There were also the Belgium Congo uranium mines that came under German control after Belgium was invaded in May 1940.

Given that the uranium isotope 235 makes up only 0.7% of natural uranium, how much of it was necessary to build a bomb? On this question, there was significant disparity in the estimates. Heisenberg was convinced that the critical mass (the amount needed to sustain a nuclear chain reaction) was tons, and his opinion counted more than anyone else's in Germany. Doing their own secret calculations, Szilard and Einstein also initially concluded that the figure was tons.[23] Then only four months after the start of the war, Otto Frisch and Rudolf Peierls (who had studied under Heisenberg) at the University of Birmingham in England reworked the nuclear calculations, coming up with the astounding figure of one single kilogram. If Heisenberg was right, the prospects of successfully building a bomb were low. If Frisch and Peierls were right, the prospects were high. The difference was—all the difference in the world. Frisch and Peierls were far nearer the mark. (The number would be revised upward to 60 kilograms.) In their report in March 1940, they wrote: "The energy liberated in the explosion of such a super-bomb is about the same as that produced by the explosion of 1000 tons of dynamite. This energy is liberated in a small volume, in which it will, for an instant, produce a temperature comparable to that in the interior of the sun. The blast from such an explosion would destroy life in a wide area. The size of this area is difficult to estimate, but it will probably cover the centre of a big city."[24]

In his memoir *What Little I Remember*, Frisch attempted to answer why he did not stop right there when he realized that it was feasible to build a bomb and that his work was pointing the way. "I have often been asked why I didn't abandon the project there and then, saying nothing to anybody. Why start on a project which, if it were successful, would end with the production of a weapon of unparalleled violence, a weapon of mass destruction such as the world had never seen? The answer is very simple. We were at war, and the idea was reasonably obvious; very probably some German scientists had had the same idea and were working on it."[25]

Just as the question of critical mass was of huge significance, so also was the question of what moderator should be used. In the USA, Fermi, Szilard, and Eugene Wigner had come to the conclusion that instead of heavy water, pure graphite (carbon) could function as a modifier. However, it was not easy to obtain. Furthermore, even the purest graphite available tended to be contaminated by elements, such as boron, that absorb neutrons. Not realizing that the graphite was contaminated, some German scientists reached the wrong conclusion about graphite's efficacy as a moderator. Based on a flawed experiment with a contaminated sample done by his colleague, Walter Bothe, Heisenberg had decided by 1940 to focus on heavy water instead, influenced by advice from Harteck and Karl-Friedrich.[26] Essentially, the Uranium Club was beginning the research on the atomic bomb on the basis of at least two wrong assumptions: That critical mass was very high—an opinion Heisenberg and his colleagues held until the USA bombed Hiroshima and Nagasaki in 1945; and that heavy water was the only possible moderator with slow neutrons, not fast ones. These were far-reaching errors. It must be said that the Allied scientists also made several wrong assumptions early on, but the unified structure of what would become the Manhattan Project, with Major General Leslie Groves (eventually promoted to lieutenant general) in command, helped in the long run to correct errors. The German structure tended to magnify them. For example, Heisenberg ridiculed Kurt Diebner because he believed him to be an inferior physicist, which was not the case. When Diebner and his colleagues made significant advances that could have led to the achievement of a chain reaction, Heisenberg dismissed the advances out of hand.

Once the war began in September 1939, Norsk Hydro in Norway saw an immediate increase in German interest in heavy water. By the beginning of 1940, before the German invasion of Norway, Norsk Hydro was still producing only ten kilograms per month, even though in January of that year I.G. Farben had placed an unfillable order of 100 kilograms per month. At that time, Norsk Hydro was using a process developed by Karl-Friedrich that had been improved by Harteck, Groth, and Clusius.[27]

In January 1940, Kurt Diebner, who was at the time the director of the German Army Ordnance (HWA) research council, held a meeting in Berlin with Heisenberg, Karl-Friedrich, and Karl Wirtz in attendance at which Heisenberg was asked whether a plant should be built in Germany for the full-scale production of heavy water. Heisenberg, however, first wanted to test neutron absorption in a small sample of heavy water. He was not interested in large-scale construction, nor was he interested in the administration and financial planning attendant thereto. He was only interested in the science. A few weeks later, Harteck asked Heisenberg if he knew anyone who was working on mass production of heavy water in Germany. According to Dahl, "Heisenberg wrote back, advising Harteck to talk it over with his Leipzig colleague, Bonhoeffer. Harteck did just that, asking Bonhoeffer if he would be interested in jointly pursuing the production of heavy water by the catalytic exchange between hydrogen and water vapor—a process quite different from the electrolytic process normally used, and one explored by Harteck and the physical chemist Hans Suess some time earlier. Bonhoeffer, in turn, passed the buck to I.G. Farben Ammoniakwerk at Merseburg, who expressed guarded interest in Harteck's suggestion."[28]

It is peculiar that Harteck would ask Heisenberg about who was working on mass production of heavy water and that Heisenberg would suggest Karl-Friedrich, given the fact that Harteck was in close touch with Karl-Friedrich and always had been. This leads to the supposition that because of Karl-Friedrich's antipathy toward the research, Harteck was casting about for someone else with whom to work.

There were other groups in Germany that were researching atomic power, including the private Ardenne Laboratory, which was under the joint leadership of the Post Office and the Reichsforschungsrat. Manfred von Ardenne and his group were researching electromagnetic mass separation of isotopes, a very important aspect of bomb development. The Ardenne Laboratory was a prime example of how disjointed research was in Germany. As has already been seen, atomic research was not united but was spread among several universities, private and public institutions, and government departments, none of which wanted to share power or information, instead fiercely guarding their own fiefdoms. The Ardenne Laboratory was one such fiefdom, unusual in that Manfred von Ardenne leaned toward the Soviet Union to which he would switch his loyalty completely in 1945. Because of this toxic atmosphere, Harteck had to be circumspect in seeking help. He did ask Karl-Friedrich, only to have him sidestep by suggesting Harteck contact I.G. Farben. However, Harteck already knew of Farben's involvement. Furthermore, he was contemptuous of the level of scientific knowledge at Farben and other German companies. "They hadn't learned anything about nuclear physics. You cannot imagine this. They didn't realize the ramifications," he declared in an interview many years after the war.[29]

Then on April 8, 1940, Germany invaded Norway and Denmark on the pretext of protecting them from an imminent attack by France and England. German troops seized control of Vemork one month later on May 3. This set the heavy water question in a new light, according to a letter that Harteck wrote to Karl-Friedrich.[30] Immediately, Karl Wirtz, a member of the Uranium Club, was directed by Army Ordnance to travel to Norway to meet with Jomar Brun, head of the Norsk Hydro's electrolysis plant, with the goal of obtaining a large amount of what the Germans called SH-200.[31] It was obvious to Wirtz that Vemork had to expand production in the long term, but there was no clear answer as to how to do that. In the short term, there seemed to be no heavy water to be had.

The Norwegians had known that an attack was imminent. Highly suspicious that heavy water would be used for something terrible, but not knowing exactly how or what, some of the Norsk Hydro staff, including Axel Aubert, general director, and Jomar Brun had arranged for the remaining supply of heavy water, about 185 kilograms, to be shipped to the laboratory of Joliet-Curie in Paris, France. The ruse involved two different planes taking off from Oslo on March 12, 1940, with French military officers dressed as civilians. On one plane was a simple wooden crate filled with Norwegian granite and on the other plane metal canisters of heavy water. To throw the Germans off, tickets were purchased for different days to different locations. Then creating an artificial disturbance on the runway, the cargos were switched before the planes took off. When one plane was forced down in Hamburg by the Luftwaffe, all the Germans found was the granite. About this incident, Dahl wrote that it was of the "greatest importance in the political affairs of the Western World." Yet it was only the beginning of the high espionage and outrageous courage that would lead to the destruction of Vemork. After the switched flight, the heavy water made its way safely to Paris, but it did not stay there long. When France capitulated in June 1940, it was loaded onboard the vessel the *Broompark*, sailing from Bordeaux to Falmouth Roads in England. Curiously, it seems that the *Broompark* was sunk by a magnetic mine. As with the two planes, this was a case of two similar vessels and switched identity. The heavy water was no more, at least that was the story British and French intelligence spread widely. But in truth, the canisters arrived safely and were transported to London by train where in a most strange twist they were stored in Windsor Castle under the care of a librarian until they were put to use in experiments at Cambridge.[32]

Heavy water was not the only item that made it to England safely. With the Gestapo rapidly closing in on him, in October 1940, Leif Tronstad slipped across the Norwegian border into Sweden on his way to England. There he began to work closely with British intelligence, planning a daring raid by Norwegian resistance fighters to put Vemork out of action. The very plant he had helped build he would help destroy.

Vemork never reached the levels of heavy water production that Germany demanded. There were *natural* causes: In the spring of 1940, a severe drought and unseasonably cold weather prevented the snowpack from melting, resulting in the drying up of the reservoir that fed the Vemork penstocks. Then, there were *unnatural* causes: Brun was not above pouring castor oil into the electrolyte, causing heavy foaming. Brun still did not know exactly why the Germans were so fixated on heavy water, but he was smart enough to realize a little sabotage was in order. Several other workers at Vemork came to the same conclusion, adding cod-liver oil to the high-concentration containers. With the single-minded efficiency for which Germans were renowned, improvements to the plant went ahead anyway, foam or no foam.[33]

Harteck made several visits to Vemork. In October 1941, he traveled with Wirtz for a meeting with Norsk Hydro executives, the upshot being the proposed construction of a new catalytic-exchange process developed by Harteck. In November 1941, a meeting was held at Army Ordnance in Berlin at which Harteck and Karl-Friedrich were in attendance on the new process. By year-end, production finally began to rise at Vemork, reaching 100 kilograms per month, but it was still not enough.[34]

By early 1942, the future of the Uranium Club was in serious doubt. Besides the shortage of heavy water, isotope separation had proven to be very difficult. Progress was so slow that in February 1942, German Army Ordnance concluded in a report that uranium would not decide the outcome of war and that research should be transferred to a civilian authority. At that point, Harteck and Weizsäcker were called up for military service, but Heisenberg and Karl-Friedrich interceded with a high-level military authority on their behalf. According to Hentschel, the "high-level" authority was none other than Karl-Friedrich's brother-in-law, Hans Dohnanyi, who was very influential "in obtaining military deferrals for members of the staff."[35]

There were other serious setbacks. In April 1942, the reactor Heisenberg was working on in Leipzig actually achieved 13% neutron generation, using 140 kilograms of heavy water and 750 kilograms of uranium metal powder, but two months later the reactor was destroyed by an explosion, which was chemical in nature not atomic, although when word of the explosion reached Britain (probably via Rosbaud), there was grave concern about its possible causes. No one was injured. Subsequently, Heisenberg became acting head of the Kaiser Wilhelm Institute of Physics in Berlin where he directed the construction of a larger reactor as well as a bombproof underground laboratory. This was a major improvement over the small shed known as the Virus House, so named to frighten people away and to dissuade possible saboteurs. To further obscure its real purpose, the Virus House was located not at the Kaiser Wilhelm Institute of Physics but down the street at the Institute of Biology and Virus Research. Obviously, the shed was more than a shed; inside and beneath was a circular pit two meters deep, lined with bricks and filled with water. Earlier efforts to achieve a chain reaction in the pit by lowering a neutron source into a uranium pile had failed, basically because the amount of material used was insufficient, hence Heisenberg's efforts to build a larger reactor. Another problem Heisenberg faced was his insistence on using uranium plates (instead of cubes as Diebner had done) that were exceedingly difficult from a metallurgical standpoint for Auer, under Riehl's guidance, to manufacture.[36]

Heisenberg's power seemed to increase at this point. Writing to Elisabeth on June 5, 1942, while returning from Berlin to Leipzig on the train, he told her of an important meeting attended by Albert Speer and Field Marshall Erhard Milch at which he and Bothe had made presentations:

> In the Harnack House truly every preeminent person of the Reich was assembled—essentially the entire armaments council and many others in power. It went far better than expected, my own and Bothe's presentations left, apparently, a strong impression; at any rate we were treated fabulously afterward, both personally and in terms of the facts. To me this is all exceedingly strange and unnerving; suddenly, I will no longer have to concern myself at all with the Lenard-Stark clique, and can push through anything I find important.[37]

However, according to statements made by Speer and Milch following the war, they were far less enthusiastic about Heisenberg's presentation. Speer recalled asking Heisenberg after the lecture about the feasibility of building a nuclear bomb. "His answer was by no means encouraging," Speer said, which was what he put in his subsequent report to Hitler. Milch remembered that at that meeting he had asked how large the bomb would have to be to destroy a large city such as London, to which Heisenberg replied "about the size of a pineapple."[38] Milch was dismayed. Did Heisenberg actually believe an atomic bomb could be that small? Did he make that comparison merely to please his listeners and to gain more Reich support? Or had he accidentally said the wrong thing, meaning that only the U-235 used in the bomb was equivalent to the size of a pineapple? Only one thing is clear: The level of confusion and misunderstanding was very high.

Karl-Friedrich continued his research on heavy water, although exactly what he was doing is nowhere spelled out. The Germans maintained good records to the point of obsession, so who received a grant, when, and for what purpose were usually carefully recorded, as were the attendees at meetings. Therefore, it is possible to prove Karl-Friedrich's continuing involvement in bomb research. For example, on April 18, 1944, Diebner sent to the president of the Reich Research Council a listing of nuclear research commissions (grants) made over the preceding year (April 1, 1943–March 31, 1944). Under the column titled "Commissions in Nuclear Physics Research" is the following: "Development of catalysts for H-D [hydrogen-deuterium] exchange. Physical Chemistry Inst. At Leipzig University, Prof Bonhoeffer. Status is secret." Also listed is a grant to Paul Harteck, noting that his work is urgent and is partly top secret. No further information is given.[39]

That the German atomic bomb project was in some respects stalled by the beginning of 1943 was known to the British because Paul Rosbaud had kept them informed. He had also informed the British of the rocketry project at Peenemünde although at the time the British were not able to grasp its scope or ramifications. Unfortunately, some of Rosbaud's information was not given sufficient weight, particularly when it came to the atomic bomb. Some people in British intelligence suspected that Rosbaud was a double agent who was intentionally downplaying German atomic advances so that the Allies would back off on their own bomb research. Frederick Seitz, who had great respect for Rosbaud, having personally witnessed the scope of his espionage, wrote about the problem:

> It was not Rosbaud's fault that the Western Allies failed to appreciate the fact that the German experimental nuclear fission program was essentially at a standstill by the end of 1942. The politically oriented leaders in the United Kingdom and the United States were inclined to distrust information that was not secured by on-site inspection made by individuals in whom they had well-justified confidence. In particular General Leslie Groves decided with support from the White House that he could be convinced about the status of German developments only when his own specially selected and trusted staff had made personal inspections. Too much was at stake to rely on a complex foreign espionage network. Although Allen Dulles's intelligence agency, linked to the U.S. embassy in Switzerland, might have been of help to Groves, its cloak-and-dagger mindset was highly oriented toward political, military and economic affairs.[40]

Rosbaud became a scientific spy in the 1930s at a time when the need for such a specialized individual was not fully recognized by military, intelligence, and political leaders. The First World War had been, from a scientific standpoint, principally in the domain of the chemists (first and foremost being Haber), but Rosbaud knew that the next war would be in the domain of the physicists. Up-to-date on the latest scientific developments, Rosbaud was able to appreciate the advances being made while simultaneously imagining the darkest consequences.

Meanwhile, things were about to get worse for the Germans, because in England, Leif Tronstad was working hard on plans to destroy the heavy water cells at Vemork. The difficulties in doing so were enormous. First, there was the rugged terrain; second, the tight security at Norsk Hydro; and third, the location of the cells in the basement of the eight-story electrolysis building that was built of concrete. Bombing the surface of the plant would leave the cells untouched, that is, if it were even possible for the bombs to be targeted accurately enough to drop into the narrow gorge. The first sabotage attempt took place in November 1942 when the English along with Norwegian resistance attempted to land men by means of two gliders towed by two bombers. Unfortunately, success was reliant on a moonlit night, but after takeoff the weather over the North Sea turned bad and the moon was obscured by clouds. The gliders and one of the two planes made it to Norway only to crash into mountaintops. Those men who survived the crashes were captured, interrogated, and shot.[41] Only one tow plane made it back safely.

The British and Norwegians were not deterred. A more carefully planned raid, codename Gunnerside, took place three months later on February 27, 1943. A few weeks before, a small team of thoroughly trained Norwegian commandoes had parachuted onto a snow-covered plateau well to the north of Vemork. Again, a full moon was essential, but this time the weather and the moonlight held just long enough to get the men safely on the ground. However, not long after landing, a blizzard set in and they had to make their way on skis under extreme conditions, stopping midway at cabins, before finally reaching the Vemork gorge on a moonlit night. There, they took off the skis and their white camouflage suits. In British battle dress, they descended the precipice in the darkness, crossed the icy Maane River, and scaled the south side, finally reaching the railroad tracks that ran into the plant. This was the only unguarded entry, the main suspension bridge being protected by soldiers with submachine guns, and the ladders by the steep penstocks being protected by mines and trip wires. Fortunately, the deafening thrum of the huge turbines covered any noise the commandoes made as they cut chains and gained entry to the electrolysis building by crawling through a cable duct. Once inside, they packed the explosive around the cells, hurried a startled Norwegian night watchman out of the building, set the fuses, and bolted for the rim of the gorge as the blast went off behind them, destroying from 500 to 900 kilograms of heavy water.[42] Descending, almost plummeting, down the gorge, they crossed the river, scrambled up the north side, grabbed their skis and gear, and headed toward the Swedish border.

Although the Gunnerside raid had done major damage to Vemork, the Germans set about repairing and reinforcing it immediately. By the summer of 1943, production was returning to normal. That rapid restoration was too much for the Americans who were very concerned about what the Third Reich was doing with heavy water, and so on November 16, 1943, they sent in 140 Flying Fortresses that dropped over 800 bombs.[43] The plant was devastated. This time the Germans had to admit that Vemork should not be rebuilt, at least not the heavy water part. There is one more incident worthy of mention. In February 1944, the Germans tried to ship the last of the heavy water still at Vemork to Berlin. It was loaded on the ferry *SS Hydro*. But the resistance got word of what was happening and planted a bomb on the ferry, sending it to the bottom of Lake Tinn at its deepest point, 430 meters down, with considerable loss of life.

Germany had begun to look at alternate suppliers and to develop new heavy water technologies well before the sabotage of Vemork and the sinking of the *Hydro*. One other possible source of heavy water was I.G. Farben's Ammonia Werk (Leuna Works) near Leipzig. Nothing had come of Karl-Friedrich's recommendation of I.G. Farben to Harteck made in 1940, basically because Heisenberg was not interested in pursuing it, much to Harteck's annoyance. Now the recommendation was reconsidered. A pilot plant could be built using what was called the hot-cold principle, which was a more efficient, but technically complex, catalytic water-hydrogen exchange process engineered by Harteck. Per Dahl explains the process as follows: "By a sequence of circuits, the equilibrium balance is progressively adjusted; as a result, the water is correspondingly enriched in deuterium in the cold section and depleted in the warm section, but with a net gain in concentration." The pilot plant, called Stalin-Organ, so named for its numerous pipes and columns similar to a church organ, had been proposed in February 1943 by a physical chemist at Farben following a major conference at the Kaiser Wilhelm Institute for Physics, and after discussions with Karl-Friedrich and Hans Suess. By September 1943, it was ready to go on line. "Harteck was sufficiently satisfied with the design to push for a full-scale effort – that is, a plant 100 times as large capable of the five or six tons of heavy water deemed adequate to sustain a viable uranium-power project."[44] But again, there were major disagreements about the process and the cost, with I.G. Farben (hoping to make a profit) insisting on control, even patent rights. It was yet another example of German disunity. After the war in the article he wrote for *Nature*, Heisenberg stated that the pilot plant made slow progress: "It was proving, in fact, barely possible, in view of air raids and the overall strain on German production, to undertake such big building projects."[45] He was right about the air raids. On July 28, 1944, Farben's Leuna plant was bombed by the Allies in a huge daylight raid. When Harteck and Dieber arrived to survey the smoking remains, they were accosted by a furious Farben executive who charged that the Allies had somehow been alerted to the existence of the pilot plant and that henceforth Farben was not interested in the project. Yet again, Harteck was going to have to look elsewhere for his heavy water.[46]

Notes

1. Shirer, *Third Reich*, 598–599.
2. Harteck, Ermenc interview, PHP, RPI.
3. Dahl, *Heavy Water*, 137. See footnote 9-32. Harteck to D. Irving, April 19, 1966; PHP, Acc. 85-22, Box 1, Folder 1, RPI.
4. Hirsch-Heisenberg, *My Dear Li*, 76.
5. Of Speer, Heisenberg wrote on June 5, 1942, "I do not envy him the fact that his words decide the work of many millions of people, the care for the hundred thousand homeless from Cologne and the whole human misery." Ibid., 176.
6. Begun, *Isotopes*, n.p.
7. In a letter to Harteck dated January 24, 1943, Bonhoeffer mentioned both Heisenberg and Weizsäcker. PHP, Box 1, Folder 7, RPI. See also Carl Friedrich von Weizsäcker, *Pioneer of Physics, Philosophy, Religion, and Peace Research*, ed. Ulrich Bartosch (Springer ebook, 2015), 3, ch. 1.
8. Klaus Hentschel, ed., *Physics and National Socialism: An Anthology of Primary Sources* (Basel: Birkäuser Verlag, 1996), lxxxii.
9. In a letter to Harteck dated November 11, 1941, Karl-Friedrich wrote that he had sent Diebner information on deuterium to help distinguish between "experiment and theory" on issues related to separation. He also wrote that he had gone to Berlin to hear a lecture by Planck and to talk with Wirtz at the Kaiser Wilhelm Institute. PHP, Box 1, Folder 7, RPI.
10. Hentschel, *Physics and National Socialism*, lxviii.
11. Many historians have placed on this group of scientists the derogatory label "fellow travelers," because they saw themselves as apolitical. Mark Walker writes: "Moreover, as far as science in the Third Reich was

concerned, fellow travelers were often far more dangerous than either Hitler's true believers or his bitter opponents." Walker, *Nazi Science*, 271.

12. Ibid., 195.
13. Among his many positions after the war, Seitz was director at the Oak Ridge National Laboratory, chair of the physics department at the University of Illinois, and president of The Rockefeller University.
14. Nikolaus Riehl and Frederick Seitz, *Stalin's Captive: Nikolaus Riehl and the Soviet Race for the Bomb* (Washington, DC: American Chemical Society, 1996), 18.
15. Ibid., 91.
16. Hentschel, *Physics and National Socialism*, ixix.
17. Ruth Lewin Sime, *Lise Meitner: A Life in Physics* (Berkeley: University of California Press, 1996), 307. See endnote 153.
18. DBW, Vol. 8, 334.
19. Cornwell, *Hitler's Scientists*, 234.
20. Hentschel, *Physics and National Socialism*, 365.
21. Ibid., lxxxiv, see footnote 304. See also an excerpt of a report by Heisenberg, February 26, 1941. Samuel A. Goudsmit Papers, Series IV. Alsos Mission. Subseries A. Misc. loose paper from the 1989 accession, 1939–1944. Box 27, Folder 40, Niels Bohr Library and Archives.
22. Dahl, *Heavy Water*, 138, 329 (endnotes).
23. Jonothan Logan, "The Critical Mass," *American Scientist* 84, no. 3 (May–June 1996), 263.
24. Frisch-Peierls memorandum. https//en.wikipedia.org/wiki/Frisch-Peierls_memorandum.
25. Frisch, *What Little I Remember*, 126.
26. Dahl, *Heavy Water*, 139.
27. Hentschel, *Physics and National Socialism*, 369. See footnote 39.
28. Dahl, *Heavy Water*, 149. See endnote 10-6: "Harteck-Bonhoeffer correspondence, January–March 1940: PHP. Acc. MC 17, Box 1, Folder 6. Bonhoeffer to I.G. Farben, February 28, 1940, ibid., and IMC, DJ 29-632."
29. PHP, Ermenc interview, 28, RPI.
30. Dahl, *Heavy Water*, 149. See footnote 10-9. PHP, Acc. MC 17, Box 1, Folder 6, RPI.
31. Ibid., 115.
32. Dahl, *Heavy Water*, 127–129.
33. Ibid., 158.
34. Ibid., 153.
35. It is likely that Karl-Friedrich approached his brother-in-law, a member of the Abwehr, to get deferrals because Dohnanyi had gotten deferrals for others, including Dietrich. During the winter of 1941–1942, the German

military had suffered major loses in Russia, so conscription meant being sent to the eastern front, which had high casualty rates. Years later, Weiszäcker was asked to present an essay and speech on the 70th anniversary of Dietrich Bonhoeffer's birthday hosted by the World Council of Churches. In the opening of that speech, he mentioned he came to know the Bonhoeffer family "in most cases only after the war," never mentioning his relationship with Karl-Friedrich. Susanne Helm, Carola Sachise, and Mark Volker, *The Kaiser Wilhelm Society Under National Socialism* (Cambridge: Cambridge University Press, 2009), 353–354. See Hentschel, *Physics and National Socialism*, biographical profiles, xx. See also Weizsäcker, "Thoughts of a Non-Theologian on Dietrich Bonhoeffer's Theological Development" (Geneva: *The Ecumenical Review*, April 1976), 156.
36. Dahl, *Heavy Water*, 188, 190–191.
37. Hirsch-Heisenberg, *My Dear Li*, 175.
38. Cassidy, *Beyond Uncertainty*, 331.
39. Hentschel, *Physics and National Socialism*, 322–323. See also Samuel A. Goudsmit Papers. Series IV, Alsos Mission. Subseries A. Alsos mission material. German Research Council correspondence on nuclear physics 1942–1945 (Box 27, Folder 30). Niels Bohr Library and Archives.
40. Riehl and Seitz, *Stalin's Captive*, 50.
41. Dahl, *Heavy Water*, 196–197.
42. Hentschel and Dahl present different numbers for the amount of heavy water destroyed. Hentschel states 900 kilograms; Dahl, 500 kilograms. Ibid., 197–205, 207. See also Hentschel, 375, footnote 63.
43. Ibid., 215.
44. Ibid., 234.
45. Hentschal, *Physics and National Socialism*, 375.
46. Dahl, *Heavy Water*, 237–238.

CHAPTER 12

Steadfast to the End

On April 5, 1943, Dietrich was arrested at the home of his parents on charges that his exemption from military service was invalid and, therefore, a crime. On the same day, Christel and Hans Dohnanyi, Josef Müller, and his wife, Marie, were arrested on suspicion of treason. A Catholic, Müller had connections to the Vatican with which the resistance was attempting to work, just as Dietrich had connections to the Church of England.[1]

As soon as she was allowed to write from prison, Christel sent a letter to her two teenage sons, Klaus and Christoph, about the arrest: "Now I want to tell you one more thing. Don't carry any hate in your heart against the power that has done this to us. Don't fill your young souls with bitterness; that has its revenge and takes from you the most beautiful thing there is, trust . . . it is after all only a really small and meager part of the human being that one can put in jail."[2] Christel was released in two weeks but continued to play a major role in passing information to the participants in the resistance and misinformation to the Gestapo interrogators. For Christel, Hans, and Dietrich, their job was no longer to take an active part in the conspiracy but, as Bethge phrased it, to throw "the bloodhounds off their scent."[3]

Klaus Bonhoeffer and Rüdiger Schleicher were not arrested until eighteen months later in October 1944. Until then, they remained involved in the resistance albeit from a different direction than Hans and Dietrich who were linked in through the Abwehr. Both Klaus

K. L. Housley, *The Scientific World of Karl-Friedrich Bonhoeffer*,
Palgrave Studies in the History of Science and Technology,
https://doi.org/10.1007/978-3-319-95801-9_12

and Rüdiger were specialists in aviation law as were their friends, the two brothers, Otto and Hans John. Otto John worked with Klaus at Lufthansa, which had ceased to operate as a private company in 1935, becoming a part of the Reich aviation ministry. Hans John taught at the Institute of Aviation Law at the University of Berlin under Rüdiger's leadership. Rüdiger was also head of the legal department of the Reich aviation ministry, a position that gave him the ability to intercede occasionally on behalf of family members to get coveted prison visitation passes.[4] It was also significant that Rüdiger and Ursula lived next door to Karl and Paula Bonhoeffer in Berlin. A garden path connected their homes on Marienburger Allee, to which Karl and Paula had moved in the mid-1930s from their spacious home on Wangenheimstrasse where the children had grown up. The Bonhoeffer home functioned as a base of operations for a component of the resistance. According to Christian Gremmels, editor of *Letters and Papers from Prison*, regular discussions took place there, with participants being Paula and Karl, Christel Dohnanyi, Rüdiger and Ursula Schleicher, Susanne Dreß (when she was in Berlin), Klaus, Otto and Hans John, Hans Oster, and other trusted friends: "The family sat together deciphering encoded news reports that came from the prisons, assessing what was to be done, and seeking advice from various quarters on how to implement their strategy."[5] In regard to Klaus, Dietrich occasionally complained to Bethge that he had not heard from him, nor had he come to visit him in prison. However, there were compelling reasons for Klaus not to write and for Dietrich to make it seem in his letters as if Klaus were uncaring and distant.[6]

At the time of Dietrich's arrest, the Gestapo suspected him of something more serious than a military exemption, as indicated by his incarceration on the top floor of Tegel prison where prisoners condemned to death were kept. The Gestapo knew that Dietrich was exempt from military service because he was working as a civilian agent for the Abwehr. They also knew of his long-standing opposition to Nazism and his tireless work for the Confessing Church, which had led to his being banned from publishing and public-speaking because of his "activities subverting the people." Straining credulity, Dietrich had explained that he had finally come to his senses and was proud to serve Hitler. What the Gestapo did not yet know, but strongly suspected, was that under cover of his ecumenical church work (which took him to several countries, including Switzerland, Sweden, Norway, and Italy), Dietrich had passed on information in an effort to win Allied support for the German resistance.

Furthermore, the Gestapo did not know of prior attempts on Hitler's life. What spurred on the Bonhoeffers, the Dohnanyis, and the Schleichers to such an extreme action was the implementation of Hitler's Final Solution. No longer were Jews to be driven out of Germany, they were to be murdered with technical efficiency. Wearing the yellow star, men, women, and children were herded onto trains and deported to the east where the gas chambers were working non-stop. There was also discussion within the Reich leadership of sterilizing all the Slavic people whom they had enslaved. The Nazi plan was subjugation followed by extermination. Lebensraum—that beautiful pristine space Hitler conjured up as the divine right of all pure Germans—was to be purged of its inferior inhabitants. According to Elisabeth Sifton and Fritz Stern in their essay "The Tragedy of Dietrich Bonhoeffer and Hans von Dohnanyi":

> As German Jews began to be deported to the East and almost certain extinction there, Dohnanyi embarked on an extraordinary project to use his offices to save at least a few of them, including some to whom he had earlier promised protection. The complicated, desperate plan—it became known as Operation 7—involved having seven people whose names appeared on the deportation lists (the number grew to fourteen) designated as Abwehr agents and sent with the Gestapo's consent to Switzerland. By September 1942 fourteen Jews had found safe haven, and were further helped by Hans's insistence that they receive funds to sustain themselves.[7]

Operation 7 involved the irregular transfer of Abwehr funds, resulting in currency violations, which led the Gestapo to Dohnanyi. There had long been bad blood between the Gestapo and the Abwehr because the Gestapo wanted to control all military intelligence, and they did not trust the Abwehr leaders. Suspecting Dohnanyi of being a traitor to the Reich, they had tapped his phone and intercepted his mail.

Despite the importance of Operation 7, Dohnanyi knew that helping fourteen Jews to safety was not enough. It would not stop the trains to Auschwitz and Buchenwald. It would not turn off the gas. Only the death of Hitler might do that. In March 1943, less than a month after the emaciated and frostbitten remains of the German army in Stalingrad surrendered to the Russians, Dohnanyi and Admiral Wilhelm Canaris flew to Smolensk, carrying an unusual British explosive from which a bomb could be constructed after arrival. The plan was for another conspirator, Colonel Henning von Tresckow, to plant the bomb on Hitler's plane. Bethge played a part in the plan, although he did not realize it at the time. "I drove Dohnanyi, in a car specially permitted to doctors that belonged to Bonhoeffer's father, to the night train that was to take him to East Prussia. There he took the plane to meet Canaris." Von Tresckow put the bomb on Hitler's plane disguised as two bottles of Cointreau, but it failed to detonate. Undeterred, they tried again on March 21 with a conspirator carrying two bombs in his pocket, intending to blow himself up, but this time a last-minute change in plans by Hitler foiled the effort.[8]

Bethge remembered that on the Sunday the assassination was to occur almost the entire family, including the children, were gathered at the Schleicher home. They were practicing the cantata *Lobe den Herren* (Praise to the Lord) written by the contemporary composer and organist Helmut Walcha. The performance was to be a special gift for Karl's upcoming seventy-fifth birthday. Within the cantata were the comforting words of the well-known hymn: "Praise to the Lord, who o'er all things so wondrously reigns; shields you under his wings, yea, so gently sustains." Dietrich was at the piano, Klaus played cello, Rüdiger and Emmi violin, and Hans and Christel were in the choir along with the children. Bethge, who was betrothed to the Schleichers's daughter, Renate, played violin and also conducted. One reason for the music-making was that the Bonhoeffers considered performing classical music (not just listening to it) an essential part of life; so it had always been, with every holiday and birthday celebrated in song. The ulterior reason was to provide a vocal smoke screen in case they were being spied on.[9] While the children were concentrating on the music, the adults had their minds elsewhere as they waited for a phone call about the assassination. "It must start any moment," Christel whispered anxiously to her sister. But the telephone never rang.[10]

Karl's birthday celebration took place on March 31 and was attended by the whole family. Ironically, well-wishers from Charité Clinic arrived at the Bonhoeffer home to award Karl the Goethe medal, which was presented with a written greeting from Hitler: "In the name of the German people I bestow on Professor Emeritus Bonhoeffer the Goethe medal for art and science, instituted by President Hindenburg. The Führer Adolf Hitler." It was the last full gathering of the family. Five days later, the arrests occurred.[11]

There may well have been one other reason why it was imperative to kill Hitler. It is possible that Karl-Friedrich kept not only Rosbaud but also Dietrich, Klaus, and Hans informed about progress on building a bomb so powerful that with it Hitler could conquer the world. They had to stop Hitler *before* the Uranium Club succeeded. There would be no *after*.

Paul Rosbaud was also appalled by the Final Solution. Early on, he alerted the British to the existence of the concentration camps. However, as with the information he passed along on the bomb and rocketry, he was grievously disappointed that the British failed to comprehend the magnitude of the catastrophe. He had first come face to face with the horrors of the concentration camps when he and his friend, Ruth Lange, had visited a camp in which Ruth's brother-in-law, Dr. Georg Benjamin, was an inmate.[12] Disguised as workers, they had managed to make contact with Benjamin and some of the other inmates, sneaking them food. Despite their efforts, Benjamin subsequently committed suicide by throwing himself against an electrified fence. After Rosbaud's home in Berlin was bombed, he and Lange were forced to move out of the city to Teltow where Lange continued to play an important role in helping numerous victims of Nazism. Rosbaud wrote:

> In 1944 when I moved to Teltow she came and shared the house, which was something like an illegal HQ, with me, my old housemaid, and a French POW, and it was there that she led in organizing help for Russian, French and Yugoslav people in concentration camps, the few still existing Jews, Dutch and Norwegian slave workers, and so on. Whether it was food for half-starved people, cigarettes, medicine for sick POWs, she arranged for it. It was there that she met my allies, Laue, Hahn, Mattauch, Bonhoeffer, and many others.

In his biography of Rosbaud, Kramish commented that this was the most Rosbaud himself ever said about this dangerous and very personal work. It is also the only known record of Karl-Friedrich being in Rosbaud's home during the war.[13]

After the war, colleagues and surviving concentration camp inmates related many stories of Rosbaud's compassionate courage. One such story was told by the nuclear physicist Arnold Flammersfeld who had received his doctorate under Lise Meitner and was a colleague of Otto Hahn, as well as an acquaintance of Karl-Friedrich. Flammersfeld recalled:

> After the heavy bomb damage of Hahn's institute in Berlin-Dahlem on February 14, 1944, I participated in the cleaning-up work. One of the employees of the institute, Ludwig Gille, the head of the mechanical shop, was an influential and dangerous Nazi. He had brought four or five concentration camp prisoners to help. They were working in their striped prison clothing. By chance I saw Rosbaud distribute to them rolls he had brought. This courageous deed made a deep impression on the young people. We would not have dared to do it.[14]

Rosbaud, the Bonhoeffers, the Dohnanyis, and their wide circle of resistance friends were working on multiple entangled moral levels, which helped them maintain focus on the individual human who was overwhelmed by systemic evil. For Rosbaud, this meant: *the inmate huddled before me needs food.* For Karl-Friedrich: *this half-Jewish student must be harbored in my lab.* For Dietrich: *the prison guard terrified by last night's bombing needs my care.* For Karl: *this mentally ill patient facing execution must be given a non-genetic diagnosis.* For Hans, *these fourteen Jews must be helped across the Swiss border.* By such actions, they ran the risk of bringing unwanted attention to themselves and their mission. For all of them, there could be no other way. It had been ten long years since Nazism transmogrified Germany, but they had not become callous (Fig. 12.1).

After his arrest, Dietrich and Karl-Friedrich stayed in touch by letter as they had always done, but now there was much greater danger, requiring every single word to be carefully chosen to protect themselves, their families, and the other conspirators. For the first three months after his imprisonment, Dietrich was only allowed to write to his parents every ten days. They shared the letters with other family members. Sometimes the topic was not what it seemed. Nor did an airy tone indicate that all was well. All letters were censored except for some that Dietrich smuggled out of prison via a sympathetic guard. Most of the uncensored letters went to Bethge and Dietrich's fiancée, Maria von Wedemeyer, then nineteen years old, to whom Dietrich had become engaged only three months before his arrest. In the ones to Bethge, he was truthful about the grimness of prison life and how at the beginning he had been treated as a dangerous criminal. Nevertheless, he was extremely careful with content so as to protect Bethge should the letters fall into the wrong hands. Referring to his tendency to resignation (acedia) and melancholy (tristitia), he had

Fig. 12.1 Hans Dohnanyi. Interfoto/Alamy Stock Photo

resolved not to yield to their "ominous consequences," by which he meant suicide. "I have told myself from the beginning that I will do neither human beings nor the devil this favor; they are to see to this business themselves if they wish; and I hope I can stick to it." This bluntness is in sharp contrast to the soothing tone he used in writing to his parents only ten days after his imprisonment. "Above all you need to know and indeed believe that I am doing well . . . Curiously those things that one usually imagines to be particularly unpleasant when in detention, that is the various external privations, do in fact hardly matter at all. It is quite possible to satisfy one's morning appetite with dry bread—and by the way, I am also getting all kinds of good things!—and the cot does not bother me in the least." But his parents were too wise to be fooled because they themselves were playing the same necessary game, downplaying the difficulties and dangers of their lives.[15]

It is exactly that obfuscating approach that Karl-Friedrich used in his first letter to Dietrich a few weeks after his arrest. Across the top of the letter a Nazi censor had written the words "delivery permitted." Karl-Friedrich opened by assuring Dietrich that he was thinking of him in his isolation, and then, groping for words, he wrote apologetically, "For this is all it can be, since there are so many more questions on one's mind, rather than the desire to tell you about all kinds of trivia. Of course, all of us are hoping very much that this time of testing will soon be over for you, and that you will be released soon. I have often been in Berlin the past couple of weeks. There is no need for you to worry about our parents. They are, of course, quite shaken but full of trust and confidence that the matter will soon be straightened out." He then wrote lightly about the garden at their Leipzig home being magical in the springtime, but that unfortunately his two youngest children had sore throats, while Grete was busy from dawn to dusk. He himself was working on a scientific manuscript; however, his thoughts often strayed to Dietrich and all the others. "Keep up your good spirits," he concluded, knowing there was so much he had left unsaid.[16] It troubled the whole family that they could no longer be open and truthful. They had no choice. Torture and death could result from a careless word.

In his powerful essay "After Ten Years," written in late 1942, four months before his arrest, Bonhoeffer wrote to his co-conspirators with regret: "We have been silent witnesses of evil deeds. We have become cunning and learned the arts of obfuscation and equivocal speech.

Experience has rendered us suspicious of human beings, and often we have failed to speak to them a true and open word. Unbearable conflicts have worn us down or even made us cynical. Are we still of any use? We will not need geniuses, cynics, people who have contempt for others, or cunning tacticians, but simple, uncomplicated, and honest human beings." He concluded with the troubling question about how they would act after secrecy was no longer necessary: "Will our inner strength to resist what has been forced on us have remained strong enough, and our honesty with ourselves blunt enough, to find our way back to simplicity and honesty?"[17] The imperative for secrecy was so strong that in the attic room of his parents' house where his books were kept, Dietrich had stashed a copy of the essay over a rafter. It was not found until after the war.

On November 21, 1943, Karl-Friedrich informed Dietrich that he had applied for permission to visit him in prison. "As you know, I had been playing with this idea for a while, except I didn't know whether I would be preventing someone else's visit that you would perhaps desire *more*." He then described a weekend visit to Friedrichsbrunn where his family, as well as the family of his youngest sister, Susanne Dreß, were staying for the winter because of the heavy bombing of Berlin and Leipzig.

Covered in snow, the Harz Mountains were serenely beautiful, and Susanne found the area to be enchanting, never having been there in the winter before. A few weeks after Karl-Friedrich's letter, she wrote to Dietrich that, "On the bare beeches the hoarfrost is hanging in four-centimeter-long crystals." To her, it seemed that a person could almost forget about the war while in Friedrichsbrunn were it not for the roar of planes passing overhead. She admitted sadly that Christmas preparations had been completely abandoned. "I have almost nothing even for the children."[18]

Unfortunately, the winter beauty did not make daily life easy, as both Susanne and Karl-Friedrich made clear in their letters. The large airy house was not built for year-round living. Most of the rooms were unheated (coal being in short supply), so everyone had to squeeze into the kitchen, one of the few rooms that had both heat and electricity. Despite all these drawbacks, the house was not without its crowded periods of normalcy, as Karl-Friedrich described it to Dietrich in his letter, "this frenzied commotion, everything happening in the kitchen—cooking, eating, dishes, homework, play, sewing on the machine, ironing, silliness and bickering, and at the end of the day getting ready for bed."[19]

In his letter, Karl-Friedrich wrote sorrowfully that not only his home life but also his professional life at the University of Leipzig had changed for the worse. His good friend and anti-Nazi colleague, Wilhelm Jost, had left for the University of Marburg.[20] "This is a great loss for me. I miss him at every turn." Even worse, Wolfgang Ostwald, the son of Nobel laureate, Wilhelm Ostwald, was dying of cancer. "So now everything depends on me. Furthermore, two old gentlemen have died: first, the physiologist here, Gildemeister, with whom I had a great deal to do in my new area of responsibility and whom I miss terribly. The other, my predecessor [Max] Le Blanc, for whom I delivered a eulogy yesterday in the public meeting of the academy." Anti-Nazi, Martin Gildemeister was the one who had helped Bernard Katz, a brilliant Jewish student, earn his degree and then get out of Germany to safety in England in 1935. This meant that virtually all the professors who had been working with Karl-Friedrich on the interconnections between physiology and physical chemistry were either dead or had left Leipzig. Alone, Karl-Friedrich wrote to Dietrich from his nearly empty and unusually quiet home in Leipzig.

> I used today to think through a couple of papers I have been wanting to write for some time now. I have the feeling that time is pressing, yet I make no real progress. It's not only the empty stomach that tickles a person a bit at times and makes one get up from the desk at ill-timed intervals. But it is simply not at all easy at present to concentrate on pure science. On the other hand, one ought to do so as long as one still has unbroken windows, which is still the case for us at home. At the institute, my auditorium is already pretty well blown out and unusable. We can still publish our papers more or less without restriction. Hopefully, your time in prison will truly soon be over! It is quite inconceivable for us what it means to be forced to be alone for so long. All the best and perhaps we will see each other soon. Your Karl-Friedrich.[21]

Karl-Friedrich's longing to return to pure science was expressed yet again in a letter written on February 4, 1944, while he was visiting the University of Göttingen to lecture in the physics colloquium on "the mechanism of rhythmic reactions," about which there was interest among the biologists because it was related to neurology. "I also got to see a whole lot of people again for whom pure science is simply their whole life—and I met some new ones too. I really felt 'transported into a better world.'" The last words were quoted from Franz Schubert's

An die Musik: "You, noble Art, in how many grey hours, when life's mad tumult wraps around me, have you kindled my heart to warm love, have you transported me into a better world." For Karl-Friedrich, it was not music that was the transporting art but pure science, which was in marked contrast to the reality of his bombed-out life. "Here in Leipzig I'm fighting to get workers to put up a temporary roof to save what's left of my institute, running from one office to the next to get the plumbing and electric wiring finally repaired for my apartment, writing my fingers raw ordering replacements for the library books and chemicals that burned, and negotiating with Osram on adding an advisory position. All the scientific work that really has my heart in it is being left undone."[22]

Ulf Messow, a professor at the Wilhelm Ostwald Institute of Physical and Theoretical Chemistry at the University of Leipzig, wrote in a brief biography of Karl-Friedrich that most scientists had to subordinate their research to the war effort: "[Karl-Friedrich] was ordered to work on ice, fluorescent tubes, fluorescent lamps, defoamers for aircraft, and engine lubricating oil. But because it was secret, this work was not published."[23] The mention of fluorescent tubes and lamps is a reference to Osram, which was a lighting company in Berlin. Yet even that work—Osram had secret military contracts and was using slave labor—required something from Karl-Friedrich he was not willing to give. He wrote to Dietrich in August 1943 in regard to Osram, "Despite promises of absolute freedom of research and considerable material enticements, I don't quite trust the situation entirely. Under no circumstances will the pitiful remains of my freedom be up for sale."[24]

As to the hoped-for trip to Tegel prison, Karl-Friedrich visited Dietrich in early December 1943, much to Dietrich's pleased surprise.[25] Probably Karl-Friedrich brought books and food with him, for no one visited Dietrich or Hans empty-handed. The sparse prison diet consisted mostly of potatoes, turnips, cabbage, and bread, hardly enough to live on. As for books, the prison library in Tegel was of very poor quality. Furthermore, Dietrich was a prodigious reader in a broad range of subjects and genres. Besides theology, he read fiction, history, sociology, even a medical handbook so he could help in the prison infirmary, tending to injuries from bombings that blew out the windows.

All the Bonhoeffers used a secret code in books they sent to each other, putting dots under certain letters in reverse order, then signing their names in the front to indicate there was a message within.[26] Rosbaud used a similar system, sending coded messages to the British in scientific articles and books, which is why his job in publishing was the perfect cover. Often Rosbaud used Norwegian students studying in Germany as carriers of his messages. Arvid Brodersen, who was involved in the Norwegian resistance, recalled that: "Whenever one of them went home for vacation, he carried the current reports with him. They occasionally got some help from Swedish diplomatic couriers, who would take an envelope with them in the bag, and then hand it to one of the student group who 'happened' to travel north on the same train, as soon as they reached Swedish territory. But more often they had to carry the material all the way themselves."[27]

So also, messages to and from Karl-Friedrich and other scientists, including Meitner in Sweden, were handled in the same way. Meitner was not directly involved in spying, although she was willing to pass on information. For Rosbaud, Meitner served as his conduit to his wife, Hilde, and daughter, Angela, in England whom Meitner knew well. He would send Meitner his personal letters and she would forward them. Meitner's biographer Ruth Lewin Sime believes that Rosbaud probably did not want to jeopardize Meitner's function as an intermediary to his family by sending vital data through her, plus he had more direct routes to his British contacts. "Unsurprising, there is no written corroboration: except for occasional veiled references to a mutual friend in a prison or concentration camp, there is no obvious clandestine undercurrent to their wartime correspondence, and after the war Meitner expressed only gratitude to Rosbaud for the books he had 'so kindly selected.'"[28]

Of the books that Karl-Friedrich sent to Dietrich, some may have carried secret messages, but probably most did not. Continuing to study and write in prison, Dietrich asked for specific books as often as he asked for socks, tobacco, pipe cleaners, and numerous other personal items. Since his return from the USA in 1939 until his arrest, he had been working intermittently on his masterwork, *Ethics*, interspersing his trips for the Abwehr with lengthy stays at the homes of friends, including the estate of Ruth von Kleist-Retzow, and a Benedictine monastery in Ettal where the sympathetic monks provided him with the peace and quiet necessary to write.[29] *Ethics* was not an arcane intellectual exercise. Written under extreme pressure, it focused concretely on the disintegration of the world and the potential for peaceful reconstruction after the war by means of God's grace. In the chapter "Natural Life," Dietrich grappled with euthanasia, sterilization, and murder, referring not to the Third Reich but to "worldly authorities" in case the manuscript was discovered.

To the Nazi claim that a disabled individual was a "worthless eater," and should be killed, Dietrich responded that the value of life was not in social utility: "There is no worthless life before God, because God holds life itself to be valuable."[30] By the time of his arrest, he had completed a draft of *Ethics*, but many sections needed clarification; by no means was it ready for publication. Furthermore, Dietrich had plans to write other books and essays, one on the nature of truth, a topic that troubled him greatly given the necessity of lying, and another on taking stock of Christianity and the nature of Christian faith.[31]

Dietrich was determined to continue writing in his cramped prison cell as long as his family could keep him supplied with books, paper, and pens, which they tried hard to do. On September 13, 1943, Dietrich wrote to his parents requesting several newly published books. He remarked, "Klaus too makes lovely book discoveries from time to time." For Dietrich's birthday in 1944, Klaus sent him a Moravian community hymnal, published in 1778, as well as *Von Deutscher Dichtung und Musik* (*Of German Poetry and Music*) by Wilhelm Dilthey.[32] However, it was not always possible to obtain the books Dietrich requested. For example, in August 1943, Dietrich asked his mother to send him the novel *Witiko* by Adabert Stifter, a nineteenth-century Austrian writer, which was in his attic library. After searching for it, Paula replied disconsolately on September 20, 1943, that she had failed. Intense night bombings of Berlin had begun, forcing Paula and Karl to take protective measures. "Your bookshelves upstairs have now been completely emptied, and everything has been brought downstairs, also the pictures and contents of your cupboards. When one sees the many damaged roofs, it is clearly better this way, even though naturally everything has become quite mixed up in the process." That was not the end of it, however. Instead of sending him *Witiko*, she sent a breviary by Stifter in which was hidden a coded message about Dietrich's lawyer. Paradoxically, Bonhoeffer had managed to find a copy of *Witiko* on his own in a most unlikely place: the prison library.[33]

Dietrich had also expressed interest in obtaining books on science, and on this topic Karl-Friedrich was helpful.[34] In the field of theology, Dietrich knew exactly what books he wanted, but in the field of science he had to rely on his brother's suggestions. The result was some fascinating book selections that began to stir Dietrich's thinking in unique ways. In his 1955 essay, "Challenge of Dietrich Bonhoeffer's Life and Theology," Bethge wrote that much of Dietrich's theological reflections in prison "came from Karl-Friedrich's reservations concerning Christian dogmatics and his ethical convictions and actions. They became creatively fruitful in Dietrich's mind."[35]

On June 12, 1943, Karl-Friedrich wrote that he was reading a little book by Gerhard Ritter, a contemporary historian.[36] "Next time I'm in Berlin, I will try to find out if you're familiar with it and then include it in the package, and perhaps also a freshly published collection of lectures on modern physics, or rather the philosophy of nature; but I first have to read it a little more carefully myself in order to decide whether you would get something from it." The book he was referring to was *The World View of Physics* by Carl Friedrich von Weizsäcker, although he did not mention that he knew him.

For some reason, Karl-Friedrich did not send *The World View of Physics* to Dietrich until the following year. In the meantime, he sent him *Microbe Hunters* by Paul de Kruif, published in 1926. Dietrich wrote to his parents on August 24, 1943, that since Karl-Friedrich had sent it, "I assume it is scientifically accurate. It certainly is a very impressive piece of research history."[37] Actually, the book was not very accurate, however, it was as close to being a page-turner as a medical history could get. For dramatic effect, De Kruif hyped the lives and work of thirteen men, beginning with Leewenhoek and his discovery of microbes called "wretched beasties," through Pasteur, Koch, and ending with Paul Ehrlich and his breakthrough on syphilis. A bacteriologist, De Kruif was also an atheist who was exhilarated by what scientists had achieved against disease and death without the help of God, relying on their own intelligence and ingenuity.[38]

Nothing showed up in Dietrich's correspondence that indicated he was influenced by De Kruif, at least not in 1943. Only when he began to read Weizsäcker's book a year later can the influence be seen. On May 24, 1944, he wrote to Bethge, "I'm now reading with great interest Weizsäcker's book on the 'world view of physics' and hope to learn a great deal from it, even for my own work." The book gave Dietrich fresh insights into the problem of a God-of-the-gaps approach to the relationship between science and religion. Weizsäcker had quoted Pierre-Simon Laplace, the eighteenth-century French scientist, who when asked whether there was still room for God replied, "I do not need this hypothesis." Then Weizsäcker had turned to Johannes Kepler for whom positive knowledge of science pointed to God, and to Isaac Newton for whom the opposite was true. Weizsäcker's conclusion was that as science filled in the gaps in knowledge, God and "the faded, half-religious concepts which have often been substituted for him in recent times, were in continuous and dishonorable retreat."[39] This was in agreement with De Kruif who had made the point in *Microbe Hunters* that diseases were being cured not by miracles but by the ever-expanding knowledge of humans. For example, leprosy was not caused by God's curse for sinful behavior but by a bacillus. So also a person with syphilis did not have to wait for a miracle, he could be treated with Salvarsan. After five days of intensive reading and study, Dietrich wrote again to Bethge:

> Weizsäcker's book on the *World View of Physics* continues to preoccupy me a great deal. It has again brought home to me quite clearly that we shouldn't think of God as the stopgap for the incompleteness of our knowledge because then—as is objectively inevitable—when the boundaries of knowledge are pushed ever further, God too is pushed further away and thus is ever on the retreat. We should find God in what we know, not in what we don't know; God wants to be grasped by us not in unsolved questions but in those that have been solved. This is true of the relation between God and scientific knowledge, but it is also true of the universal human questions about death, suffering, and guilt. Today, even for these questions, there are human answers that can completely disregard God. Human beings cope with these questions practically without God and have done so throughout the ages, and it is simply not true that only Christianity would have a solution to them. As for the idea of a 'solution,' we would have to say that the Christian answers are just as uncompelling (or just as compelling) as other possible solutions. Here too, God is not a stopgap. We must recognize God not only where we reach the limits of our possibilities. God wants to be recognized in the midst of our lives, in life and not only in dying, in health and strength and not only in suffering, in action and not only in sin. The ground for this lies in the revelation of God in Jesus Christ. God is the center of Life and doesn't just 'turn up' when we have unsolved problems to be solved.[40]

Weizsäcker had assumed the double role of physicist and philosopher early on, over which he wrapped the mantle of the theologian. He had always been religious, placing great value on the Sermon on the Mount, as had Dietrich who focused on it in his book *Discipleship*. Weizsäcker wrote in an autobiographical essay that at the age of twelve he had looked up into the night sky and thought, "God is present in the stars of the sky, and they are gaseous spheres; how does that fit together?

Science and religion had met. The problem of (theoretical) philosophy was posed." A teenager when he was introduced to Heisenberg in Denmark, Weizsäcker became Heisenberg's protégé and lifelong friend. His principal research before joining the Uranium Club was on stellar energy and the formation of the solar system, one of his most important early papers being on the process of fusion in stars, which he coauthored with the physicist Hans Bethe, who had been forced to leave Germany in 1933. Simultaneously, Weizsäcker pursued his interest in philosophy although Heisenberg discouraged him, advising him to concentrate on physics when young and only when old to take up philosophy. Ignoring Heisenberg's advice, Weizsäcker never stopped mixing the two, becoming an expert on Immanuel Kant.[41]

In March 1941, Weizsäcker had lectured on "Is the world infinite in time and space?" in occupied Denmark and had been invited back by Nazi bureaucrats to lecture on "The Relationship between Quantum Mechanics and Kantian Philosophy." Due to the influence of Weizsäcker's father, a Nazi diplomat, Heisenberg had received permission to go with him in September 1941 to lecture on cosmic radiation, perchance to meet with Bohr. From the German standpoint, it was the equivalent of a goodwill tour meant to impress the staff at Bohr's institute. From the Danish standpoint, having been overrun by Germany in April 1940, the tour reeked of coercion and false camaraderie. Bohr had no intention of setting foot in the same room with Heisenberg and Weizsäcker, and only diplomatic subterfuge made the meetings happen.[42] Nor was Bohr about to divulge to his German visitors that sitting on the shelf in the laboratory of his colleague, George de Hevesy, was an unremarkable bottle of aqua regia in which were dissolved two gold Nobel Prize medals that had been entrusted to Bohr's safekeeping by Max von Laue and James Franck. Because it was illegal to take gold out of Germany, Bohr and de Hevesy knew there would be trouble if the medals were found. Their scientifically creative answer was to dissolve them in the hope that a time of peace would come when the gold could be precipitated out and the medals recast. As for Bohr's gold medal, it had been sold to raise money for the Fund for Finnish Relief.[43]

An aristocrat, Ernst Heinrich von Weizsäcker was in the diplomatic service, serving in Denmark and Switzerland during the 1920s and early 1930s. After Hitler rose to power, he became a state secretary. From 1943 to 1945, he served as ambassador to the Vatican. He was convicted after the war of war crimes and spent a year in prison.

He defended himself by claiming he was anti-Nazi and was in league with the resistance. However, the historical record stands: If Ernst was anti-Nazi, he was still very effective in bringing about Nazi goals. Repeatedly, he opened political doors for his son and Heisenberg, the door to Bohr's institute in Copenhagen being one.[44] His son, Carl, was not a member of the Nazi party, but as has already been shown, he was caught up in nationalistic fervor as was Heisenberg. Carl was politically adept and scientifically brilliant, but until well after the war when he became a pacifist, he was also a shape-changer.

None of this affects his book *The World View of Physics* in which Weizsäcker looked at the structure of the atom, the effectiveness of mathematics, the scientific method, the law of the conservation of energy, and finally quantum mechanics. Beginning with the Greeks (he himself was a Platonist), Weizsäcker discussed a long line of philosophers and scientists, including Kant, Kepler, Locke, and Newton. Other than the fact that Weizsäcker explained the theory of relativity without mentioning Einstein, there was no hint in his book of Aryan physics or a Nazified approach to history in which facts were fitted to ideology. Nor was it too technical for a theologian to read. Weizsäcker's main point was that modern physics had led to the loss of the sense of wholeness that was found in the Christian cosmos of the Middle Ages. However, there was value in that loss:

> A "world view" has been destroyed, more thoroughly than could have been expected. That is no misfortune, but a wholesome lesson. A world view is more than a scientific theory; it is meant at least symbolically to comprehend the whole of reality. Looked at from the point of view of the particular demonstrable items of knowledge, this regard for the whole is always a faith; it is the faith which is the condition of our life. The old physical world view also attempted to represent the whole, but with inadequate means; and it was bound to founder. But physics was the first science of nature to possess a complete system, and so, independently of the "world view," it played the role, as it were, of a model among the sciences. The new physics is the first complete system of natural knowledge beyond the boundaries of the mechanical world view that is expressible with mathematical exactitude. Let us do what we can to make it worthy of this role.[45]

These ideas rang true with Dietrich who knew too well "the faded, half-religious concepts" that had so weakened Christianity and which had led him to conceive of a religionless Christianity in which the outward trappings, what Dietrich referred to as "religious clothes," were stripped away and the person of faith turned toward the broken world wherein God suffered.[46] He was especially intrigued with Weizsäcker's discussion of whether the universe is finite or infinite (both infinitely small and infinitely large). No matter how it began, an infinite universe is self-subsisting, *etsi deus non daretur*—even if there is no God. "The real use which science has made of the idea of infinity," Weizsäcker wrote, "lies in the potential infinity of the known world. We can always push further the limits of the known world and need not fear that a halt will be called to this advance by anything but the slackening of our powers."[47] Weizsäcker saw a connection between the infinity of God and the nature of Christianity, in which a new dimension of the human soul was disclosed. "As against the previous world, it must be conceived as infinite in at least three things: in its longing, in its moral demand, and in its religious fulfillment." God is also infinite in grace; therefore, salvation comes when a man no longer shuts himself off from grace. According to Weizsäcker, "Every attitude which man can assume of himself limits him; it is God's gift that boundless love can come to power in him."[48]

From his study of Weizsäcker, Dietrich began to shape the idea of "this world-come-of-age," which was among his most daring. He was influenced as well by Wilhelm Dilthey's use of the term in his historical analysis of religion, which Dietrich was also reading.[49] The idea did not mean an ethical growing up. It meant a maturation of the believer away from the concept of a powerful God who was a guardian and a gap-filler and toward the concept of a God who was powerless by choice. On July 16, 1944, he wrote to Bethge, "Thus our coming of age leads us to a truer recognition of our situation before God. God would have us know that we must live as those who manage their lives without God. The same God who is with us is the God who forsakes us (Mark 15:34!). The same God who makes us live in the world without the working hypothesis of God is the God before whom we stand continually. Before God, and with God, we live without God."[50]

It is doubtful that *The World View of Physics* ever received a closer, more intense reading than that given it by Dietrich alone in his prison cell. Unfortunately, Dietrich never had time to develop his ideas on science and religion further because the last failed attempt on Hitler's life occurred shortly after he finished the book. The brevity in Dietrich's letters about his idea of the world-come-of-age makes it open to misinterpretation. It is worthy of note that Dietrich wrote to Bethge that he wished they could have a discussion on Weizsäcker's ideas, sharing their thoughts.[51] Years later in an essay that Weizsäcker wrote about Dietrich, he took that wish personally, as if Dietrich had wanted to talk to *him*, regretting that it had never occurred.[52] A case can be made that over the shifting and shifty course of Weizsäcker's long life—from an enthusiastic member of the Uranium Club to a nuclear disarmament pacifist—Dietrich had more influence on him than he had on Dietrich.

Everything changed on July 20, 1944, when the resistance led by Claus von Stauffenberg planted a bomb (the same kind used in the 1943 attempts but with a thinner detonating wire) in the room at the Wolf's Lair in East Prussia where Hitler was holding a meeting with his staff. It went off, killing four people and wounding several others. But Hitler was only bruised and slightly burned by the blast. His sadistic rage knew no bounds. He wanted revenge and he wanted it to be brutal. In the aftermath, as many as six thousand people were rounded up and tortured to death. Rommel, who was part of the plot, was allowed to take poison. Some, including Beck and Stauffenberg, were executed or were forced to commit suicide that night. Paula's cousin, Paul von Hase, commandant of Berlin, was executed on August 8, 1944, along with eight others, meat-hooked to the ceiling of the prison and strangled with piano wire, their execution filmed for Hitler's pleasure. Others, including Klaus, Rüdiger, Dietrich, Hans, and Max Planck's son, Erwin, lived until 1945 under horrific conditions.[53]

Hans's physical condition had been dire since November 1943, eight months before the assassination attempt. Suffering a severe brain embolism when the prison was hit by an incendiary bomb, he had been taken first to Charité Clinic and then to a military hospital. Sometime after that, he caught diphtheria. On July 30, 1944, Karl wrote, from the viewpoint of a physician, to Bethge about a visit to Hans:

> He is in a pitiful state as a result of the dreadful paralysis from his diphtheria. It has improved somewhat in his face and the roof of his mouth, but it has progressed further to his arms, legs, and torso, making him almost immobile. To be sure, one can expect everything to turn out all right, but it is nevertheless a terrible burden and a test of patience for him and for Christel.[54]

As Hans began to improve, he had to find a way to fend off further interrogation that might force him by means of torture to reveal resistance secrets. Hans was particularly worried about his files on Nazi atrocities that he had meticulously kept since 1933. The files were stored at the Zossen military base, which Ludwig Beck considered a safe location. Then on September 22, 1944, two months after the assassination attempt, the Gestapo found some of the Zossen files and showed them to Hitler. Shortly thereafter Klaus and Rüdiger were arrested. Bethge, who was serving in the military, was arrested in November. Knowing what was to come, Klaus considered committing suicide but his sister, Ursula, dissuaded him, which she came to regret bitterly for the rest of her life.[55] It was also the beginning of the end for Dietrich. In October, he was moved from Tegel prison, where he had been incarcerated for eighteen months, to the infamous prison in the cellar of the Gestapo headquarters, which was already crowded with other participants in the resistance.[56]

Early in 1945, Hans begged Christel to infect him with a serious disease so that he would be too sick to reveal information under torture. Getting a pathogen from her father, Christel put it in the food that she brought Hans, color-coding the infected parcel. Karl had access to pathogens through his neurological work at the Charité Clinic, which included the study of how certain diseases affected psychosis. Writing to Christel in a letter that was smuggled out of the prison, Hans told her that the parcel had been received and he had eaten the food. He told her of the immense misery around him, explaining, "That is why I'm not frightened of any infectious illness. I know very well that I would lie down with the feeling that this would not only save my life, but that of many others whose situation is bound up with mine, certainly Dietrich's."[57]

As to any direct connection to the resistance on the part of Karl-Friedrich, the evidence is circumstantial and thin. However, he would have been in danger even if he knew nothing. All the Bonhoeffers, including Karl and Paula, were in danger. There was more than enough information, if tortured out of one of the conspirators, to implicate them. Karl's car had been used in the first assassination attempt; he had obtained the pathogen to infect Hans. The Bonhoeffer women had served as couriers of secret information. Karl-Friedrich was in danger not just because of the activities of his family but also because of his having been denounced for harboring half-Jews in his laboratory.[58]

Furthermore, Karl-Friedrich had attempted to intercede on behalf of a Jewish relative of Michael Polanyi who had been sent to a concentration camp in Poland, although he had failed to win her release (Fig. 12.2).[59]

Karl-Friedrich was struggling to provide for his parents and the families of his brothers, most of them suffering from malnutrition and extreme stress. In the late winter of 1945, with the unstoppable advance of Soviet troops toward Berlin, Karl-Friedrich tried desperately to see his brothers and brothers-in-law one last time. He succeeded in seeing only Klaus not long after he had received the death sentence. On entering Klaus's cell, he saw that the score of Bach's *St. Matthew's Passion* was lying open on a little folding table. Karl-Friedrich later related to his sister, Sabine, that he told his brother how impressed he was that Klaus could hear the music just by reading the score. Klaus replied, "but the words too, the words!" Those words were on the

Fig. 12.2 Flossenbürg concentration camp in May 1945 after liberation. United States Holocaust Memorial Museum, courtesy of National Archives and Records Administration, College Park

crucifixion of Christ.[60] Shortly thereafter in the last days of the war, the executions began. On April 9, 1945, Dietrich was hanged and his body burned at Flossenbürg concentration camp near the Czech border. On the same day, Hans was hanged in Sachsenhausen concentration camp, carried to the scaffold on a stretcher. On April 23, Klaus and Rüdiger were executed by an S.S. firing squad near a train station, their bodies hastily buried in a bomb crater.[61] And on April 30, Hitler committed suicide.[62]

In June 1945, Karl-Friedrich wrote to his children in Friedrichsbrunn that he had not been in Berlin since late March and had heard nothing about the fate of his family. On that last visit, Hans, Klaus, and Rüdiger were still alive, but no one knew where Dietrich was. Karl-Friedrich told his children about the supreme effort made by his sisters, sisters-in-law, and Maria von Wedemeyer to get food parcels into the prisons. "They often made the trip in vain, they often had to listen to insults from contemptible officials, but sometimes, too, they found a friendly guard who thought like a human being and delivered a greeting, accepted a parcel outside the prescribed hours, or gave food to the prisoners in spite of the rules." Since then, Karl-Friedrich had heard that the Nazis had executed four thousand political prisoners just before the Russians captured Berlin. Overcome with worry, he asked:

> Is everyone still alive? Did your grandparents manage all right through those difficult days? Both were already on the verge of exhaustion even before this. Grandma had suffered frequent bouts of weakness and loss of memory in recent years as a result of overwork, anxiety and malnutrition. They have no proper household help. Uncle Dietrich spoke to someone at some length somewhere near Passau as late as April 5. From there he was supposed to be taken to the Flossenbürg concentration camp, near Weiden. Why isn't he here yet?[63]

Notes

1. Bethge states that Müller's secretary Annie Haser was arrested, but the Index of Names in DBW lists his wife, Marie, who was released on April 25. Bethge, *Bonhoeffer*, 785–786; DBW, Vol. 8, 700.
2. Elisabeth Sifton and Fritz Stern, "The Tragedy of Dietrich Bonhoeffer and Hans von Dohnanyi," *New York Review of Books* (October 25, 2012): 5.

3. Müller made it alive through the war, escaping from a concentration camp, and crossing the Alps into Italy. Bethge, *Bonhoeffer*, 750.
4. Schleicher was Bethge's father-in-law, his daughter, Renate, having married Bethge on May 15, 1943. In December 1943, Schleicher arranged for a pass for Bethge to visit Dietrich in prison. He also attempted to get a pass for Ursula. DBW, Vol. 8, 355, 372–373.
5. DBW, Vol. 8, 594–595.
6. On February 1, 1944, Dietrich wrote to Bethge, "I've seen nothing of Klaus, nor heard from him; apart from everything else, it could be that he's too sensitive by nature to subject himself to the sight of this place." That Dietrich knew of Klaus's involvement can be gleaned in a coded letter Dietrich wrote to Bethge on July 16, 1944, alluding to the upcoming attempt on Hitler's life: "I am very glad that Klaus is in such good spirits! He was so depressed for quite some time. Well, I think that all his worries will soon be over; I very much hope so, for his own and the whole family's sake." DBW, Vol. 8, 283, 474, 581.
7. Sifton and Stern, "Tragedy," 4.
8. Bethge, *Bonhoeffer*, 779–780.
9. Walcha's cantata was an unusual selection for the birthday. He was born in Leipzig in 1907, losing his eyesight in 1919 from a smallpox inoculation. An organist and composer, he was a master interpreter of Bach. en.wikipedia.org/wiki/Helmut_Walcha.
10. Bethge, *Bonhoeffer*, 780.
11. Ibid., 785.
12. Georg Benjamin was the brother of the philosopher, Walter Benjamin, who killed himself in 1940, having crossed the border from France into Spain in an effort to escape the Nazis. Georg's wife and Ruth's sister, Hilde Lange Benjamin, became a notorious judge in East Germany after the war.
13. Kramish, *Griffin*, 220.
14. Ibid., 221.
15. DBW, Vol. 8, 180, 56.
16. Ibid., 58, see footnote 1.
17. "After Ten Years" was written to Hans Dohnanyi, Hans Oster, and Bethge. Ibid., 52.
18. The letter from Susanne Dreß is dated December 14, 1943. In November, the air raids had begun to occur on dark nights and in bad weather. DBW, Vol. 8, 216–217.
19. Ibid., 192.
20. Jost wrote the essay "The First 45 Years of Physical Chemistry in Germany." In a moving eulogy for Karl-Friedrich in 1958, Jost wrote that he was "upright and independent in the formation and expression of

his opinions" and that he had an "unflinching sense of truth and justice." Wilhelm Jost, "Friedrich Bonhoeffer," *Die Naturwissenschaften*, 44; Jahrgang, Heft 24 (Zweites Dezemberheft), 1957.

21. DBW, Vol. 8, 191–193.
22. Ibid., 287–288.
23. Ulf Messow, "Karl Friedrich Bonhoeffer," in *Saxon Biography*, herausgegeben vom Institut für Sächsische Geschichte und Volkskunde e.V. (December 21, 2015). http://www.isgv.de/saebi.
24. Osram was conducting research for the Luftwaffe, including heat-seeking sensors and an experimental substance for the motors of jet aircraft. Kramish, *Griffin*, 150; DBW, Vol. 8, 141.
25. Ibid., 212–213.
26. Ibid., 175.
27. Kramish, *Griffin*, 228.
28. Ruth Lewin Sime. *Lise Meitner: A Life in Physics* (Berkeley: University of California Press, 1996), 305. See endnote 143, "Meitner to Rosbaud," 4 August 1946 (MC).
29. The connection to Ettal was made by Josef Müller. Bethge, *Bonhoeffer*, 701.
30. DBW, Vol. 6, 193.
31. DBW, Vol. 8, 499–450.
32. Ibid., 155, 291.
33. Ibid., 136, 164, 175. For the importance of *Witiko* to Dietrich Bonhoeffer, see Kathleen L. Housley, "Daring to Do the Good: The Knight and the Theologian," *Image* no. 89 (Summer 2016): 72–84.
34. The first book that Dietrich read on science and economics was *Science Breaks Monopolies* by Anton Zischka who in the 1930s presented a positive image of German technology. The cause of war was conflict over land and raw materials. Dietrich wrote to Karl-Friedrich that he was greatly impressed by the book. DBW, Vol. 15, 291–292.
35. DBW, Vol. 8, 22. See footnote 29.
36. Ritter was connected to the resistance. It is possible the reference had hidden meaning. DBW, Vol. 8, 104.
37. Ibid., 140.
38. De Kruif displayed the racial prejudice of his age and did not condemn dangerous procedures and experiments done on colonized people. Paul de Kruif, *Microbe Hunters* (San Diego, New York, and London: Harcourt Brace, Jovanovich, 1926, 1954), 21.
39. *The World View of Physics* was first published under the title *Zum Weltbild Der Physik*. There were four German editions prior to the English edition published by the University of Chicago Press in 1952. At least the 1949 edition was revised, with an ending added on human suffering

that sounded exculpatory. He added an introduction to the 1952 English version. Carl Friedrich von Weizsäcker, *The World View of Physics* (Chicago: University of Chicago Press, 1949), 157.

40. DBW, Vol. 8, 401, 405–406.
41. It was common for European scientists to be well versed in philosophy. Carl Friedrich von Weizsäcker, "Self-Portrait," in *Pioneer of Physics, Philosophy, Religion*, ed. Ulrich Bartosch (Springer ebook, 2015), 6–25. https://doi.org/10.1007/978-3-319-13446-8.
42. Much has been written about the Copenhagen meetings. Besides what was said, there was disagreement about when and in what settings the meetings occurred. These disagreements were not just among later historians but also among the participants themselves, specifically Bohr, Heisenberg, Weizsäcker, and members of Bohr's staff. Then in 2002, unpublished documents were released by the Niels Bohr Archives that showed Bohr remembered clearly Heisenberg's confidence that Germany would win the war and that he was leading the development of atomic weapons. Weizsäcker's response to the release of the papers was that Bohr's memory was flawed. See www.nbi.dk/NBA/papers/docs/cover.html. See www.spiegel.de/wissenschaft/0,1518,181179.oo.html. Also see Cornwall, *Hitler's Scientists*, 301.
43. Because the names of the recipients were engraved on the medals, Bohr was worried about burying them. After the war, de Hevesy reversed the dissolution process and sent the gold to the Nobel Prize committee, which recast the medals and presented them to Laue and Franck. Birgitta Lemmel, "The Nobel Medals and the Medal for the Prize in Economic Sciences," www.nobelprize.org/nobel_prizes/aboutmedals.
44. Indicative of the connection between the Reich foreign office and German science is Ernst Weizsäcker's election in 1942 to the Kaiser Wilhelm Society's advisory board and senate. Kristie Macrakis, "The Ideological Origins of Institutes at the Kaiser Wilhelm Gesellschaft in National Socialist Germany," in *Science, Technology and National Socialism*, ed. Monika Renneberg and Mark Walker (Cambridge: Cambridge University Press 1994), 159.
45. Weizsäcker, *World View*, 35–36.
46. Letter to Bethge, June 25, 1942. DBW, Vol. 16, 329.
47. Weizsäcker, *World View*, 157.
48. Ibid., 147–148.
49. Dietrich's idea of the world-come-of-age is in a letter to Bethge, dated June 8, 1944. In *Weltanschauung und Analyse*, Dilthey wrote that the needs of seventeenth-century society gave rise "on the new ground of a science come of age . . . to a scientific system of knowledge providing generally valid principles for the conduct of life and leadership in society." DBW, Vol. 8, 426, 587.

50. Mark 15:34 refers to the crucifixion text in which Jesus cries, “My God, my God, why have you forsaken me?” Dietrich included the exclamation point. Dietrich loaned *The World View of Physics* to Gaetano Latmiral, an Italian officer imprisoned in Tegel. During their exercise period, they talked about political, religious, and scientific problems. Bethge, *Bonhoeffer*, 851; DBW, Vol. 8, 479.
51. DBW, Vol. 8, 401.
52. Carl Friedrich von Weizsäcker, “Thoughts of a Non-Theologian on Dietrich Bonhoeffer’s Theological Development,” *The Ecumenical Review* 28, no. 2 (April 1976): 168. https://doi.org/10.1111/j.1758-6623.1976.tb01205.x.
53. DBW, Vol. 17, 302–303.
54. DBW, Vol. 8, 496.
55. Schlingensiepen, *Bonhoeffer*, 360.
56. Death sentences for Klaus Bonhoeffer and Rüdiger Schleicher were handed down in February 1945. Hitler ordered more executions of conspirators, including Dietrich’s execution, on April 5, 1945. DBW, Vol. 16, 702–703.
57. In a letter to Christel dated March 8, 1945, Hans mentioned diphtheria, but given that he had had diphtheria the preceding year, Christel may have delivered some other diseases. Bethge, *Bonhoeffer*, 913.
58. Letter from Rosbaud to Goudsmit, June 22, 1947. Paul Rosbaud correspondence and manuscripts, 1946–1949 (Box 28, Folder 43). Samuel A. Goudsmit Papers, 1921–1979, Series IV. Alsos Mission. Subseries B. Niels Bohr Library & Archives with the Center for History of Physics.
59. Scott and Moleski, *Polanyi*, 189, see endnote 88.
60. Leibholz-Bonhoeffer, *Bonhoeffers*, 28.
61. Execution dates vary. Bethge states that Klaus and Rüdiger were killed on the night of April 22. The notes to *Letters and Papers from Prison* indicate that Klaus was executed on April 23, and Rüdiger on April 24 at Moabit prison (Lehrter Straße). Bethge, *Bonhoeffer*, 929–930; DBW, Vol. 8, 675, 708.
62. Bethge was released on April 25. DBW, Vol. 17, 120.
63. DBW, Vol. 8, 561–563. See also Dietrich Bonhoeffer, *Letters & Papers from Prison*, ed. Eberhard Bethge (New York: Touchstone, Simon & Schuster, 1997), 409–410.

CHAPTER 13

Rebuilding the World

There had been four Bonhoeffer brothers. Walter was killed in World War I; Klaus and Dietrich in World War II; and only Karl-Friedrich was left. Besides the two revered brothers-in-law, Hans Dohnanyi and Rüdiger Schleicher, numerous cousins, uncles, aunts, and close friends lost their lives. Some were executed for their part in the resistance (among them, General Paul von Hase and Mildred and Arvid Harnack). Some were killed in the bombings (Paula's sister, Elizabeth, died in the firebombing of Dresden). Others were killed in the brutal aftermath (Justus Delbrück died of diphtheria in a Soviet prison camp).[1] The surviving children and grandchildren were hungry, their homes damaged or destroyed, and their education disrupted. In front of the advancing Soviet forces, a tidal wave of terrified refugees from the eastern areas of Germany had swept westward trying to get to safety, making an intolerable situation even worse. There was little food, no means of communication, inoperable transportation systems, and ruined hospitals and schools.

Among the refugees from the east was Dietrich's fiancée, Maria von Wedemeyer, who had fled her family's estate near the Polish border in late January in a wagon pulled by three plow horses and driven by a Polish driver. In her care were several siblings and family friends. In severe winter weather as Russian guns began firing in the village behind them, they headed west for a relative's home. Just after their departure, Maria's mother, Ruth von Wedemeyer, wrote to Karl and Paula Bonhoeffer: "Join me in

K. L. Housley, *The Scientific World of Karl-Friedrich Bonhoeffer*, Palgrave Studies in the History of Science and Technology, https://doi.org/10.1007/978-3-319-95801-9_13

praying that she proves equal to her difficult task. They should be there in two weeks if all goes well, but the snow and wind have been very severe since then. I was strongly advised against Berlin. We are profoundly grateful to you for your offer to take the children." She informed them that if Maria managed to complete the trip, she would stay just long enough to settle the children into their new surroundings and then would attempt to reach the Bonhoeffers in Berlin. Maria did just that, spending part of February in a futile search for Dietrich, writing to her mother that her efforts had come to nothing: "Dietrich simply isn't here. Who knows where he is. In Berlin they won't tell me and at Flossenbürg they don't know."[2]

Maria heard of Dietrich's death in June. Due to the communication lines to Berlin being cut, his parents did not hear until July. At the beginning of August, Karl-Friedrich wrote to Sabine and Gert Leibholz in England to tell them of the deaths of Dietrich, Klaus, and Rüdiger (about which they had already heard) and that Hans's fate was still unknown, although Karl-Friedrich suspected that he too was dead. "We have no absolute proof as yet, but really I cannot see any possible alternative." Then, he gave a synopsis of what had transpired with the resistance, knowing that Sabine and Gert had been completely cut off from all but the most general war news, save for what Bishop Bell could tell them, which was very little. Above all else, he wanted them to know about the family's "courage, prudence, caution and endurance," and that all the members were in complete agreement on political matters, which gave them strength.[3]

Preparing to return to the practice of medicine at Charité Clinic because he had no other means of income and his services were badly needed, Karl wrote to his former assistant, Paul Jossmann, expressing exactly the same sentiments as Karl-Friedrich. He admitted that the deaths had taken "its toll on us old folk," due to the constant tension and anxiety. "But since we all agreed about the necessity of action, and my sons were also fully aware of what they could expect if the plot miscarried, and had resolved if necessary to lay down their lives, we are sad, but also proud of their straight and narrow attitude. We have fine memories of both sons from prison . . . that move both of us and their friends greatly."[4]

Beyond letting family and close friends know what had occurred, the Bonhoeffers could not speak openly about their role in the resistance because many Germans continued to venerate Hitler, believing that he would have won the war were it not for treachery. Interviewed by Dorothea von Meding for the book *Courageous Hearts: Women and the*

Anti-Hitler Plot of 1944, Emmi Bonhoeffer told of a typical postwar incident that happened to her daughter, Cornelia. Returning from school and having missed the train, she hitched a ride. The driver asked her who her father was and she replied that he was dead. "Did he die at the front?" the driver asked. Cornelia answered that he had been killed by the Nazis for working against Hitler, to which the driver replied "Poor little traitor's kid".[5]

When Dietrich Bonhoeffer's friend, Pastor Martin Niemöller, who had survived for eight years in concentration camps, dared to raise the issue of Germany's collective guilt in a speech at the University of Erlangen in January 1946, he was roundly booed and was called a "tool of the Allies."[6] There was neither sympathy for what Niemöller had tried to achieve through his involvement with the Confessing Church, nor recognition of what he had sacrificed. There was only condemnation (Fig. 13.1).

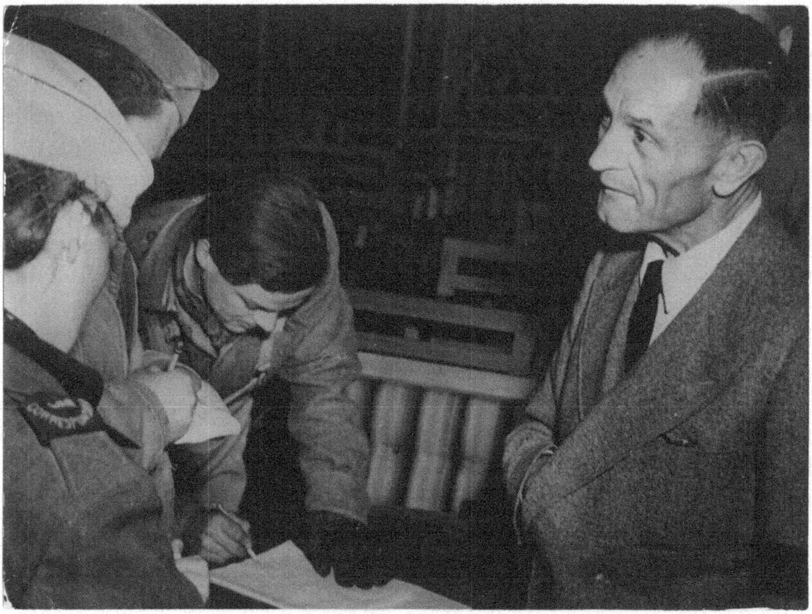

Fig. 13.1 Martin Niemöller after his release from concentration camp. United States Holocaust Memorial Museum, courtesy of Joseph Eaton

Paul Rosbaud, who had gone to live in England after the war, remarked with great heaviness of heart that he was in the same position as the Bonhoeffers. He could not tell anyone in Germany that he had wished for the Third Reich to be defeated because they would not understand. In fact, it was dangerous to do so because people were out for revenge. Rosbaud was aghast at the number of Nazis who were attempting to regain power, in the process deforming the recovery efforts. In a letter to Samuel Goudsmit, he vowed to keep fighting the "evil spirit" while admitting that at times it seemed hopeless. "By the way, you can also hear sometimes, not from scientists but from the man in the street, that the Germans would have had the A-bomb long before you [the Allies] if the Norsk Hydro plant would not have been destroyed in time," Rosbaud wrote. "Somebody told me even, that a German 'traitor' had given away the secret of the heavy water plant. I was very impressed indeed."[7]

In the closing months of the war, the Uranium Club had been busier than it had ever been. Because of the bombing of Berlin and the destruction of the Kaiser Wilhelm Institute of Chemistry in 1944, most of the nuclear testing operations had been moved to a safer location near Hechingen in the Jura Mountains of southern Germany. In January 1945, as the Soviet army fought its way across Poland, the only Uranium Club scientists who remained in Berlin were Karl Wirtz and Kurt Diebner. Yet they were not just sweeping up broken glass from blown-out windows. They were assembling one last heavy water experiment, for which purpose Harteck had made a quick trip to Vemork to secure any low-concentration heavy water that might still be available. But then Walther Gerlach (who had assumed leadership of the nuclear project through the Reich Research Council in late 1943) made the decision that everything and everyone had to retreat south to the mountainous protection of Hechingen and the neighboring village of Haigerloch, where several other members of the Uranium Club, including Heisenberg, Weizsäcker, Hahn, and Laue, had moved the prior year. There was also a slave-labor camp nearby, so any required construction could be done as fast as possible without regard for human welfare. On January 30, a convoy left Berlin with Gerlach, Wirtz, and Diebner in the front cars, followed by overloaded military trucks in which jostled canisters of heavy water, uranium, and equipment. That the British knew of their departure was, as usual, thanks to Rosbaud.[8]

In Heisenberg's letter to Elisabeth dated February 1, 1945, he wrote that the weather in Hechingen had finally turned warm and with it his vital spirits were perking up. He expected that his fingers that were badly swollen from the cold—"perhaps due to playing on the cold piano or something else"—would finally improve. "At any rate, I am feeling well and ready to take up the battle for one's existence again. In our nuclear physics group, the internal battle (Diebner vs K.W.I.) has broken out anew, probably as a result of a new wave of conscriptions and the threatening danger to the east." He gave no details of the internal battle, but it was clear that the long-standing animosity between himself and Diebner would continue to the bitter end.[9]

It was in a cave previously used as a wine cellar in remote Haigerloch that the scientists tried desperately to achieve a chain reaction. Last minute assembly operations were carried out at breakneck speed with no consideration for safety. Per Dahl points out in *Heavy Water* that there were no cadmium-sheathed control rods, nor was there a way to rapidly drain out the heavy water from the bottom of the reactor vessel should the reaction begin to get out of control. There wasn't time for such niceties. When the experiment took place, neutrons did multiply, but not to the level needed for a self-sustaining chain reaction, but they came close. Even if there had been more heavy water and uranium available, it was too late to try again. On April 23, the Allies reached Haigerloch. They found the cave, gathered evidence, and began to round up the scientists. Heisenberg was at his summer home in Urfeld, and Harteck was in Hamburg. By May 3, all were in custody.[10]

There is no adequate explanation why the Uranium Club tried so hard in the closing months of the war to achieve a chain reaction. If it were true that most of the members did not want to build a bomb for Hitler (as many of them vociferously maintained after the war), then why did they not quit, or at least make a token effort, when it was obvious the war was lost? Some historians have speculated that they were jockeying for preferential treatment either from the Soviets or the Allies by being able to offer their expertise in a successful endeavor. (This is exactly what von Braun and his rocket scientists did.) According to Mark Walker in *Nazi Science*, there was no defeatism among the Uranium Club scientists who assumed they were ahead of their rivals in other countries, an assumption they staunchly held until the day the atomic bomb was dropped on Hiroshima. "For them reaching their goal was also being the *first* to do so, an accomplishment which would have obvious professional rewards, no matter who won the war." Walker also states that the greater the threat of impending doom, the more they buried their heads in their work, which made them oblivious to the moral overtones.[11]

In July, the ten rounded-up scientists were sent to an estate in England called Farm Hall near Cambridge. A pleasant place, it was in actuality thoroughly bugged so that the British could record everything that was said. The ten were: Heisenberg, Gerlach, Diebner, Weizsäcker, Hahn, Laue, Harteck, Wirtz, Bagge, and Korsching. Under comfortable conditions (there was even a grand piano for Heisenberg), the men stayed at Farm Hall for six months until January 1946. When the Farm Hall tapes were finally declassified in 1992, forty-seven years after they were recorded, they revealed an enormous amount about the workings of the Uranium Club and the personalities of its members. Uppermost among the revelations was the effort led by Weizsäcker to morally justify the German scientists. Not long after they had heard about Hiroshima, Weizsäcker drafted the following: "History will record that the Americans and the English made a bomb, and that at the same time the Germans, under the Hitler regime, produced a workable engine. In other words, the peaceful development of the uranium engine [nuclear reactor] was made in Germany under the Hitler regime, whereas the Americans and the English developed this ghastly weapon of war."[12]

This statement became, for Heisenberg especially, a core belief repeated in various forms for decades. By careful word-choice, Weizsäcker had presented the German scientists as more moral than the American and British scientists. He had made it seem as if the Germans had succeeded in being the first to develop a nuclear reactor useable for peaceful purposes. This was a lie on two levels: They never succeeded and their primary goal was not for peace but as a breeder reactor for plutonium. Weizsäcker's explanation came to be known as the *Lesart*, which in German means an interpretation that does not align with reality. It was reinforced with additional arguments that the members of the Uranium Club had intentionally undercut their atomic research by not asking for sufficient money and resources and not pushing as hard as they could.

Self-delusion was typical of German scientists in all fields. Overwhelmingly, they wanted to deny what had happened during the twelve-year period from 1933 to 1945. If that proved impossible, at least they wanted to deny that they willingly participated. Part of the impetus for their denial was that in order to be allowed to work after the war, many of them had to go through a process of denazification in which they proved by means of a questionnaire and testimony from colleagues before tribunals that they had not been Nazis, or that they had cooperated under coercion. But the process was easily manipulated, with misinformation listed on the questionnaires and testimony provided by close cronies.[13] According to historian Mark Walker, by 1948 when the tribunals ended, they had caught some "little Nazis" but few big ones.[14] Rosbaud was so appalled by the process that he saw it as part of *re*-nazification of Germany and Austria. He raged in a letter to Goudsmit that the Nazi Martin Kröger, who had denounced Karl-Friedrich in Leipzig, had made it through the process and been cleared and would probably be appointed a professor at Munster University.[15]

Having closed their eyes for so long, many German scientists resolved to keep them closed, if possible for the rest of their lives. Again, it was Lise Meitner who realized what was happening and cut through the lies. Even before the war ended, Meitner was dismayed by the lack of moral responsibility and absence of strong inner resistance she saw in her former German colleagues, including those who claimed in private to be anti-Nazi. She wrote to her friend, Eva von Bahr-Bergius, on March 1945, a month before Hitler committed suicide:

> The letters from German friends sound very depressed, yet I do not think they comprehend just what sort of fate has befallen Germany through their passivity. And they understand even less that they share responsibility for the horrible crimes Germany has committed. How shall the world trust a new Germany when its best and intellectually most prominent people do not have the insight to understand this and do not have a burning desire to make whatever amends are possible.[16]

In a furious letter to her old colleague, Otto Hahn, barely three months later in June 1945—after the war had ended in Europe and the gruesome details of the death camps were being revealed—she hammered home the point of moral bankruptcy: "You all worked for Nazi Germany and you did not even try passive resistance. Granted, to absolve your consciences you helped some oppressed person here and there, but millions of innocent people were murdered and there was no protest." Meitner went on to suggest that Heisenberg and millions of others like him should be forced to look at the horrors of the concentration camps. Unfortunately, Otto Hahn never received Meitner's letter because he was in custody with the other members of the Uranium Club. Instead, the letter made its way to the Allies' Office of Strategic Services (OSS). Eventually, in the histories of Nazi science, the letter found its rightful place as a passionate condemnation of the egregious moral laxity of German scientists.[17]

That was absolutely not how Meitner felt about the Bonhoeffers as evidenced by the fact that she took the time to prepare a booklet about Dietrich's murder, sending a copy to Paul Rosbaud, who she knew was in contact with the family.[18] So also several scientists and doctors who had been forced to leave Germany, including Franz (Francis) Simon, James Franck, and Rudolf and Käthe Stern sent care packages to the Bonhoeffers who were suffering from malnutrition after the war.[19] Franck's letter to Karl-Friedrich written in March 1947 is both poignant and revelatory:

> You really ought to be the last to have reason to reproach yourself in any way. On the contrary, we all owe you our thanks. And more so your family. It gives us confidence in the future to know that people like you continue to exist in Germany, who have kept a clear head in a sea of madness and moral degeneracy and guarded their knowledge of what is right and what is wrong. Believe me that almost all of your colleagues know about you and your family and that many are ready to help, not just those who used to live in Germany and even now cannot feel indifferent to the German fate and especially to the fates of proven friends and colleagues.[20]

Wilhelm Jost, who had worked with Karl-Friedrich closely in the 1930s and early 1940s before going to the University of Marburg, also received much needed help from Franck. Jost had tuberculosis, made worse by malnutrition and lack of medical care. Franck sent him money so that he could go to a sanatorium in Switzerland to recover. It was Jost who wrote about Karl-Friedrich's own severe health problem: "After the collapse he felt an obligation to contribute his part to rebuild, and he did so without sparing his person." Suffering from angina pectoris, Karl-Friedrich had a heart attack in 1946 from which he only partially recovered. However, he would not quit in his efforts on behalf of his family and his staff and students at Leipzig where his institute had been totally

Fig. 13.2 Soviet tanks roll through Leipzig in March 1945. United States Holocaust Memorial Museum, courtesy of National Archives and Records Administration, College Park United States

destroyed during the final months of the war. Jost recalled that Karl-Friedrich was particularly concerned in those "straitened times" with attempting to broaden educational opportunities in physical chemistry which the deprivations of war and the conscription of young men had severely diminished (Fig. 13.2).[21]

It was not only the buildings that needed to be reconstructed. So also did the entire mind-set toward learning and the nature of healthy, collegial human relationships. About this last, Karl-Friedrich's old friend, the Dutch physicist Bartel van der Waerden, who had been at the University of Leipzig, wrote to him on July 30, 1946, complaining that life in Germany had become so tumultuous he could not even get permission to return to Leipzig to retrieve his bicycle and papers. "Everywhere there are walls of division, mistrust, and hate, and not much constructive work. Indeed, as you write, it is very difficult to make it clear to people everywhere that there are still decent Germans."[22]

In January 1947, after spending a grueling year-and-a-half rebuilding the bombed-out institute of physical chemistry at Leipzig while trying to recover from his recent heart attack, Karl-Friedrich accepted a professorship in the Department of Physical Chemistry at the University in Berlin. Heinz Gerischer, whom Karl-Friedrich had harbored at Leipzig and who was working on his doctorate, went with him as his assistant.[23] There were several reasons for the move: first, to be nearer his parents, and second, Leipzig was totally under Russian control, whereas Berlin was divided between the British, Americans, French, and Russians. The severely damaged Kaiser Wilhelm institutes were in the American sector; unfortunately, the university was in the Russian sector. Karl-Friedrich clearly expressed his reasons and reservations about Berlin in two letters written in August and September, 1945, to Paul Harteck in which he also informed him of the murders of his family members: "I feel like I must take care of my parents and siblings. It would be a thousand times better if I had a place to work in a West German city, which I want to do. Going to Moscow I don't have in my mind. My life and my work are in Germany." Harteck apparently did not receive the first letter because he was incarcerated in Farm Hall along with the other members of the Uranium Club. Fearing that the letter had not gotten through, Karl-Friedrich wrote again, reiterating the news about Dietrich, Klaus, Rüdiger, and Hans, who was still missing. Karl-Friedrich described a visit to his parents, finding them to be composed in their grief, having prepared themselves for the murders. "But naturally they are still very sad. I am not sure if they will ever get over that. My mother is very often forgetful, which is probably a relief for them. It is really sad to see all that and to be helpless."[24]

Karl-Friedrich wrote similarly to Michael Polanyi whom he had last seen in 1939. Since then, Polanyi had shifted from physical chemistry to philosophy and the social sciences at the University of Manchester. Because Polanyi's sister and brother-in-law had been murdered in the concentration camps, his reply to Karl-Friedrich, dated May 21, 1946, was heart-felt:

> From your letter that I received three days ago, I got the news about your personal losses, which were more tragic than I had heard from rumors. In the middle of total destruction, one must say that you have suffered beyond all measure. You have the comfort that your dead belong to the martyrs whose sacrifice carries all our hopes. Dietrich Bonhoeffer's conduct has already become a part of the world's heritage.

To Polanyi, Dietrich would be remembered as a person whose love of freedom, for which he heroically fought, was founded upon a "commitment to the transcendent." Having known all the Bonhoeffers before he was forced to leave Germany in 1933, Polanyi expressed his surprise at the involvement in the resistance of Hans Dohnanyi and Klaus Bonhoeffer, whom Polanyi remembered as carefree and fun-loving. Feeling ashamed that he had not comprehended their true worth when he had known them in person, he wrote: "It shows that one must expect the greatest of every human being, regardless of the way it may represent itself. It teaches respect for the mystery of human greatness." Toward the end of the letter, Polanyi wrote that he totally understood how Karl-Friedrich and his family could have hoped for the defeat of Germany, and that one day the whole world would find inspiration in what they had done.[25]

When Karl-Friedrich's friends in the USA and England heard of his serious illness, they raised money to enable him and his wife to go to Switzerland for a month of recuperation, but they ran into difficulties getting permission from the Soviets. In April 1947, Harold Urey wrote to Karl-Friedrich that he had tried to get permission for the trip as discreetly as possible in order that there should be no difficulty. "All such efforts have proved useless, and for that reason I am writing to you directly to see whether it is possible for you to make such a trip and whether you wish to, as well as whether the authorities will allow you to do so." Karl-Friedrich wrote back in English that he was much improved, had begun to lecture at the University of Berlin, and that it was very kind of Urey to take such an interest in his health. He also mentioned that the food parcels he had received from American colleagues had helped him to recover. "With respect to the trip in Switzerland for which you collected money for my wife and me, there is at the moment therefore no urgency. Besides that, I had a month ago an invitation from the University of Manchester which I accepted and where I hope to be from the middle of May to the middle of June supposing that I get permission of the Soviet Military administration." Then, Karl-Friedrich added a paragraph indicating subtly that he was not totally recovered and was very concerned about the future. "If the food parcels from my American friends continue to come I hope to regain my health without a special recuperation trip." Without going into great length, he mentioned that his arthritis was bad, he might have his tonsils removed, and his physicians were considering sending him for a course of treatment using hot baths. As a result of his continuing health problems, he was reluctant to turn down Urey's offer, thinking that perhaps a postponement might be in order:

> May be that next winter will be for us as hard as the last one and that I would [be] very happy to be able to go abroad for a month with my wife. I would like to give you a complete insight into the situation with us in order [that] you can make your own decision. My institute in Berlin is situated in the center of the city, which is almost completely destroyed. The institute is quite empty, no books, no instruments, no gas. It is impossible to do research work, and I have to get some instruments back which are now in the British zone near Göttingen. My collection of reprints is completely burnt so I would [be] very obliged if you would send me some reprints of your papers. It is now allowed.[26]

Fortunately, Karl-Friedrich did receive permission from the Soviet authorities to travel to England in May 1947. The trip proved to be of importance for numerous personal and professional reasons. Of greatest significance was his reunion with his sister and her family after eight years of separation. The last time he had been with them was when he and Dietrich had returned from the USA in 1939 just before the outbreak of war. On seeing Karl-Friedrich, Sabine was astonished at how gaunt and pale he was. Although their reunion brought great joy, it was also "clouded with infinite pain," because of his physical decline.[27]

Karl-Friedrich also visited Polanyi, who had arranged for Karl-Friedrich to deliver a series of lectures at the University of Manchester, writing to him before his arrival, "I am profoundly happy to see you coming at last. You have many friends here who feel as I do."[28] Polanyi had nominated Karl-Friedrich for the Nobel Prize ten years before in 1937. Ironically, during his visit, Karl-Friedrich was nominated for the Nobel Prize in Physics for the fourth time, not by Polanyi but by Heisenberg, although, again, nothing would come of it.

Polanyi was writing with great fervor on the nature of the scientific enterprise, its knotty connection with politics, and threats to the autonomy of science. Polanyi shared with Karl-Friedrich his recent publications to which Karl-Friedrich had not had access. Published in 1945, *Science and the Modern Crisis* delved into the issues of science under fascism and Marxism, and the resultant fanaticism, which Polanyi defined as the willful suppression of morals and the giving of "blind support" to a supposed scientific theory promising social salvation by means of violence. "When our intellect convinces us, backed by the authority of science, that our morality is pointless, and teaches us that we can achieve everything to which morality aspires merely by letting loose our animal force—then our morality is converted into scientific bestiality."[29]

In *The Autonomy of Science*, also published in 1945, Polanyi strongly opposed control of science by the state.[30] Like Karl-Friedrich, he knew that science was not pure, that personal commitment and coordination among scientists were essential, and that reward and societal recognition played key roles not solely in what was chosen to be studied but in what questions were asked of the resulting data. Yet scientific truth still existed, as did the sudden exhilarating and *pure* revelations about nature, such as Karl-Friedrich's discovery of para and ortho hydrogen. In his major work *Personal Knowledge: Towards a Post-Critical Philosophy*, published in 1958, Polanyi attempted to explain it as follows: "Scientific discovery reveals new knowledge, but the new vision which accompanies it is not knowledge. It is *less* than knowledge, for it is a guess; but it is *more* than knowledge, for it is a foreknowledge of things yet unknown and at present perhaps inconceivable. . . . Theories of the scientific method which try to explain the establishment of scientific truth by any purely objective formal procedure are doomed to failure." Perhaps one of the reasons for the unbroken friendship between Polanyi and Karl-Friedrich was that their scientific perceptions had been formed in the same hot furnace of the hard sciences in the 1920s, while the tempering that had occurred during the 1930s and 40s had made them better able to comprehend the human component of science.[31]

Karl-Friedrich also had the opportunity to see his friend, Paul Rosbaud, who was in the process of establishing the firm Butterworth-Springer to continue scientific publishing in the UK. (Eventually, it would become Pergamon Press in partnership with Robert Maxwell.) In that role, Rosbaud frequently traveled to Germany where he was disheartened by the chaos, particularly about the people who had been in the resistance who were, as he put it, "left terribly in the lurch," while people who were pro-Nazi were being defended. "I had a long talk with Bonhoeffer who was in England for a few weeks," Rosbaud reported to Goudsmit after Karl-Friedrich had returned to Germany. "He [had] the same experience, and said to me that the English and also many of the Americans don't understand and also don't like people like him and me. They find others who may have invented Panzerfaust or rockets much more interesting and also much more decent." In the same letter, Rosbaud let Goudsmit know that he had been warned about returning to Berlin "because some of the Nazis might be interested to take revenge."[32] Rosbaud was also suffering from survivor's guilt. He wrote to his brother in November 1946, "The last years have not passed without leaving marks upon me. There were too many in the underground who could not be saved and, at the end, only I slipped through by a hairsbreadth. My hatred of the Nazis has not diminished." Those lost people, unnamed and unhonored, haunted him. However, according to Arnold Kamish in *The Griffin*, Rosbaud never spoke of them except in vague and guarded terms, perhaps because the British Secret Service had warned him to remain silent.[33]

Probably the most important meeting of the trip was with Francis Simon at which Karl-Friedrich told him that Heisenberg was lying about the German approach to bomb building. Awarded the Iron Cross in the First World War, Simon had left the University of Breslau in 1933 in great anger, subsequently taking a position at Oxford University. At the start of the war, he was a member of MAUD, the British atomic bomb group, and then worked on the Manhattan Project in Los Alamos. In a report dated March 1948 to Michael Perrin, a MAUD member who was still working on atomic energy issues, Simon gave details of a recent meeting with Heisenberg, in the process mentioning the meeting with Karl-Friedrich the previous year. Heisenberg had insisted to Simon that German scientists knew about the fast neutron reaction and the potential for plutonium, but that they had wanted to prevent Hitler from getting the bomb and so did not tell him of these things. But Simon disagreed about what Heisenberg and the rest of the Uranium Club members knew: "The German scientists thought of a bomb as a whole heavy water pile which one had to drop—this incidentally was confirmed by Bonhoeffer when I saw him in Oxford last year. Heisenberg tried to ridicule this. In particular, he emphasized that he always knew that the bomb would be the size of a pineapple." Simon was careful not to let on to Heisenberg how he knew his "story did not align to the facts." He concluded his report with the observation that he did not know if Heisenberg was deliberately lying, or if he had constructed a picture of what he would have liked to occur, and then persuaded himself it was absolutely true.[34]

In 1948, Karl-Friedrich became director of the Kaiser Wilhelm Institute of Physical Chemistry in Berlin-Dahlem, which eventually would be renamed the Fritz Haber Institute of the Max Planck Society. The name Kaiser Wilhelm was unacceptable to the Allies because he had been an ardent militarist and because of the complicity of the various institutes with Nazism. On the other hand, Max Planck was recognized as an elder statesman of physics who did what he could in his declining years to protect pure science during the Third Reich. Furthermore, he had died in October 1947, shattered in the last two years of his life by the execution of his son, Erwin, for his part in the resistance. Also appropriate was the renaming of the physical chemistry institute after Fritz Haber. At the time of his death, Karl-Friedrich had written to Haber's children that he would do everything possible to keep alive the memory of their father. The renaming was a way to keep that promise. However, renaming was the easy part; reorganization was the hard part due to numerous political and financial difficulties.

Karl-Friedrich had been offered the directorship in the summer of 1945 to replace Peter Adolf Thiessen who, because of his strong communist leanings, had voluntarily gone to the Soviet Union at the end of the war along with Manfred von Ardenne, Gustav Hertz, and Max Volmer. (When the Russians rolled into Berlin, Thiessen had ridden up to Von Ardenne's institute in a Soviet tank.) Once in the USSR, they put their considerable knowledge of atomic power to work to build a bomb. What Thiessen took with him was immense; what he left behind was a vacant building, which is why Karl-Friedrich had declined the directorship when the offer was first made.[35]

The following year, Karl-Friedrich accepted the founding directorship of the Max Planck Institute for Physical Chemistry to be established in Göttingen instead of Berlin. The institute existed only on paper, so Karl-Friedrich had to build it from the ground up: laboratories, staff, and equipment. Again, one of the reasons why he accepted this challenge was the Soviet Union, which on June 24, 1948, imposed a blockade on West Berlin, shutting down access by road and rail. Until May 12, 1949, air crews from the US Air Force, the British Royal Air Force, and air forces from all corners of the UK landed tons of supplies, fuel, and food, finally forcing Russia to lift the blockade. Heinz Gerischer recalled that while life went on in Berlin during the blockade, it made doing science extremely difficult.[36] Then, in 1949, the iron curtain descended and Germany was split into the German Democratic Republic (east) and the Federal Republic of Germany (west).

It was during the blockade that Karl Bonhoeffer suffered a stroke while doing rounds at the Charité Clinic, dying on December 4, 1948. He had always been considered the perfect model of the man of science and medicine, dedicated to his patients, making an effort to keep politics out of the examining room, teaching his students strict diagnostic criteria. But over and over he had been faced with horrific ethical situations: shell shock on First World War battlefields; hospitalized mental patients allowed to die of starvation, sterilization, and euthanasia of "useless eaters." On his retirement from the Charité Clinic, Karl had strongly opposed the appointment of Maximilian de Crinis, a sadistic Nazi physician, to replace him.[37] Failing in that, he began to work with Dietrich to save patients by providing non-genetic diagnoses. All of these things had led to his unwavering support of his family and the resistance. He knew well the Hippocratic oath "first do no harm." But sometimes it had been necessary to do harm to prevent greater harm, such as infecting Hans with diphtheria. Haunted by the pathological behavior he had witnessed, he took part in rebuilding psychiatry in West Berlin in his last years, literally attempting to return sanity to an insane world. He also undertook a psychological study of Hitler, publishing an essay in 1947 in which he discussed mass hysteria and induced madness. According to his daughter-in-law Emmi, "What he meant was a condition that is brought about by the transmission of a madman's delusions to his environment."[38] It was Karl's friend, the psychiatrist Robert Gaupp, who best summed up the man:

> By nature so astute and critical, Bonhoeffer was cautious and modest toward philosophy, and remained within the boundaries of the empirical world that was accessible to him. There are few research scientists who at the age of eighty, after a scholarly career of more than half a century, have had to retract so little of what they once taught. Bonhoeffer, the research scientist of 1897, could still be compared with the Bonhoeffer of 1948. He adhered to the course carved out by his life experience, cautiously, but with determination.[39]

After Karl's death, Paula, emaciated and bent, was taken care of by her family. She died three years later in 1951.

Karl-Friedrich's decision to accept the position in Göttingen was prudent because the city had survived the war with minimal damage. Furthermore, the city was totally in the British-occupied zone, instead of being split among the Allies and Soviets as was Berlin. As a result, it was rapidly becoming a magnet for scientists from destroyed universities and institutes. From the Uranium Club, Heisenberg, Hahn, Weizsäcker, Wirtz, Laue, and Korsching were all there, although Karl-Friedrich had little interaction with them because his new goal was to build an interdisciplinary institute in which physics, chemistry, and biology were brought together synergistically. Appointed a research associate, Heinz Gerischer and his wife, Renate, who was studying for her doctorate under Karl-Friedrich, returned from Berlin with him. Theodor Förster, who had been Karl-Friedrich's assistant in Leipzig, was already there, pursuing research on energy transfer in biological systems.[40] From this small core of like-minded colleagues, Karl-Friedrich began to build the institute.

Because of the serious financial and administrative turmoil in all the Max Planck institutes, Karl-Friedrich had to shuttle between Göttingen and Berlin for a while. It was during this period that an unusual string of events occurred related to the fissure between east and west, Communism and democracy, and the arms race between the USA and the Soviet Union. Karl-Friedrich and the physical chemist Robert Havemann had known each other since the early 1930s at the Fritz Haber Institute (Fig. 13.3). A highly ethical man, Havemann was a communist who was openly opposed to Hitler and Stalin, considering the policies of both men to be aberrations. As was the physicist Herbert Jehle, Havemann was the exception to the generalization that all German scientists were fellow travelers with Nazism. In 1943, he helped establish a group called the European Union with the goal of sheltering Jews and helping prisoners in forced labor camps. He and fourteen other members were arrested and sentenced to death. Unidentified friends who were scientists interceded on his behalf with Nazi leaders, claiming that Havemann was doing "research of great military importance." As a result, Havemann remained in prison but was not executed. His name appears in the Bonhoeffer correspondence in a very brief but unusual entry. In one of the last desperate letters that Hans Dohnanyi managed to sneak out of prison to his wife, Christel, dated February 25, 1945, there is a four-word sentence: "For Klaus, recommendation

Fig. 13.3 Robert Havemann was accused of being a Soviet spy. PJF Military Collection/Alamy Stock Photo

Havemann."[41] There is no way to know what Hans meant. What is known is that in the final months of the war, Havemann was preparing for a Soviet takeover of the prison and a final battle with the Gestapo. He was also keeping his fellow prisoners informed, getting news about the situation outside the prison via a radio he had somehow constructed and kept hidden from the guards.[42]

All this helps to explain why Karl-Friedrich strongly defended Havemann several times in the late 1940s and early 1950s under circumstances that could have made Karl-Friedrich himself look suspect. As Germany split in two and the USSR became an atomic power, the hatred and fear of Communism increased exponentially throughout the West. From the point-of-view of the Allies, a German who continued to be a supporter of Nazism but who was against the USSR was superior to a person who was bravely anti-Nazi but was pro-USSR, which was true of Havemann.[43] In 1949, the American authorities in West Berlin called for Havemann's removal from the Max Planck Institute. As head of the institute, Karl-Friedrich defended him on the basis of his courageous resistance to Hitler as well as his ability to bring renown to the institute in the future. Then, in 1950, Havemann wrote an article sharply criticizing the atomic weapons policy of the USA, specifically the development of the hydrogen bomb. Karl-Friedrich again interceded, but this time Havemann's anti-nuclear position was too much for the Allies and German authorities in Berlin, who suspected he was a spy.[44] Havemann lost his job. However, Karl-Friedrich's efforts on his behalf did not end there. He secured a professorship for him at Humboldt University (University of Berlin) in the Soviet sector. Havemann returned to physical chemistry, although he did not forsake his devotion to humanitarianism, speaking out as strongly against the corruption of Communism as he had against Nazism. By helping establish Havemann's reputation as the head of physical chemistry at Humboldt University, Karl-Friedrich provided him with a modicum of scientific fame in the West, which in turn shielded him from going to prison in the East for his continuing outspokenness. Havemann lived out most of the last years of his life in East Germany under house arrest.[45]

Karl-Friedrich's move to Göttingen had one major benefit besides the obvious one of being located in West Germany. It became the place where some members of the Bonhoeffer family began to re-gather. Sabine and Gert Leibholz had decided to leave England and return to Göttingen, where Gert was re-appointed as a law professor. Maria von Wedemeyer found her way there to study mathematics in 1947 before leaving for the USA to attend Bryn Mawr College the following year. Here, too, the influence of the tight circle of Bonhoeffer friends made its appearance, proving how entangled their lives were. The story is brief but worth telling. It was the physicist Herbert Jehle—who had been introduced by Karl-Friedrich to Dietrich in the early 1930s—who under Dietrich's influence had refused an academic appointment in Germany—who had been thrown in a concentration camp in France because he was a pacifist—who had made it to the USA with the help of the Quakers—who now reached back across the Atlantic to clear the way for Maria, a young woman he had never met but whom Dietrich had loved above all others, getting her a scholarship to Bryn Mawr.[46] In the last letter Dietrich sent to Maria in December 1944, he wrote that because of her, his family, and friends, he never felt lonely and forlorn even in the solitude of his cell. "I live in a great unseen realm of whose real existence I'm in no doubt. The old children's song about the angels says 'two to cover me, two to wake me.' And today we grown-ups are no less in need than children of preservation, night and morning, by kindly unseen powers."[47] In a way, that describes Jehle—a kindly unseen power.

At Göttingen, Karl-Friedrich attracted students who would go on to have stellar scientific careers. Among them was Manfred Eigen who would eventually follow Karl-Friedrich as director of the Max Planck Institute of Physical Chemistry during the period when it changed from physical chemistry to biophysical chemistry. At that time (1971), it also took on an additional name in honor of its founder: the Karl Friedrich Bonhoeffer Institute. Eigen won the Nobel Prize for Chemistry for his studies of extremely fast proton transfer reactions and the development of techniques to measure those "immeasurable" reactions to the nearest billionth of a second. In his Nobel lecture in 1967, Eigen remembered Karl-Friedrich with gratitude for turning his path from physical chemistry to biology. Later in his life, Eigen wrote, "Bonhoeffer was a wonderful scientist and man, a person." Bonhoeffer tried to persuade Eigen to study reactions. "He was especially interested in nerve conduction and made a model of that, but I always had to tell him I first wanted to study more fundamental reactions to understand better the course of complicated reactions before I would go into such complicated problems, as we know nowadays biological reactions are."[48]

Karl-Friedrich's model to which Eigen referred, published in 1948, concerned nervous excitation and conduction based on an iron-wire analog of the nerve. Max Delbrück helped Karl-Friedrich reconnect with the mathematician John von Neumann at the Institute for Advanced Study in Princeton. Prior to leaving Germany in 1933, Newmann had attended Polanyi's discussion group that Karl-Friedrich had also attended. However, time, space, and diverging research interests had separated the two men. Finally, back in touch, they carried on a lengthy correspondence about Karl-Friedrich's model. At that time, von Neumann, who was without hyperbole one of the smartest mathematicians in the world, was developing a theory of automata.[49]

Karl-Friedrich's first two sons were also beginning their university studies, with his second son, Friedrich, expressing an interest in science. In 1950, Karl-Friedrich wrote to Polanyi that his oldest son, Karl, was studying at Oxford and that he might get a chance to visit the Polanyis.

As for Friedrich, "He wants to be a scientist, which is not really original, but it is the only thing he likes besides music," Karl-Friedrich wrote. "And I think we should not discourage him. Mathematics is very easy for him. He has an excellent practical and technical talent for it. He still wavers between pursuing math and biology."[50] In 1958, Friedrich received his doctorate from Göttingen in nuclear physics, but his research focus was on biological molecules. Following in the footsteps of Max Delbrück at Caltech, he did postdoctoral work at the University of California at Berkeley on the physical chemistry of macromolecules, eventually doing important research on molecular mapping.[51]

In 1951–52, Karl-Friedrich served as chairman of the prestigious German Bunsen Society. In that position, he invited Francis Simon to participate in a meeting of the society. The invitation was part of Karl-Friedrich's unflagging effort to close the divide and to overcome the ingrained anti-Semitism and lack of responsibility still poisoning German society and science. Simon declined the invitation, writing to Karl-Friedrich that while he would like to see Karl-Friedrich, Max von Laue, and some others again, to do so would mean he would also see former Nazis who were opportunists and cowards. He wrote:

> In my opinion, the German scientists lost their honor in 1933 and have done nothing to regain it. I admit one can say that it is not everyone's business to risk his position or his life, but after the war that was not necessary. The least that could be expected after all the misfortune that has been done was that the German scientists in their entirety or through their scientific societies had stated publicly and clearly that they regretted what had happened. I have not noticed anything like this. If I am wrong, please correct me.[52]

Simon closed the letter by inviting Karl-Friedrich and his wife to Oxford where he wanted to show them his new low-temperature laboratory, the largest in the world. In his reply dated April 4, 1951, Karl-Friedrich wrote that he and his wife would be very pleased to visit the Simons in the fall. He also wrote that he was not surprised by Simon's negative response to the invitation to the Bunsen Society but that Simon misunderstood. "If the mood in Germany were as you suppose, I would not have been elected to be the first chairman and Mr. Jost would not have been elected to the standing committee." Karl-Friedrich disagreed with Simon that "the protest of professors could have changed anything" in

1933 given the level of terror, making the point that such an accusation of silence and opportunism could also be leveled, unfairly, against scientists under Soviet rule. "This is not to say that I do not understand if you prefer to stay away from a meeting wherein you might meet some real followers and maybe some real Nazis. I understand that this is unbearable; and it is not easy for me. But they vanish in the crowd of decent and sensible people, from whom you and I should not separate."[53]

In trying to convince Simon, Karl-Friedrich was up against a massive amount of denial and self-pity by German scientists who had remained in Germany, which infuriated scientists who had been driven out. An egregious, but all-too-common, example is an invitation sent out by Rudolf Smend, president of the Göttingen Academy of Science in 1946, which read, "Now that the deplorable circumstances that led to your cancellation of membership no longer exist, we request permission to list you as a member again." Both James Franck and Max Born had received the invitation. Franck minced no words in a letter he wrote to Born:

> I found Smend's remark simply outrageous, dismissing what had happened in Germany with the words "deplorable circumstances." I can very well picture a group of humanists sitting there who perhaps even consider it a great sacrifice of theirs to ask people like you and me to consider ourselves as their fellow members and that it ought to be an honor for us. I feel I cannot tolerate such an attitude.[54]

Eventually, by dogged persistence, Karl-Friedrich won Simon over about visiting Germany. This is made clear from a statement that Simon's wife, Lotte, made in a BBC interview regarding Simon's being awarded the prestigious Linde Medal in 1952 from the Low-Temperature Society. Initially, he did not want to accept the medal since it would mean returning to Germany. But then he received a letter from Karl-Friedrich urging him to accept because not to do so would only confirm the prejudicial attitude of many German scientists, resulting in their making even less of an effort at reconciliation. Lotte Simon stated that it was Karl-Friedrich's letter that changed her husband's mind. She paraphrased the letter as follows: "I know exactly what your reaction will be. But please think of us. This is the first time we are active to make a gesture to say we regret what has happened. If you refuse now, they'll always say, 'Well, there you are. They don't want it.' Please come."[55] Simon accepted, and for the first time since being forced to leave in 1933, he returned to Germany to receive the medal. The wounds remained, but the healing had begun, in part because of Karl-Friedrich's quiet persistence.

Michael Polanyi returned to visit the Bonhoeffers in Göttingen. Max Born also returned, first to visit his former physics colleagues and then permanently, moving from Scotland to West Germany in 1954, the year he finally received the much-deserved Nobel Prize. (Born died in Göttingen in 1970.) So also did Max Delbrück who came back to visit in 1947 and then for three months in 1954 to lecture at Karl-Friedrich's institute, bringing with him samples of the phototropic fungus *Phycomyces* to which he had recently turned his attention. Delbrück was impressed with Karl-Friedrich's multi-disciplinary approach that expanded the boundaries of physical chemistry. "Bonhoeffer was very liberal in what he permitted to happen at his physical chemistry institute," he recalled. "He also actually took on [Carsten] Bresch, who was a geneticist, for several years, which was a very useful thing." Bresch's research focus was the genetics of bacteriophages, as was Delbrück's. In fact, in 1947 on his first visit back to Germany, Delbrück had managed, via a complicated diplomatic maneuver, to supply Bresch with live bacteriophages, handing over the containers he had carried in his luggage to an American control officer in Berlin. Not knowing quite what to do with them, the officer reluctantly arranged to have them delivered to Bresch. With nothing available to keep the bacteriophages alive, Bresch immediately had to cobble together his own equipment, including a constant temperature bath, in a ramshackle laboratory. Subsequently, Bresch visited Delbrück and they talked for hours on end. What Delbrück saw in Bresch and other German scientists impressed him deeply. "The enthusiasm for going back to work and to research was just immense, a real explosion of enthusiasm getting back to something really worthwhile," he recalled.[56]

In 1952, Karl-Friedrich received his fifth nomination for the Nobel Prize, but again he did not win. He was awarded the prestigious Bunsen Memorial Medal in 1955 "for his contributions to the kinetics of the reactions in gases and on electrode surfaces and the relationship of passivity phenomena with the physiological problems of nerve conduction."[57] In 1956, he was made an honorary member of the New York Academy of Sciences. Then, on May 15, 1957, he died of a heart attack at the age of 58. Most of the obituaries spoke at length about his scientific achievements but only mentioned briefly (if at all) the role of his family in the resistance. However, all of the obituaries mentioned

Karl-Friedrich's integrity and dignity. One stated: "His scientific achievements have made him famous, however, there are many excellent scientists. There are few with his character."[58] Unfortunately, despite his personal efforts to reach out to Jewish scientists, Karl-Friedrich did not live long enough to see Germany begin to look behind the curtain of self-imposed amnesia about its complicity in the horrors of World War II. That would come in the following decade as a new generation came to adulthood and began to ask questions.[59]

So also Paul Rosbaud's role was barely recognized in Germany at the time of his death in 1963. He destroyed his personal papers, and his body was honorably consigned to the sea by his own wishes as if to say that he was truly a man without a country. Only in the USA did he receive any recognition before his death. Similar to Karl-Friedrich's recognition by the New York Academy of Science, Rosbaud was awarded the John T. Tate Medal of the American Institute of Physics in 1961 for the quality of his scientific publishing and for his "extraordinary courage" during the war.[60]

There is no way to sum up who Karl-Friedrich was. There is no single event that captures his essence. However, the following comes close: In Dietrich's last letter to his parents dated January 17, 1945, from the Gestapo's infamous cellar prison, he wrote, "I'm doing well. You two just see that you stay healthy. I thank you for everything."[61] Then, Dietrich, who believed that God was found in simple human goodness, requested several items, among them coffee beans, toothpaste, and some books, including one by Plutarch, the first-century Greek historian. It was Karl-Friedrich who somehow managed to obtain, amidst the bombed-out stores and boarded up libraries, a rare copy of Plutarch, sending it to Dietrich for his thirty-ninth birthday on February 4, 1945. It was delivered to the door of the Gestapo prison by Karl and Paula along with a birthday cake. That Dietrich received the book is clear because he carried it with him when he was transported to Flossenbürg concentration camp. Stopping on the way at a schoolhouse on April 6, Dietrich wrote his name in the book in three places, and then intentionally left it behind as a sign to anyone who might try to trace his steps.[62] Whether Dietrich got the chance to read Plutarch in the last two months of his life is unknown, nor does it matter. What matters is that as he journeyed toward the scaffold, Dietrich held in his hands a gift from his beloved brother who had made a final effort to reach out in simple human goodness.

Notes

1. DBW, Vol. 12, 302–303, 349.
2. Ruth-Alice von Bismarck, ed., *Love Letters from Cell 92* (Nashville: Abingdon Press, 1992), 276.
3. Leibholz-Bonhoeffer, *Bonhoeffers*, 194.
4. Jossmann came to the USA in the 1930s. Bethge, *Bonhoeffer*, 933.
5. Dorothea von Meding, *Courageous Hearts: Women and the Anti-Hitler Plot of 1944*, trans. Michael Balfour and Volker R. Berghahn (Providence and Oxford: Berghahn Books, 1997), 24.
6. Klaus Hentschel, *The Mental Aftermath: The Mentality of German Physicists, 1945–1949* (Oxford: Oxford University Press, 2007), 113.
7. Letters to Goudsmit, August 9, 1946; February 2, 1949. Samuel A. Goudsmit Papers, Series IV, Alsos Mission. Subseries B. Paul Rosbaud. Niels Bohr Library & Archives. American Institute of Physics.
8. Dahl, *Heavy Water*, 252–255.
9. Hirsch-Heisenberg, *My Dear Li*, 239.
10. Ibid., 258.
11. Walker, *Nazi Science*, 198–199.
12. See Cassidy, *Beyond Uncertainty*, for information on Farm Hall. The Lesart by Weizsäcker appears on 378, endnote 22.
13. Weizsäcker made it through the denazification process only to admit in his memoir *The Ambivalence of Progress* that he had gotten off too easily.
14. Walker, *Nazi Science*, 201.
15. Rosbaud to Samuel Goudsmit, June 22, 1947. Goudsmit Papers. Niels Bohr Library & Archives, American Institute of Physics.
16. Sime, *Meitner*, 308, endnote 155, pg. 478. "Meitner to Bahr-Bergius," 30 March 1945 (MC).
17. Ibid., 310.
18. A copy of the booklet has not been found. Rosbaud to Samuel Goudsmit, August 9, 1946. Samuel A. Goudsmit Papers. Niels Bohr Library & Archives.
19. Franz Simon changed his name to Francis when he moved to Great Britain. He signed his letters to Karl-Friedrich Bonhoeffer as F. Simon. Fritz Stern recalled that his mother sent CARE packages to Karl Bonhoeffer, and that Karl-Friedrich visited the Sterns in the USA following the war. Stern, *Five Germanys*, 176.
20. Jost Lemmerich, *Science and Conscience*, 268.
21. Wilhelm Jost, "Karl Friedrich Bonhoeffer," *Die Naturwissenschaften*, 44; Jahrgang, Heft 24 (Zweites Dezemberheft), 1957.
22. Alexander Soifer, *The Scholar and the State: In Search of Van der Waerden* (Berlin: Birkhäuser, 2014), 283–284.

23. Heller et al., "The Life and Work of Heinz Gerischer," ECS Classics, Electrochemical Society, 1.
24. Bonhoeffer to Harteck, August 5, 1945, September 9, 1945. Series 1, Subseries A, Box 1, Folder 8. PHP, RPI.
25. Polanyi to Bonhoeffer, May 21, 1946, Abtielung Rep 23, Personal Correspondence, Nachlass Karl Friedrich Bonhoeffer, Archiv der Max-Planck-Gesellschaft.
26. There are two letters from Urey about Bonhoeffer's health and efforts to get him to Switzerland. One is addressed to Fritz Karson, a US official, dated February 7, 1947. The second, dated April 1, 1947, is to Bonhoeffer. Bonhoeffer's reply is a draft with many words and phrases crossed out. Abtielung Rep 23, Personal Correspondence, Nachlass Karl Friedrich Bonhoeffer, Archiv der Max-Planck-Gesellschaft.
27. Leibholz-Bonhoeffer, *Bonhoeffers*, 197.
28. Polanyi to Bonhoeffer, April 14, 1947, Abtielung Rep 23, Personal Correspondence, Nachlass Karl Friedrich Bonhoeffer, Archiv der Max-Planck-Gesellschaft.
29. Michael Polanyi, *Scientific Thought and Social Reality*, ed. Fred Schwartz (New York: International Universities Press, 1974), 40.
30. Ibid., 15.
31. Michael Polanyi, *Personal Knowledge: Towards a Post-critical Philosophy* (Chicago: University of Chicago Press, 1958), 135.
32. Panzerfaust means cannons on tanks. Rosbaud to Goudsmit, June 22, 1947, Samuel Goudsmit Papers.
33. Kramish, *The Griffin*, 215.
34. Kenneth D. McRae, *Nuclear Dawn: F. E. Simon and the Race for Atomic Weapons in World War II* (Oxford: Oxford University Press, 2014), 206.
35. "Max Planck Institute," "Thiessen" https://en.wikipedia.org/wiki/Peter_Adolf_Thiessen and https://en.wikipedia.org/wiki/Max_Planck_Institute_for_Biophysical_Chemistry.
36. Heller et al., "Gerischer," 1.
37. Zeidman, *Neuroscience in Nazi Europe*, 834.
38. Von Meding, *Courageous Hearts*, 3.
39. Bethge, *Bonhoeffer*, 22.
40. Heller et al., "Gerischer," 1. See also Albert Weller, "In Memoriam: Theodor Förster," *EPA Newsletter*, 2.
41. Eberhard Bethge and Renate Bethge, eds., *Last Letters of Resistance: Farewells from the Bonhoeffer Family*, trans. Dennis Slabaugh (Philadelphia: Fortress Press, 1984), 66.
42. Klaus Ruthenberg, "Robert Havemann," *HYLE-International Journal for Philosophy of Chemistry* 4, no. 2: 163. http://www.hyle.org/journal/issues/4/ruthb.htm. See also "Dissidents of Totalitarianism: Robert

Havemann," *Garden of the Righteous Worldwide.* http://en.gariwo/righteous/the-righteous-biographies/exemplary-figures/dissidents-of-totalitarianism/robert-havemann-7638.html.

43. The subject of Communists and the resistance is ideologically fraught. Mildred and Arvid Harnack were Communists who were part of the group that the Gestapo called the Red Orchestra. Because they passed secrets to the USSR, after the war they were judged to be spies, not resisters, even though their intent was to bring down Hitler. But by that token, many of the resisters, including Dietrich, could be judged to be spies as well because they were passing information to the West. Since the fall of the Berlin wall, the end of the USSR, and access to archival records, scholars have begun to reassess this important subject. See Brysac, *Resisting Hitler: Mildred Harnack and the Red Orchestra.*
44. The espionage case against Klaus Fuchs was in 1950, igniting concern in the West about communism and scientific spying.
45. James et al., *One Hundred Years*, 142–143.
46. Bismarck, *Love Letters*, 349.
47. Ibid., 268–269.
48. Manfred Eigen, Noble Prize speech. See also Eigen, *Bonhoeffer*, Web of Stories. http://www.webofstories.com/play/manfred.eigen/25.
49. The letters between Bonhoeffer and von Neumann are in the Library of Congress. Bonhoeffer's paper was titled, "Activation of passive iron as a model for the excitation of nerve." See James G. Glimm, John Impagliazzo, and Isadore Singer, eds., *The Legacy of John von Neumann* (Providence: American Mathematical Society, 2006), 299.
50. Bonhoeffer to Polanyi, Polanyi Papers, Special Collections Research Center, University of Chicago Archives.
51. "Dedication to Friedrich Bonhoeffer," Wiley Periodicals, Inc, 2004, na, pg. 1, https://doi.org/10.1002/nen.20026.
52. Letter from Simon to Bonhoeffer, March 22, 1951, Abtielung Rep 23, Personal Correspondence, Nachlass Karl Friedrich Bonhoeffer, Archiv der Max-Planck-Gesellschaft. See also Ute Deichmann, "The Expulsion of German-Jewish Chemists and Biochemists and Their Correspondence with Colleagues in Germany After 1945: The Impossibility of Normalization?" in *Science in the Third Reich*, ed. Margit Szöllösi-Janze (Oxford and New York: Berg, 2001), 271.
53. Letter from Bonhoeffer to Simon, April 4, 1951, Abtielung Rep 23, Personal Correspondence, Nachlass Karl Friedrich Bonhoeffer, Archiv der Max-Planck-Gesellschaft.
54. Hentschel, *Mental Aftermath*, 154–155, see footnote 387.
55. McRae, *Nuclear Dawn*, 232.
56. Delbrück interview, 104, 106.

57. Messow, "Karl Friedrich Bonhoeffer," *Saxon Biography*. Institut fur Sächsische Geschichte und Volkskunde (April 2014), 4. http://www.isgv.de/saebi.
58. Jaenicke, "Bonhoeffer," *Physikalische*, 1957, 369.
59. In a letter from Herbert Jehle to Karl-Friedrich Bonhoeffer dated April 1952, Jehle complained that the connection between German science and the resistance movement was virtually unknown. He blamed reports, such as the Alsos reports, for the misconception among Americans. Abtielung Rep 23, no. 30, 7. Personal Correspondence, Nachlass Karl Friedrich Bonhoeffer, Archiv der Max-Planck-Gesellschaft.
60. "Dr. Paul Rosbaud to Receive First John T. Tate Medal of American Institute of Physics" News Release American Institute of Physics (September 27, 1961), 2.
61. DBW, Vol. 8, 553.
62. Ibid.

Appendixes

Appendix I: Chronology of Main Events

1898 Marriage of Karl Ludwig Bonhoeffer and Paula von Hase in Breslau.

1899 January, birth of Karl-Friedrich Bonhoeffer in Breslau; December, birth of Walter Bonhoeffer in Breslau.

1900 Physicist Max Planck sets forth quantum theory.

1901 Birth of Klaus Bonhoeffer in Breslau.

1902 Birth of Ursula (Ursel) Bonhoeffer in Breslau.

1903 Birth of Christel (Christine) in Breslau.

1905 Einstein publishes paper on Special Theory of Relativity.

1906 February 4, birth of the twins, Sabine and Dietrich, in Breslau.

1909 Birth of Susanne in Breslau; the physical chemist Fritz Haber achieves nitrogen fixing and synthesizes ammonia, revolutionizing worldwide food production.

1912 Bonhoeffers move to Berlin where Karl Ludwig becomes head of neurology and psychiatry at Charité Clinic and University of Berlin.

1914 Einstein moves to Berlin as director of the Kaiser Wilhelm Institute for Physics and professor at University of Berlin; August, World War I starts; December, first meeting held at Charité Clinic on shell shock.

K. L. Housley, *The Scientific World of Karl-Friedrich Bonhoeffer*, Palgrave Studies in the History of Science and Technology, https://doi.org/10.1007/978-3-319-95801-9

1915 Fritz Haber develops poison gas and directs its first use at the Battle of Ypres; Einstein publishes paper on the General Theory.

1916 Bonhoeffers move to Grunewald section of Berlin; friendship with Delbrücks and Harnacks begins.

1918 April, Walter is killed in First World War; October, Karl-Friedrich is wounded; November, war ends; Haber awarded Nobel Prize, which is presented in 1920; worldwide influenza epidemic breaks out, lasting three years.

1919 Unrest in Germany; Spartacist uprising; formation of Weimar Republic; Karl-Friedrich begins to study at University of Tübingen and then goes to University of Berlin.

1920 First meeting in Munich of National Socialist German Workers' Party (NSDAP), known as Nazi party.

1922 Karl-Friedrich earns Ph.D., becomes assistant to Fritz Haber; Rathenau shot in Berlin.

1923 Hitler attempts coup in Munich; writes *Mein Kampf* while in prison in 1924.

1925 Heisenberg publishes paper on quantum mechanics.

1927 Heisenberg publishes paper on uncertainty principle; Dietrich earns Ph.D.; Karl-Friedrich habilitates and teaches physical chemistry and atomic physics.

1929 Karl-Friedrich and Paul Hartech discover ortho- and para-hydrogen; Karl-Friedrich travels with Michael Polanyi to USA; New York Stock Exchange crashes; Nazis show strength at huge rally in Nuremberg.

1930 Dietrich attends Union Theological Seminary in NYC; Karl-Friedrich takes chair of physical chemistry at University of Frankfurt; Karl-Friedrich marries Margarete Dohnanyi, sister of Hans Dohnanyi.

1931 Harold Urey discovers deuterium; wins Nobel Prize in 1934.

1932 James Chadwick discovers neutron; wins Nobel Prize in 1935.

1933 Hitler becomes Reich chancellor; following Reichstag fire, the Enabling Act grants Hitler dictatorial power; Aryan clause takes effect and Jewish scientists lose jobs; Haber resigns; Pastor's Emergency League established to support non-Aryan pastors; Dietrich serves two churches in London; Karl-Friedrich begins heavy water research with samples from Norsk Hydro plant in Vemork, Norway.

1934 Death of Haber in exile; Karl-Friedrich becomes chair of physical chemistry at University of Leipzig where Heisenberg is chair of physics.

1935 Dietrich directs Confessing Church seminary at Finkenwalde; Haber memorial service held in Berlin; Arnold Berliner and Julius Springer lose publishing positions; Paul Rosbaud remains on Springer staff.

1937 Finkenwalde seminary closed by Gestapo.

1938 Munich Agreement leads to the abandonment of a coup attempt by German resistance; annexation of Austria; Leibholz family flees to England; Lise Meitner escapes to Sweden; Otto Hahn and Fritz Strassmann fission atom; Karl-Friedrich shifts research focus to biochemistry.

1939 Rosbaud publishes article on fission in *Naturwissenschaften*; race to build atomic bomb begins; Czechoslovakia invaded; Uranium Club forms; Dietrich, Karl-Friedrich, and Heisenberg visit USA before war begins; Szilard and Einstein write President Roosevelt on potential of an atomic bomb.

1940 Germany invades Norway, Denmark, Luxembourg, and France; in June, Dunkirk falls; Germany gains control of Norsk Hydro plant that produces heavy water; involved in resistance, Dietrich also works on *Ethics.*

1941 Heisenberg and Weizsäcker make a controversial visit to Niels Bohr in Copenhagen; Dietrich continues work on *Ethics.*

1942 Enrico Fermi achieves first sustained atomic chain reaction at University of Chicago; German military suffers severe losses on eastern front; USA enters the war; Dietrich writes essay "After Ten Years."

1943 Dietrich and Maria von Wedemeyer become engaged; British and Norwegian raid on Vemork; in March, an attempt on Hitler's life fails; in April, Dietrich, Hans and Christel Dohnanyi are arrested; Christel is released; Dietrich put in Tegel prison; Allied bombing of Vemork causes severe damage.

1944 Hans Dohnanyi falls ill with diphtheria in prison; *SS Hydro* carrying heavy water is blown up in Norway; on June 6, Allied invasion of Normandy takes place; on July 20, an attempt on Hitler's life, led by Claus von Stauffenberg, fails; September, some of Dohnanyi's files on Nazi crimes are discovered; October, Klaus Bonhoeffer, Rüdiger Schleicher, and Eberhard Bethge are arrested.

1945 Maria von Wedemeyer escapes from family estate as Soviets advance; April 9, Dietrich is hanged in Flossenbürg concentration camp, and Hans Dohnanyi is hanged in Sachsenhausen; on April 23, Klaus and Rüdinger Schleicher are shot; on April 30, Hitler commits suicide; July, twelve members of Uranium Club, including Heisenberg, are imprisoned at Farm Hall in England for six months.

1946 While attempting to take care of his family and rebuild the destroyed physical chemistry institute at Leipzig, Karl-Friedrich suffers a heart attack.

1947 In poor health, Karl-Friedrich becomes professor at University of Berlin to enable him to be closer to his parents.

1948 Soviet blockade of West Berlin; Karl-Friedrich becomes director of the Fritz Haber Institute of the Max Planck Society; Karl Ludwig Bonhoeffer dies at the age of 80 while working at the Charité Clinic.

1949 Iron Curtain descends and Germany is divided into the German Democratic Republic (east) and the Federal Republic of Germany (west); Karl-Friedrich becomes director of the Max Planck Institute for Physical Chemistry in Göttingen.

1951 Paula Bonhoeffer dies; Karl-Friedrich serves as head of German Bunsen Society; he attempts to convince Jewish scientists to return to Germany.

1957 On May 15, Karl-Friedrich dies of a heart attack at age 58.

1971 Max Planck Institute of Physical Chemistry takes on additional name, the Karl Friedrich Bonhoeffer Institute.

Appendix II Biographical Profiles

The following brief profiles provide basic information on people mentioned in this book. Besides birth/death dates and academic affiliations, the profiles give a glimpse into what occurred in their lives after World War II. Data was gleaned from numerous sources, among them the *Indexes and Supplementary Materials, Vol. 17*, of *Dietrich Bonhoeffer Works*, and *Physics and National Socialism: An Anthology of Primary Sources* by Klaus Hentschel. For medical colleagues of Karl Ludwig Bonhoeffer, the online Charité Memorial Site, University of Berlin, provided valuable information. Entries were cross-referenced with Wikipedia and other internet sites.

A

Adorno, Theodor: (1903–69) German-Jewish philosopher and sociologist; professor at Frankfurt University at same time as Karl-Friedrich Bonhoeffer; collaborated with Max Horkheimer; 1933, lost position at Frankfurt; 1934, to England; 1938, to USA; 1949, returned to Frankfurt, focusing research on nature of prejudice and psychology of fascism.

Anderson, Carl: (1905–51) American physicist; spent career at Caltech; 1932, discovered positron by observing its track through a cloud-chamber; 1936, discovered muon; 1936 Nobel Prize in Physics.

Angell, James W.: (1898–1986) American economist; taught at Columbia University; 1929, wrote *The Recovery of Germany* on World War I reparations; 1930–31, Dietrich Bonhoeffer heard Angell debate with Hjalmar Schacht, a German economist, while studying at Union Theological Seminary.

Antropoff, Andreas von: (1878–1956) Estonian-born German physical chemist; professor at Bonn University; member of the Nazi party; nominated Karl-Friedrich Bonhoeffer for Nobel Prize in Chemistry in 1934.

Ardenne, Manfred von: (1907–97) German physicist; ran private laboratory funded in part by the German Post Office during World War II; researched electromagnetic mass separation of isotopes, also radar; at end of war, went to USSR where he worked on atomic bomb project; 1955, returned to Dresden.

Aubert, Axel: (1873–1943) Norwegian chemical engineer; 1895, Ph.D. Basel University; 1926, general director of Norsk Hydro; chairman of company at time of his death.

B

Bagge, Erich: (1912–96) German nuclear physicist; 1938, Ph.D. under Heisenberg at Leipzig University; member of the Uranium Club researching enrichment of U-235 and isotope separation; on staff at Kaiser Wilhelm Institute for Physics in Berlin; 1945–46, detained at Farm Hall, England; 1948, professor at Hamburg University; 1957, Kiel University.

Barth, Karl: (1886–1968) Swiss theologian; 1934, drafted Barmen Declaration leading to formation of Confessing Church; 1935, lost his position at Bonn University for political reasons; major influence on Dietrich Bonhoeffer with whom he corresponded.

Beck, Ludwig: (1880–1944) German military leader; during 1930s, chief of German general staff; 1938, resigned due to disillusionment with Hitler; participant in resistance; executed following failure of plot to kill Hitler.

Bell, George: (1883–1958) Bishop of Chichester, England, from 1927–1957; ecumenical leader; met Dietrich Bonhoeffer in the early 1930s and maintained contact with him; known to Bonhoeffer family members in England as Uncle George; 1942, met Dietrich in Sweden; July 1945, three months after Dietrich's death, organized a memorial service at Holy Trinity Church in London.

Benjamin, Georg: (1895–1942) German-Jewish physician; younger brother of the philosopher Walter Benjamin; brother-in-law of Ruth Lange, Rosbaud's friend; died in concentration camp.

Benjamin, Walter: (1892–1940) German-Jewish philosopher; went into exile in Paris; when Paris fell to Germany in 1940, attempted to reach Portugal via Spain, but fearing he would be captured and deported back to France, committed suicide.

Berg, Otto: (1873–1939) German chemist; with Noddack and Tacke, discovered the element rhenium.

Berve, Helmut: (1896–1979) German historian; 1927, professor Leipzig University; 1933, joined Nazi party; dean at Leipzig University; 1943, Munich University; 1954, Erlangen University.

Bethe, Hans: (1906–2005) German-Jewish nuclear physicist; 1928, Ph.D. under Sommerfeld at Munich University; 1932, lost position at Tübingen University; 1935, to USA, professor Cornell University; research on astrophysics and quantum electrodynamics; took part in Manhattan Project calculating critical mass needed for chain reaction; after war worked on research into hydrogen bomb; 1967 Nobel Prize in Physics.

Bethge, Eberhard: (1909–2000) German minister; student of Dietrich Bonhoeffer at illegal Finkenwalde Seminary; extensive correspondence with Dietrich; 1943, married Renate Schleicher, daughter of Rüdiger and Ursula (Bonhoeffer) Schleicher; 1944, arrested but not executed; after Dietrich's death, edited writings, getting them published; author of *Dietrich Bonhoeffer: A Biography*, as well as numerous papers.

Birge, Raymond: (1887–1980) American physicist; entire career at University of California, Berkeley, where he studied molecular spectra and championed the establishment of standard values for constants; later work focused on statistics.

Bodenstein, Max: (1871–1942) German physical chemist; 1893, Ph.D. Heidelberg University; known for his work on chemical kinetics; director of the Kaiser Wilhelm Institute for Physical Chemistry in Berlin; encouraged Karl-Friedrich's research on photochemistry and kinetics during the 1920s.

Bohr, Niels: (1885–1962) Danish physicist; famous for the Bohr model of the atom as well as major contributions to quantum theory; 1922 Nobel Prize in Physics; his Institute of Physics in Copenhagen served as a way-station for exiled Jewish scientists during the 1930s; 1943, because his mother was Jewish, he and his wife fled to Sweden; peripheral participant in Manhattan Project; after war, called for international cooperation on nuclear energy and helped establish CERN.

Bonhoeffer, Christel (Christine): (1903–65) second daughter of Karl and Paula; 1925, married Hans Dohnanyi; was an active participant in the resistance; 1943, imprisoned and released.

Bonhoeffer, Dietrich: (1906–45) German theologian; twin of Sabine; 1927, Ph.D. theology at Berlin University; pastor and director of Finkenwalde, an illegal seminary of the Confessing Church; participant in the resistance chiefly by maintaining ecumenical contacts outside Germany, including Bishop George Bell in England, and staff at the World Council of Churches in Switzerland; 1943, became engaged to Maria von Wedemeyer; April 1943, arrested and sent to Tegel prison in Berlin; April 9, 1945, executed at Flossenbürg concentration camp.

Bonhoeffer, Emilie (née Delbrück): (1905–91) also spelled Emmi; daughter of the historian Hans Delbrück and Lina Delbrück; sister of Max; grew up in Berlin near Bonhoeffers; 1930, married Klaus Bonhoeffer.

Bonhoeffer, Julie (née Tafel): (1842–1936) mother of Karl Ludwig Bonhoeffer; important ethical role model for Bonhoeffer children; April 1933, defied boycott of Jewish businesses by shopping in a Jewish store.

Bonhoeffer, Karl-Friedrich: (1899–1957) physical chemist; first son of Karl and Paula; 1922, Ph.D. Berlin University; through the 1920s worked with Fritz Haber; 1929, discovered para- and ortho-hydrogen with assistance of Paul Harteck; 1930, married Margarete (Grete) Dohnanyi, sister of Hans Dohnanyi; their children were Karl, Friedrich, Martin, and Katharina; after war, helped rebuild science in Germany; at the time of his death he was director of the Max Planck Institute of Physical Chemistry, Göttingen.

Bonhoeffer, Karl Ludwig: (1868–1948) neurologist and psychiatrist; March 1898, married Paula von Hase; father of four sons and four daughters; career began at Breslau Psychiatric Clinic; from 1912–1936, director of neurology and psychiatry at the Charité Clinic in Berlin and professor at Berlin University; research was dedicated to establishing a solid neurobiological foundation to psychiatry; supported voluntary sterilization but opposed euthanasia; continued to work as a clinician until his death.

Bonhoeffer, Klaus: (1901–45) third son of Karl and Paula; doctorate in law; corporate lawyer for Lufthansa; 1930, married Emilie (Emmi) Delbrück; involved in resistance; April 1945, executed.

Bonhoeffer, Margarete (née Dohnanyi): (1903–92) known as Grete; daughter of composer Ernst von Dohnanyi; sister of Hans Dohnanyi; 1930, married Karl-Friedrich Bonhoeffer, children were Karl, Friedrich, Martin, and Katharina.

Bonhoeffer, Otto: (1864–1932) brother of Karl Ludwig Bonhoeffer; chemist at Bayer; lost two sons, Fritz and Werner, in the First World War; his third son, Lothar, was severely injured.

Bonhoeffer, Paula (née von Hase): (1876–1951) daughter of Clara and Karl Alfred von Hase; grew up in Breslau; earned teaching certificate before marrying Karl Ludwig on March 5, 1898; supported her children's involvement in resistance.

Bonhoeffer, Sabine: (1906–99) twin sister of Dietrich; 1926, married Gerhard Leibholz, a legal scholar whose family was Jewish; 1938, fled to England; returned to Germany after the war.

Bonhoeffer, Susanne: (1909–91) youngest child of Karl and Paula, known as Susi; 1929, married Walter Dreß, a Lutheran minister and a professor of church history.

Bonhoeffer, Ursula: (1902–83) oldest daughter of Karl and Paula; 1923, married Rüdiger Schleicher, a lawyer and legal scholar who participated in the resistance; their daughter, Renate, married Eberhard Bethge.

Bonhoeffer, Walter: (1899–1918) second son of Karl and Paula; killed in the First World War.

Born, Max: (1882–1970) German physicist; 1925, at Göttingen University where he and Heisenberg formulated matrix mechanics; 1933, exiled from Germany; went to Cambridge and then in 1936 to Edinburgh University; 1954 Nobel Prize in Physics for his work on quantum mechanics shared with Walther Bothe.

Bosch, Carl: (1874–1940) German chemist and engineer; developed at BASF the methodology to scale up Haber's synthesis of ammonia from nitrogen and hydrogen gas; 1925, helped found I G Farben; 1931 Nobel Prize in Chemistry.

Bothe, Walther: (1891–1957) German physicist; 1914, Ph.D. Berlin University under Planck; 1913–30, worked with Hans Geiger on radioactivity at Reich Physical and Technical Institute; 1925, professor Berlin University; 1932, director Heidelberg University; opposed to Deutsch Physik; director of Kaiser Wilhelm Institute for Medical Research; 1939, attended initial meeting of Uranium Club; 1943–44, built Germany's first cyclotron with Wolfgang Gentner; 1954 shared Nobel Prize in Physics with Max Born.

Braun, Wernher von: (1912–77) German physicist and rocketry engineer; 1932, graduated Technical University of Berlin; 1934, Ph.D. in physics from Berlin University with thesis on rocketry; Nazi party member and member of SS; head of German rocketry project including V1 and V2; after war, came to USA as part of Operation Paperclip; developed rocketry for USA.

Bresch, Carsten: (1921) German physicist and geneticist: worked with Karl-Friedrich Bonhoeffer in Göttingen University after the war; influenced by Max Delbrück.

Broglie, Louis de: (1892–1987) French physicist; worked on wave-particle duality in quantum theory; 1929 Nobel Prize in Physics; after war, advised French government on nuclear energy and worked to bring about a greater understanding of science.

Brun, Jomar: (1904–93) Norwegian chemical engineer; 1929, employed by Norsk Hydro; head of electrolysis plant at Vemork that produced deuterium during 1930s; went to England during World

War II where he participated in plan to sabotage Vemork; returned to Norway after war; 1951, professor Norwegian Institute of Technology.

Buber, Martin: (1878–1965) Austrian-born Jewish theologian; 1930–1933, honorary professor Frankfurt University; 1933, lost position; 1938, to Palestine where he became a professor at the Hebrew University.

Bumke, Oswald: (1877–1950) German neurologist and psychiatrist; 1901–1913, head of psychiatric clinic in Freiburg; 1916, to Breslau Clinic following the death of Alzheimer, who had been Karl Ludwig Bonhoeffer's successor; 1924, to Munich, replacing Emil Kraepelin with whom he disagreed on schizophrenia; critic of psychoanalysis.

C

Canaris, Wilhelm: (1887–1945) German admiral and chief of Abwehr; initially supported Hitler and then turned against him; became part of resistance; executed on April 9 along with Dietrich Bonhoeffer at Flossenbürg concentration camp.

Chadwick, James: (1891–1974) English physicist who worked with Rutherford at Cavendish lab; 1932, discovered neutron; 1935 Nobel Prize in Physics; 1935, Liverpool University; 1941, wrote final draft of MAUD report that convinced USA to proceed with Manhattan Project; head of the British team on Manhattan Project; remained active in nuclear power issues after war.

Clusius, Klaus: (1903–63) German physical chemist; member of the Uranium Club; worked on heavy water production and isotopes during war; after war, professor at Zurich University.

Cockcroft, John: (1897–1967) British physicist; 1928, Ph.D. under Rutherford; built accelerator with Ernest Walton and Marcus Oliphant; 1932, with Walton disintegrated atomic nucleus for which the two shared Nobel Prize in Physics in 1951; head of MAUD, Great Britain's committee on nuclear research; also worked on radar; after war remained active in peacetime use of atomic energy.

Compton, Arthur: (1892–1962) American physicist; 1927 Nobel Prize in Physics for discovery of particle nature of electromagnetic radiation (Compton effect); professor Chicago University; worked on Manhattan Project during war; after war became chancellor at Washington University in St. Louis.

Condon, Edward: (1902–74) American nuclear physicist; Ph.D. California University at Berkeley; studied at Göttingen University with Max Born; worked with James Franck, resulting in Franck-Condon principle; worked with Arnold Sommerfeld in Munich; in USA held many positions including director of the National Bureau of Standards, and president of the American Association for the Advancement of Science.

Coster, Dirk: (1889–1950) Dutch physicist; studied under Bohr with whom he published on X-ray spectroscopy and the periodic system; with de Hevesy discovered hafnium; 1924, Groningen University; 1938, helped Lise Meitner escape from Germany; hid Jews during war.

Creutzfeldt, Hans: (1885–1964) German neurologist; best known for identification of Creutzfeldt-Jakob disease; for 14 years colleague of Karl Ludwig Bonhoeffer at the Charité Clinic in Berlin; 1938, director of neuropsychiatry in Kiel at Christian Albrechts University; attempted to save patients from euthanasia and concentration camps during the war; his wife was imprisoned for anti-Nazi statements.

Crinis, Maximilian de: (1889–1945) German psychiatrist; Nazi who succeeded Karl Ludwig Bonhoeffer as director of the Charité Clinic upon Karl's retirement; 1939, wrote euthanasia decree that Hitler signed; May 1945, committed suicide.

Curie (Joliet-Curie), Irène: (1897–1956) French physicist and chemist; daughter of Marie and Pierre Curie; 1935 Nobel Prize in Chemistry shared with husband Frédéric Joliet for discovery of artificial radiation, which led to Hahn, Strassmann, and Meitner fissioning the atom in 1938; died of leukemia caused by exposure to radiation.

Curie, Marie: (1867–1934) Polish-born French physicist and chemist; famous for discovery of radioactivity; 1903 Nobel Prize in Physics shared with husband Pierre Curie and Henri Becquerel; 1911 Nobel Prize in Chemistry; died of anemia caused by exposure to radiation.

D

Debye, Peter: (1884–1966) Dutch physicist and physical chemist; doctorate under Sommerfeld; director of KWI of Physics during 1930s; 1936 Nobel Prize in Chemistry; 1940, to Cornell University where he spent the rest of his career.

Dehn, Max: (1878–1952) German-Jewish mathematician; lost position at Frankfurt University; 1939, escaped to Denmark, then to Norway and finally, via Siberia and Japan, to USA where he taught at a series of colleges, first in Idaho and then at Black Mountain College, North Carolina.

Delbrück, Emilie: (1905–91) daughter of the historian Hans Delbrück and Lina Delbrück; sister of Max and Justus; 1930, married Klaus Bonhoeffer.

Delbrück, Hans: (1848–1929) German military historian; professor at Berlin University; 1884–1890, member of Reichstag; after First World War, member of German delegation at Versailles; neighbor of the Bonhoeffers and Harnacks; his ideas were influential and at times controversial.

Delbrück, Justus: (1902–45) German lawyer; son of Hans and Lina Delbrück; brother of Max and Emilie; participant in the resistance; October 1945 died in a Soviet prison camp.

Delbrück, Max: (1906–81) German biophysicist; grew up near the Bonhoeffers in Berlin; Karl-Friedrich was his mentor and friend; 1937, to USA on Rockefeller fellowship at Caltech; during the war taught physics at Vanderbilt University; 1947–77, professor of biology Caltech; 1969 Nobel Prize in Physiology or Medicine.

Dennison, David: (1900–76) American physicist whose work focused on molecular structure and quantum mechanics; Ph.D. University of Michigan; 1924–27, studied with Bohr and many other European scientists; spent career at Michigan.

Diebner, Kurt: (1905–64) German nuclear physicist; 1932, Ph.D. Halle University; 1934, Reich Physical and Technical Institute; scientific advisor to HWA; director of the Nuclear Research Council; organized September 1939 meeting of Uranium Club; during war, he and Heisenberg were frequently at odds; after war was interned at Farm Hall until January 1946; employed in industry.

Dirac, Paul: (1902–84) English theoretical physicist; known for insights into quantum mechanics and electrodynamics; 1926, Ph.D. Cambridge followed by studies in Copenhagen and Göttingen; 1932–69, professor Cambridge; 1933, shared Nobel Prize in Physics with Erwin Schrödinger; 1970, professor Florida State University.

Dohnanyi, Hans von: (1902–45) 1925, German lawyer; married Christel Bonhoeffer; employed by the Reich Ministry of Justice; 1939, appointed to Military Intelligence Office known as Abwehr; one of the leaders of the conspiracy against Hitler; executed on April 9, 1945.

Dohnanyi Margarete: (1903–92) known as Grete; sister of Hans Dohnanyi; 1930, married Karl-Friedrich Bonhoeffer; children were Karl, Friedrich, Martin, and Katharina.

Döpel, Georg Robert: (1895–1982) German nuclear physicist; colleague of Heisenberg at Leipzig University; member of the Uranium Club; conducted experiments on heavy water and on spherical layers of uranium surrounded by water; 1942, his experimental uranium machine exploded; near end of war his wife, Clara, who worked with him in Leipzig, was killed by a bomb; he went to the USSR and worked on atomic bomb.

Dreß, Walter: (1904–79) German pastor and theologian; 1929, married Susanne Bonhoeffer; during 1930s, aligned with Confessing Church.

Drucker, Peter F.: (1909–2005) Austrian-born American economist, management consultant, and prolific writer; 1933, lecturer in international and public law at Frankfurt University; attended faculty meeting at which Jews were thrown out; went to England and then to USA, professor at Bennington College and New York University; 1971, developed first MBA program at Claremont Graduate University.

E

Eddington, Arthur: (1882–1944) English astrophysicist and Quaker; led expedition on solar eclipses in May 1919 that confirmed theory of relativity; director of Cambridge Observatory.

Ehrenfest, Paul: (1880–1933) Austrian-born Dutch theoretical physicist; Jewish; friend of Einstein and Bohr; spent career at Leiden University researching phase transitions and the relationship between statistical mechanics and quantum mechanics; his students included Samuel Goudsmit, Dirk Coster, and Enrico Fermi; 1933, suffering from depression, he shot his son who had Down Syndrome and then shot himself.

Ehrlich, Paul: (1854–1915) German physician and scientist who helped revolutionize medicine; developed first effective treatment for syphilis; 1908 Nobel Prize in Medicine or Physiology.

Eigen, Manfred: (1927) German biophysical chemist; succeeded Karl-Friedrich Bonhoeffer as director of the Max Planck Institute of Physical Chemistry; 1967 Nobel Prize in Chemistry shared with Ronald Norrish and George Porter.

Einstein, Albert: (1879–1955) German-born theoretical physicist who was culturally Jewish; 1914, returned from Switzerland to Germany to become director of the Kaiser Wilhelm Institute for Physics and professor at Berlin University; 1921 Nobel Prize in Physics; major influence on Karl-Friedrich Bonhoeffer during 1920s; strong opponent of Hitler; 1933, took position at the Institute of Advanced Study in Princeton; 1940, became U.S. citizen.

Emeléus, Harry Julius: (1903–93) British chemist; did postdoctoral studies with Hugh Taylor at Princeton University; after war, served as president of the inorganic division of the International Union of Pure and Applied Chemistry, the Chemical Society, and the Royal Institute of Chemistry.

Esau, Abraham: (1884–1955) German physicist, president of National Bureau of Standards; leader of Reich Research Council, which took control of atomic research in 1942; professor at Berlin University; 1939, helped establish Uranium Club; despite power struggles, remained influential; 1945–48, prisoner of Dutch; 1949, visiting professor Aachen Polytech; 1953, head of institute in Mülheim focused on high-frequency engineering.

Eucken, Arnold: (1884–1950) German chemist and physicist; 1915–1930, professor at Technische Hochschule Breslau, then at Göttingen University; Paul Harteck worked as his assistant in Breslau before coming to the Haber Institute; joined NSDAP in 1933.

Ewald, Paul: (1888–1985) German physicist, X-ray crystallography; 1912, Ph.D. at Munich University under Sommerfeld; assistant to Hilbert at Göttingen University; 1921–33, professor and rector at Stuttgart Polytechnic; 1933, resigned because he was opposed to Nazism although he was not Jewish; 1937, to Cambridge; 1939, to Queens University, Belfast; 1949–57, Polytechnic Institute of Brooklyn.

Eyring, Henry: (1901–81) Mexican-born American theoretical chemist whose work focused on chemical reaction rates; 1931–45, at Princeton University and then at Utah University; worked with Polanyi in Berlin on a fellowship in 1929–31 following up mathematically on Karl-Friedrich Bonhoeffer's work and quantum mechanics of chemical bonds.

F

Farkas, Adalbert: (1906–95) Hungarian-Jewish physical chemist, received Ph.D. Berlin University under Karl-Friedrich Bonhoeffer; worked at Haber Institute; went with Bonhoeffer to Frankfurt; 1933, he and Ladislaus, his brother, went to England, publishing a paper together on heavy water; 1936–1941, to Palestine and then to USA where he spent the rest of his career.

Farkas, Ladislaus: (1904–48) Hungarian-Jewish physical chemist; received Ph.D. Berlin University under Karl-Friedrich Bonhoeffer with whom he published scientific papers; 1933, went to England on fellowship at Cambridge; published paper on heavy water with Adalbert, his brother; 1935, to Palestine, professor of physical chemistry at Hebrew University of Jerusalem; 1948, killed in an airplane crash on way to USA to buy laboratory equipment.

Fermi, Enrico: (1901–54) Italian physicist; 1938 Nobel Prize in Physics; 1938, left Italy because his wife was Jewish; came to Chicago University where he created the world's first nuclear reactor; participant in Manhattan Project; after war remained at Chicago for rest of career.

Fischer, Erich: (1910–69) German physicist; 1937, assistant to Debye and Heisenberg at Kaiser Wilhelm Institute for Physics; member of the Uranium Club; at Haigerloch at end of war; 1948, Tübingen University; 1951, Ankara University.

Flammersfeld, Arnold: (1913–2001) German physicist; 1938, received Ph.D. under Lise Meitner; worked with Hahn at Kaiser Wilhelm Institute for Chemistry; a member of the Uranium Club working with Bothe; after war, at Tübingen, Mainz, and Göttingen.

Flügge, Siegfried: (1912–97) German theoretical physicist; 1933, Ph.D. under Max Born at Göttingen University; 1936–37, assistant to Heisenberg at Leipzig University; 1937–42, worked with Hahn at Kaiser Wilhelm Institute for Chemistry; 1939, wrote influential paper on nuclear power; member of the Uranium Club; 1956–84, editor of prestigious multi-volume *Handbook of Physics.*

Foley, Francis Edward (Frank): (1884–1958) officer British Secret Intelligence Service; 1930s, passport control officer for the British in Berlin; saved many Jews by stamping their passports and issuing visas clandestinely for England or Palestine; meanwhile was covertly running MI6 station in Berlin; contact for Rosbaud.

Förster, Theodor: (1910–74) German physical chemist; Karl-Friedrich Bonhoeffer's assistant at Leipzig University; 1947, Göttingen University; 1951, Stuttgart University.

Fowler, Ralph: (1889–1944) British physicist, mathematician, and astronomer; 1919, at Trinity where he worked on thermodynamics; mid-1920s, research supervisor of Paul Dirac; 1932, chairman of theoretical physics at Cavendish Laboratory; 1939, worked with British Ordnance and MAUD Committee.

Franck, James: (1882–1964) German-Jewish physicist; served in First World War; worked with Fritz Haber; 1925 Nobel Prize in Physics shared with Gustav Hertz; at Göttingen, worked with Max Born; 1933, resigned in protest of Aryan clause; 1935, came to USA, first to Johns Hopkins and then to University of Chicago; joined the Manhattan Project; in June 1945, drafted with Eugene Rabinowitch the *Franck Report*, which stated that the atomic bomb should not be dropped on Japan but instead should be used as a publicized demonstration on a desert island, the demonstration alone being enough to force Japan to surrender; report warned of nuclear arms race.

Freundlich, Herbert: (1880–1941) German-Jewish physical chemist; head of colloid chemistry at Haber Institute and organizer of the Haber colloquia; 1933, left Germany, eventually came to USA where he was professor at Minnesota University.

Frisch, Otto: (1904–79) Austrian-born, British nuclear physicist; nephew of Lise Meitner; worked under Otto Stern at Hamburg University; 1933, left for England; spent five years at Bohr's laboratory in Copenhagen; 1939, he and Meitner realized that Otto Hahn and Fritz Strassmann had split the atom, for which Frisch used the term fission; 1940, collaborated with Rudolf Peierls on the theoretical mechanics of atomic bomb detonation; participated in Manhattan Project; after war, spent most of career at Trinity College, Cambridge.

Frumkin, Aleksander N.: (1895–1976) Russian physicist and electrochemist; known for the fundamental theory of electrode reactions; studied at Strasbourg and Bern Universities; 1922, to Karpov Institute in Moscow; 1930, professor at Moscow University where he spent the rest of his career.

G

Gaupp, Robert: (1870–1953) German neurologist and psychiatrist; colleague and friend of Karl Ludwig Bonhoeffer under whom he had studied in Breslau; professor at Tübingen University; researched shell shock during First World War.

Geiger, Johannes (Hans): (1882–1945) German physicist; co-inventor of Geiger counter; 1907, worked with Rutherford; 1912, Physical Technical Institute Berlin; 1929, Tübingen University; 1936, Berlin University; member of Uranium Club; at initial meeting in 1939.

Gentner, Wolfgang: (1906–80) German nuclear physicist; 1930, Ph.D. Frankfurt University; 1932–35, worked at Curie's laboratory in Paris; 1936–45, worked with Bothe at Kaiser Wilhelm Institute for Medical Research, focusing on cyclotron; 1938, visited University of California, which had a cyclotron; 1940, after capitulation of France, went to Curie's laboratory to work on cyclotron; 1944, German cyclotron operational; after the war, he had a long career that included positions at CERN and the Max Planck Institute.

Gerischer, Heinz: (1919–94) German-Jewish physical chemist; received Ph.D. under Karl-Friedrich Bonhoeffer at Leipzig University; sheltered by Bonhoeffer, he remained with him after the war at Berlin and Göttingen Universities, where he was research associate at the Max Planck Institute for Physical Chemistry; 1970, director of the Fritz Haber Institute, part of the Max Planck Institute.

Gerlach, Walther: (1889–1979) German physicist; 1912, Ph.D. Tübingen University; served in First World War; 1920, professor Frankfurt University where he and Otto Stern collaborated; 1925, Tübingen University; 1929–45, Munich University; 1937–45, advisory board KWG; member of Uranium Club, assuming leadership of the nuclear project through the Reich Research Council in late 1943, replacing Abraham Esau; 1945, interned at Farm Hall; after war taught at Bonn and Munich Universities and held leadership positions in scientific organizations.

Gildemeister, Martin: (1876–1943) German physiologist; anti-Nazi colleague of Karl-Friedrich Bonhoeffer at Leipzig University with whom he researched the interconnection between physiology and physical chemistry; helped protect Bernard Katz.

Goebbels, Paul Joseph: (1897–1945) German politician, Reich minister of propaganda; 1921, Ph.D. Heidelberg University; 1924, joined Nazi party; highly effective propagandist using film, radio, and print; advocated for total war; killed himself and family after Hitler's suicide.

Goerdeler, Carl Friedrich: (1884–1945) German lawyer and politician; 1930–1937, mayor of Leipzig; participated in resistance with Ludwig Beck; 1944, arrested; February 1945, executed.

Goering, Hermann: (1893–1946) German military leader and politician; 1922, joined Nazi party; 1933, founded Gestapo; 1935, commander-in-chief Luftwaffe; in charge of four-year plan of economic development leading up to World War II; 1940, named Reichsmarshall, second-in-command to Hitler; at Nuremberg trial convicted of war crimes; committed suicide.

Goldschmidt, Victor: (1888–1947) Swiss-born mineralogist and geochemist who grew up in Norway and spent most of his academic career at Oslo University; 1929–1933, at Göttingen University where he was a colleague of Franck; forced to leave Germany, returned to Norway; arrested during the war; fled to Sweden and then to England.

Goudsmit, Samuel: (1902–78) Dutch physicist of Jewish descent; 1927, Ph.D., Leiden University under Ehrenfest; 1927, to USA; 1927–46, Michigan University; during war worked at MIT; his parents in Netherlands were murdered by Nazis; became head of the Alsos Mission, which had as its goal the assessment of Nazi progress on atomic bomb; 1948–1980, Brookhaven National Laboratory.

Groth, Wilhelm: (1904–77) German physical chemist; 1932, Hamburg University studied under Otto Stern, then under Paul Harteck who took Stern's position: he was assistant to Harteck when the two men wrote the letter, dated April 24, 1939, to German Army Ordnance on the possibility of an atomic bomb; member of the Uranium Club; helped develop centrifuges for uranium enrichment; 1945, Hamburg University; 1950, Bonn University.

Groves, Leslie Richard: (1896–1970) American military officer, Army Corp of Engineers; 1918, graduated from West Point; during war, in charge of Manhattan Project; 1948, retired from military with rank of lieutenant general and became vice president of Sperry Rand Corp.

H

Haber, Fritz: (1868–1934) German-Jewish chemist; 1918 Nobel Prize in Chemistry for the Haber-Bosch process for synthesis of ammonia from nitrogen and hydrogen gas; head of the Kaiser Wilhelm Institute for Physical and Electrochemistry, known as the Haber Institute; enormous influence on Karl-Friedrich Bonhoeffer who was his assistant during the 1920s; 1933, forced to resign; 1934, died in exile.

Hahn, Otto: (1879–1968) German chemist; 1944 Nobel Prize in Chemistry for the discovery of nuclear fission; 1938, with Fritz Strassmann (incorporating work of Lise Meitner), bombarded uranium with neutrons; Meitner and Otto Frisch realized fission had occurred; anti-Nazi but circumspect in his opposition; during war, loosely connected with Uranium Club; 1945, incarcerated in Farm Hall, England; helped rebuild German science after the war; peace activist.

Harnack, Adolf von: (1851–1930) German theologian and professor of church history; neighbor of the Bonhoeffers; first president of the Kaiser Wilhelm Gesellschaft (KWG).

Harnack, Arvid: (1901–42) German economist; nephew of Adolf von Harnack; friend of Klaus Bonhoeffer; he and his wife, Mildred, were communists who were opposed to Hitler; helped to establish a resistance and spy circle known as the Red Orchestra; 1942, executed.

Harnack, Mildred (née Fish): (1902–43) American literary historian and German translator; wife of Arvid Harnack whom she had met at Minnesota Univ.; communist and resistance participant; 1942, arrested; 1943, executed.

Harteck, Paul: (1902–85) Austrian physical chemist; 1926, Ph.D. Berlin University; 1926–28, assistant to Arnold Eucken at Breslau University; 1929, assistant to Karl-Friedrich Bonhoeffer at Haber Institute; 1929, discovered molecular isotopes of hydrogen with Bonhoeffer; a member of the Uranium Club during World War II; 1951, to USA, Rensselaer Polytechnic Institute.

Hase, Benedikt von: (1890–1979) Paula Bonhoeffer's youngest brother; known to the Bonhoeffer children as Uncle Bubi; World War I prisoner of war.

Hase, Paul von: (1885–1944) cousin of Paula Bonhoeffer; military commander of Berlin during World War II; took part in the July 1944 plot to kill Hitler; executed August 8, 1944.

Hassel, Odd: (1897–1981) Norwegian physical chemist; 1924, Ph.D. under Fritz Haber at Berlin University; professor at Oslo University; 1930s, researched deuterium; best known for work related to three-dimensionality of molecular structure; during World War II was arrested by Norwegian pro-Nazi group and spent a year in prison; 1969 Nobel Prize in Chemistry shared with Derek Barton.

Havemann, Robert: (1910–82) German physical chemist; communist who was opposed to both Hitler and Stalin; participant in the resistance; attempted to save Jews; arrested by Nazis; after the war, with Karl-Friedrich Bonhoeffer's help, obtained professorship at Berlin University in Soviet sector; protested development of the hydrogen bomb by USA; in East Germany, he remained an idealistic dissident.

Heidegger, Martin: (1889–1976) German philosopher; focused on the nature of being; taught at Freiburg University for most of his career; a Nazi, he strongly encouraged students to join the Nazi party.

Heisenberg, Werner: (1901–76) German theoretical physicist; 1923, Ph.D. Munich University; 1926–27, Göttingen University; 1927, Leipzig University; colleague of Karl-Friedrich Bonhoeffer at Leipzig during late 1930s; colleague and mentor of Carl von Weizsäcker; 1932 Nobel Prize in Physics for quantum mechanics; leader of the Uranium Club during World War II; strongly patriotic, his stance toward Nazism was ambiguous; 1946–58, director Max Planck Institute of Physics, Göttingen; 1958, professor Munich; 1960–70, director Max Planck Institute of Physics and Astrophysics, Munich.

Held, Robert: (n.d.) German lawyer; friend of the Bonhoeffer family; 1926, traveled to Italy with Dietrich and Karl-Friedrich Bonhoeffer.

Herkner, Heinrich: (1863–1932) German economist and sociologist; taught at Freiburg, Karlsruhe, Zurich, and Berlin Universities; 1913, moved to Berlin and became one of the circle of friends around the Harnacks, Delbrücks, and Bonhoeffers.

Hertz, Gustav: (1887–1975) German physicist; shared 1925 Nobel Prize in Physics with James Franck for research on mercury vapor; because of Jewish lineage, he was at risk after Hitler came to power;

he found a measure of protection by becoming a director at Siemens, a position he held until 1945 when he went to the USSR where he worked on the nuclear bomb.

Herzog, Fritz: (1902–2001) German mathematician; studied at Berlin University; 1934, to USA, earned Ph.D. at Columbia; spent most of career at Michigan State University.

Heuss, Theodor: (1884–1963) German journalist and editor, sporadic involvement in politics until after World War II when he served as president of the Federal Republic of Germany (West Germany) from 1949–1959; during 1920s, he was one of the circle of friends around the Harnacks, Delbrücks, and Bonhoeffers.

Hevesy, George de: (1885–1966) Hungarian chemist of Jewish descent; professor of physical chemistry at Freiburg University; 1933, lost his position; went to Bohr's institute in Copenhagen; 1943, fled to Sweden and was awarded the Nobel Prize in Chemistry.

Hildebrandt, Franz: (1909–85) Lutheran pastor; friend of Dietrich Bonhoeffer; worked with Martin Niemöller; part of the Confessing Church; arrested for his activities with Niemöller and because his mother was of Jewish descent; 1937, after release, went to England; changed to Methodism; came to the USA; 1953–67, professor at Drew University, Madison, New Jersey; 1968, went to Scotland.

Hindenburg, Paul von: (1847–1934) German military leader and politician; First World War General Field Marshall; 1925–1934, president of Germany; played key role in Hitler's rise to power by appointing him chancellor; on his death in 1934, the office of president and chancellor were combined in the Führer.

Hinshelwood, Cyril: (1897–1967) English physical chemist; early research focused on molecular kinetics; 1956 Nobel Prize in Chemistry.

Hitler, Adolf: (1889–1945) born in Austria; served and was injured in First World War; 1920, founded National Socialist German Workers Party (NSDAP, known as Nazi party); 1923–24, while in prison for attempted coup in Munich, wrote *Mein Kampf* (*My Struggle*); 1933, became chancellor; 1934, on death of Hindenburg, took title Führer and consolidated power; built up military, crushed political opposition, and enforced policies on Aryan purity and superiority; invaded Poland in September 1939, starting World War II; committed suicide April 30, 1945.

Hoffmann, Gerhard: (1880–1945) German physicist; 1906, Ph.D. Bonn University; 1908, Königsberg University; 1928–37, Martin-Luther-University Halle-Wittenberg; researched radioactivity and cosmic rays; 1937–45, professor Leipzig University; member of the Uranium Club working on cyclotron.

Horkheimer, Max: (1895–1973) German-Jewish philosopher and sociologist; colleague and collaborator with Theodor Adorno on critical theory at Frankfurt University; 1933, lost job; went to Switzerland and then to USA where he continued to work with Adorno who had also emigrated; both men returned to Frankfurt in 1949.

Horney, Karen: (1885–1952) German psychoanalyst; 1914, Ph.D. Berlin University under Karl Ludwig Bonhoeffer; went into psychoanalysis; 1920–32, taught at Berlin Psychoanalytic Institute which she helped found; 1932 to USA with her daughters, Chicago Institute of Psychoanalysis; 1934, to Brooklyn, New School for Social Research and New York Psychoanalytic Institute; deviated from Freud on psychology of women; New York Medical College.

Hund, Friedrich: (1896–1997) German physicist; 1925, Ph.D. Göttingen University under Max Born; taught at several universities including Leipzig, Jena, Frankfurt, and Göttingen; worked with Heisenberg, Bothe, Born, and Schrodinger among others; colleague of Karl-Friedrich Bonhoeffer; protested firing of Jewish colleagues at Leipzig; after war, professor at Jena, Frankfort, and finally Göttingen.

Hutton, Robert: (1876–1970) British metallurgist; 1931–1942, Cambridge; during 1930s helped Jewish academics get out of Germany and find jobs in England.

J

Jaenicke, Johannes: (1888–1984) German chemist; assistant to Fritz Haber and department head at Haber Institute; 1920s, worked on failed project to extract gold from seawater; father of Walther Jaenicke who studied under Karl-Friedrich Bonhoeffer.

Jaenicke, Walther: (1921–2010) German physical chemist; son of Johannes Jaenicke; received Ph.D. under Karl-Friedrich Bonhoeffer at Leipzig University; 1952, professor Karlsruhe University of Technology; 1963, professor Erlangen-Nuremberg University.

Jander, Gerhart: (1892–1961) German chemist; 1925, professor Göttingen University; 1933–35, temporary director of Haber Institute after Haber was forced to resign; appointed by the Nazis with the intent of shifting research to armaments; unclear as to what Jander accomplished; 1935, Greifswald University.

Jászi, Oszkár: (1875–1957) Hungarian political scientist, social scientist, and historian; author of *The Dissolution of the Hapsburg Empire*; 1925, to USA, Oberlin; friend of Michael Polanyi.

Jehle, Herbert: (1907–83) German physicist and Quaker pacifist; knew Karl-Friedrich and Dietrich Bonhoeffer; with the help of Quakers, was released from a concentration camp in France and came to USA; taught at Harvard, Pennsylvania, Nebraska, and George Washington Universities; after the war, helped Dietrich's fiancée Maria von Wedemeyer attend Bryn Mawr in USA.

John, Hans: (1911–1945) German lawyer; brother of Otto; member of the resistance; August 1944, arrested; April 1945, executed.

John, Otto: (1909–1997); German lawyer; brother of Hans; member of resistance; worked with Klaus Bonhoeffer at Lufthansa; 1944, escaped from Germany; 1954, critical of Adenauer and politics in West Germany, went to East Germany; 1955, returned to West Germany, charged with treason; 1958, released from prison; attempted to clear himself of charges.

Johnson, James Weldon: (1871–1938) African American writer, lawyer, and civil rights activist, leader of NAACP; his *Book of American Negro Spirituals* was presented to Dietrich Bonhoeffer by Frank Fisher, a fellow student at Union.

Joliot (Joliot-Curie), Jean Frédéric: (1900–58) French physicist; husband of Irène Curie with whom he researched the structure of the atom; the couple won the Nobel Prize in 1935 for the discovery of artificial radioactivity.

Joos, Georg: (1894–59) German theoretical physicist; 1920, Ph.D. Tübingen University; 1924, professor at Jena University; 1935, took Franck's post at Göttingen University; 1941, left academia and took position at Zeisswerke Jena; peripheral member of the Uranium Club who, early on, called attention to the potential of atomic power.

Jossmann, Paul: (1891–1978) German-Jewish neurologist and psychiatrist, who came to the United States with the help of Karl Ludwig Bonhoeffer; worked at the Veterans Administration Outpatient Clinic in Boston.

Jost, Friedrich Wilhelm: (1903–88) German physical chemist; close friend and colleague of Karl-Friedrich Bonhoeffer at Leipzig University; 1944–51, professor at Marburg University; then at Göttingen University; wrote essay "The First 45 Years of Physical Chemistry in Germany" (1966).

K

Katz, Bernard: (1911–2003) German-Jewish biophysicist; 1933, received degree from Leipzig University; he was denied a major academic prize because he was Jewish; defended by Martin Gildemeister, head of the Physiological Institute; went to England and worked with Archibald Hill; 1938, Carnegie fellowship to Sydney Medical School in Australia; after war, returned to University College London; 1970, Nobel Prize in Physiology or Medicine shared with Julius Axelrod and Ulf von Euler.

Kistiakowsky, George: (1900–82) Ukrainian-American physical chemist; 1925, Ph.D. under Bodenstein at Berlin University; 1926, to USA, taught at Princeton; 1930, professor at Harvard; during war, participated in Manhattan Project.

Kleist-Retzow, Ruth von: (1867–1945); German; wife of Jürgen von Kleist; mother of Hans; widowed grandmother of Maria von Wedemeyer; supporter of the Confessing Church; helped Dietrich Bonhoeffer by providing support for Finkenwalde Seminary, and shelter for him at her estate near Kieckow so he could write.

Koeth, Joseph: (1870–1936) German military officer and politician in charge of economic demobilization after First World War; interacted with industrialists and scientists, including Fritz Haber; spoke at Haber memorial service in 1935.

Kopfermann, Hans: (1895–1963) German physicist; 1925, Ph.D. Göttingen under James Franck; 1920s, worked with Rudolf Ladenburg at Haber Institute; 1933, assistant to Gustav Hertz; 1937, professor Kiel University; a peripheral member of the Uranium Club working on isotope separation; 1942, professor Göttingen; 1953, Heidelberg University.

Korsching, Horst: (1912–98) German physicist; 1938, Ph.D. Berlin University; 1935, joined staff of Kaiser Wilhelm Institute for Physics; member of the Uranium Club; 1945–46, incarcerated at Farm Hall; 1946, Göttingen University; 1958, Munich University.

Kraepelin, Emil: (1856–1926) German psychiatrist; recognized the biological and genetic causes of mental illness, as did Karl Ludwig Bonhoeffer; campaigned for more humane treatment of mentally ill; supporter of eugenics.

Kramer, Franz: (1878–1967) German-Jewish neurologist and psychiatrist; 1907, habilitated at Breslau University under Karl Ludwig Bonhoeffer; 1912, went with Bonhoeffer to the Charité Clinic in Berlin; specialized in child psychology and neurology; 1933, lost position but remained in private practice; 1938, fled to Netherlands with Bonhoeffer's help.

Krieck, Ernst: (1882–1947) German educator; graduate of a teacher's college, honorary Ph.D.; 1933, appointed Nazi president of Frankfurt University, a position for which he was unqualified; published a magazine during the 1930s about Nazi educational ideas; eventually left party; during war, professor Heidelberg University.

Kröger, Martin: (1894–1980) German physical chemist; member of the Nazi party; professor at Leipzig University; during war, denounced his colleague Karl-Friedrich Bonhoeffer; after war, professor at Munster University.

L

Ladenburg, Rudolf: (1882–1952) German-Jewish physicist; in 1920s, replaced Franck as head of atomic physics at Haber Institute; colleague of Karl-Friedrich Bonhoeffer; 1930, went to Princeton; played a key role in helping Jewish exiles find employment in the USA.

Landsberger, Benno: (1890–1968) German-Jewish Assyriologist; 1935, lost position at Leipzig University; went to Turkey; 1945, to USA, professor Chicago University.

Langmuir, Irving: (1881–1957) American chemist and physicist; 1909–1950, career spent at General Electric; 1929, invited Karl-Friedrich Bonhoeffer to address annual meeting of the American Chemical Society; 1932 Nobel Prize in Chemistry.

Lasserre, Jean: (1908–83): French Reformed pastor and theologian; friend of Dietrich Bonhoeffer at Union Theological Seminary; influenced Dietrich on pacifism.

Laue, Max von: (1879–1960) German physicist; 1903, Ph.D. Berlin University; 1914 Nobel Prize in Chemistry for diffraction of X-rays in crystals; director Kaiser Wilhelm Society; 1919–1943, professor Berlin University; anti-Nazi; 1945, one of twelve scientists held at Farm Hall, UK, although his involvement in Uranium Club was not great; after war, went to Göttingen University and helped rebuild science in Germany; 1951–1958, director of Max Planck Institute for Physical Chemistry and Electrochemistry.

Lawrence, Ernest: (1901–55) American nuclear physicist; 1928, career started at California University where he built cyclotron with Stanley Livingston; 1939 Nobel Prize for development of cyclotron; during war, worked on Manhattan Project; founded Lawrence Berkeley National Laboratory and Lawrence Livermore National Laboratory.

Le Blanc, Max: (1865–1943) German chemist; head of physical chemistry at Leipzig University where he specialized in electrochemistry; after his retirement, Karl-Friedrich Bonhoeffer took his position.

Lehmann, Paul: (1906–94) American theologian; 1930–31, studied at Union Theological Seminary where he became friends with Dietrich Bonhoeffer; 1933, with his wife, visited the Bonhoeffers in Berlin; professor at Princeton University.

Leibholz, Gerhard: (1901–82) German-Jewish lawyer and law professor; known by name Gert; 1926, married Sabine Bonhoeffer; 1938, fled with family to England; 1947, returned to Germany and resumed legal and academic career.

Leiper, Henry Smith: (1891–1975) American missionary and advocate for world peace; associate general secretary World Council of Churches, and foreign secretary of the Federal Council of Churches based in New York City; 1939, offered a position to Dietrich Bonhoeffer.

Lenard, Philipp: (1862–1947) Hungarian-born German physicist; 1905 Nobel Prize in Physics for his work on cathode rays; a Nazi, he was a strong supporter of Hitler; promoted Deutsche physik movement, stating that all Jewish scientists and their ideas were fraudulent, first among them being Einstein and relativity; clashed with Heisenberg over quantum mechanics, which he considered to be Jewish science.

Levi, Friedrich: (1888–1966) German-Jewish mathematician, known for work on abstract algebra; served in First World War; 1935, lost position at Leipzig University; went to Calcutta University in India; 1952, returned to Germany.

Lewis, Gilbert: (1875–1946) American chemist; spent most of career at University of California, Berkeley, serving as dean; best known for work on covalent bonds and electron pairs; 1926, named the photon; 1933, worked on isotope separation and deuterium research.

Liebknecht, Karl: (1871–1919) German socialist and Marxist who cofounded Spartacus League with Rosa Luxemburg; both were killed in the 1919 uprising in Berlin.

Lind, Samuel: (1879–1965) American chemist; 1904–05, studied at Leipzig University under Ostwald and Bodenstein; spent most of career at University of Minnesota; known for research on radiation chemistry.

Livingston, Milton Stanley: (1905–86) American physicist; 1928, University of California where he worked with Ernest Lawrence on cyclotron; 1934, Cornell, worked with Hans Bethe; 1938, MIT; after war, while maintaining MIT affiliation, became chairman of the Brookhaven Accelerator Project; later served as associate director of the National Accelerator Laboratory at Batavia, Illinois.

London, Fritz: (1900–54) German-Jewish physicist, assistant to Erwin Schrödinger at Berlin University; an expert in chemical bonding and intermolecular forces; 1928, suggested to Karl-Friedrich Bonhoeffer that to facilitate transition, hydrogen should be at a low temperature; 1933, forced out of Berlin University; 1939, to U.S. after holding positions in England and France; eventually became professor at Duke University.

Lubbe, Marinus van der: (1909–34) Dutch communist who was executed for setting the Reichstag fire, although there was great controversy about his guilt; Karl Ludwig Bonhoeffer did psychological assessment and attended the trial.

Ludendorff, Erich: (1865–1937) German general who along with Hindenburg lead the German war effort during the First World War.

Luxemburg, Rosa: (1871–1919) German socialist and Marxist who founded Spartacus League with Karl Liebknecht; both were killed in the 1919 uprising in Berlin.

M

Mattauch, Josef: (1895–1976) German physicist; known for his work on mass spectrometry; succeeded Lise Meitner as head of the physics department at Kaiser Wilhelm Institute for Chemistry after she fled to Sweden; member of the Uranium Club although he was anti-Nazi; 1947, Max Planck Institute of Chemistry, Mainz; subsequently at Tübingen, Berne, and Mainz Universities.

Meinecke, Friedrich: (1862–1954) German intellectual and cultural historian who held anti-Semitic and anti-Slavic views and believed in German expansionism; was an attendee at the discussion group that included the Harnacks, Delbrücks, and Bonhoeffers.

Meitner, Lise: (1878–1968) Austrian-Jewish physicist; spent most of career at Kaiser Wilhelm Institute for Chemistry as director of physics working with Otto Hahn; professor of physics Berlin University; mentor to Max Delbrück; 1938, fled to Sweden; several months later, Hahn and Fritz Strassmann fissioned the atom, sending their puzzling results to Meitner who, with her nephew, Otto Frisch, realized what had occurred; after war, professor Nobel Institute of Physics, Stockholm Polytech.

Mendelssohn, Franz von: (1865–1935) German-Jewish banker and economist; treasurer and member of the governing senate of the Kaiser Wilhelm Society.

Menzel, Donald: (1901–76) American astrophysicist and theoretical astronomer; 1924, Ph.D. Princeton; 1926, University of California Lick Observatory; 1932, Harvard where he remained except for a period in World War II when he was in the Navy, code-breaking for military intelligence.

Minkowski, Hermann: (1864–1909) Russian-born German mathematician and physicist; 1885, Ph.D. Königsberg University; taught at Bonn, Königsberg, Zurich, and finally Göttingen; mathematically formalized Einstein's special theory of relativity.

Mitscherlich, Alexander: (1908–82) German neurologist and psychologist; 1941, Ph.D. Heidelberg University; during 1930s jailed several times for anti-Nazi stance; after war, observer at Nuremberg trials of Nazi doctors; used psychoanalysis to understand the causes for the rise of Hitler and Nazism.

Müller, Josef: (1898–1979) German lawyer; Catholic member of the resistance with connections to Vatican; reserve lieutenant in Military Intelligence Office in Munich; arrested; April 1945, liberated by Allies.

Murphy, George Moseley: (1903–68) American chemist; 1932–1936, assistant to Harold Urey at Columbia University, helping Urey build a spectrograph; co-discoverer of deuterium; 1936, Yale; during war, worked on Manhattan Project; 1946–68, New York University.

N

Nachmansohn, David: (1899–1983) Russian-born German-Jewish physical chemist and biochemist; 1920s, at the Kaiser Wilhelm Institute for Biology; attended Haber colloquia; 1933, left Germany, going to Sorbonne in Paris; 1939, to USA, teaching at Yale and Columbia.

Nernst, Walther: (1864–1941) German chemist; 1920 Nobel Prize in Chemistry; famous for work on thermodynamics; director of the Kaiser Wilhelm Institute for Physical Chemistry in Berlin; served as Karl-Friedrich Bonhoeffer's doctoral advisor at Berlin University; lost two sons in the First World War.

Neumann, John von: (1903–57) Hungarian-Jewish mathematician and physicist; 1926, after receiving Ph.D., ETH Zurich, worked with David Hilbert at Göttingen before going to Berlin University where his lectures became famous; briefly at Hamburg University; 1929, to Princeton University; 1933, joined the Institute for Advanced Study in Princeton; during war, part of the Manhattan Project.

Niebuhr, Reinhold: (1882–1984) American theologian and ethicist; professor at Union Theological Seminary; Dietrich attended his classes; 1939, when Dietrich was in imminent danger, he helped arrange employment for him in USA.

Niemöller, Martin: (1892–1984) German pastor; First World War naval officer and submarine captain; after war, studied theology, became pastor in Berlin-Dahlem; initially supported Hitler, believing he would not interfere with churches; 1933, established the Pastors' Emergency League to support pastors who were of Jewish descent; 1937, arrested; 1938–45, imprisoned.

Noddack, Walter: (1893–1960) German chemist; supervised Karl-Friedrich Bonhoeffer's thesis work; 1926, with his wife Ida Tacke, discovered element 75 rhenium and element 43 (masurium/technetium) about which there was much controversy; during war, the Noddacks worked at the Nazified University of Strausbourg.

Nonne, Max: (1861–1959) German neurologist; spent most of career in Hamburg; early research was on syphilis; advocate for use of hypnosis on soldiers during the First World War as a cure for male hysteria.

Nordheim, Gertrud (née Poschl): (n.d.) German-Jewish physicist; worked with Lothar Nordheim at Göttingen without pay; 1934, came to USA where Lothar took a temporary position at Purdue University; 1935, married Lothar; 1937, to Duke University where she worked with Lothar and also with Hertha Sponer, a German physicist and chemist.

Nordheim, Lothar: (1899–1985) German-Jewish physicist; 1923, Ph.D. Göttingen University; research on field electron emission and quantum theory; 1934, forced to leave Germany, went to Purdue University where he and Gertrud worked on cosmic rays; 1935, married Gertrud; 1937, to Duke; research on electron band theory; took part in Manhattan Project; 1956, Johns Hopkins Laboratory in San Diego; joined General Dynamics.

O

Oliphant, Marcus: (1901–2000) Australian physicist; 1922, graduated Adelaide University; 1927, to Ernest Rutherford's Cavendish laboratory in England; 1934, discovered tritium with Rutherford and Harteck; 1937, professor Birmingham University, researching radar; part of MAUD; after war, to Australian National University; 1970s, became involved in politics.

Oppenheim, Hermann: (1858–1919) German-Jewish neurologist; colleague of Karl Ludwig Bonhoeffer; conducted research on shell shock and what he termed traumatic neurosis.

Oppenheimer, J. Robert: (1904–67) American theoretical physicist; undergraduate degree Harvard; 1924, to Cambridge; 1926, to Göttingen University where he met Heisenberg, Teller, and Franck; 1927, Ph.D. under Max Born; professor at University of California, Berkeley, and Caltech; during war, head of Los Alamos Laboratory, Manhattan Project; 1947, Institute for Advanced Study in Princeton.

Oster, Hans Paul: (1887–1945) German military officer; deputy head of Abwehr; initially in support of Hitler but became disillusioned, joining resistance; hanged with Canaris and Dietrich Bonhoeffer at Flossenbürg concentration camp.

Ostwald, Carl Wilhelm Wolfgang: (1883–1943) German chemist; son of Friedrich Wilhelm Ostwald; Karl-Friedrich Bonhoeffer's colleague at Leipzig University.

Ostwald, Friedrich Wilhelm: (1853–1932) physical chemist; helped found field of physical chemistry; chair of physical chemistry Leipzig University making it the leading center in Germany; 1909 Nobel Prize in Chemistry.

P

Pauli, Wolfgang Ernst: (1900–58) Austrian-born physicist and mathematician; 1923–28, Hamburg University where he worked on quantum mechanics; 1928, ETH Zurich; 1938, after Hitler took over Austria, attempted to get Swiss citizenship which was denied; 1940, to Institute for Advanced Study in Princeton; 1945 Nobel Prize in Physics, nominated by Einstein; 1946, returned to Zurich.

Peierls, Rudolf: (1907–95) Austrian-born, Jewish, British physicist; worked with Otto Stern at Hamburg University; 1933, at Cambridge on a fellowship; did not return to Germany; worked on nuclear physics; 1940, with Otto Frisch, calculated theoretical mechanics for detonation of atomic bomb using uranium-235; during war, played major role in British nuclear effort, taking part in Manhattan Project.

Perrin, Francis: (1901–92) French physicist; son of Nobel Prize laureate Jean Perrin; during the 1930s worked on fission of uranium and, with Joliot-Curies, on chain reactions and nuclear energy.

Perrin, Jean: (1870–1942) French physicist; father of Francis; 1926 Nobel Prize in Chemistry.

Perrin, Michael: (1905–88) British chemist; born in British Columbia; Ph.D. Oxford; 1930s, worked for Imperial Chemical Industries; during war, member of MAUD committee; coordinated British role in Manhattan Project; after war, continued to work on nuclear issues.

Phillip, Lee Cato: (n.d.) American minister; student at Howard University whom Dietrich met in 1930 on a trip to Washington D.C.; attended Union Theological Seminary; dean of the chapel at Prairie View A & M University in Prairie View, Texas, for more than forty years.

Planck, Max: (1858–1947) German theoretical physicist; his discovery of energy quanta led to a new understanding of physics; 1918 Nobel Prize in Physics; son Karl was killed at Verdun in the First World War; president of Kaiser Wilhelm Gesellschraft; influenced generations of scientists; 1945, son Erwin was executed for role in resistance; after Planck's death in 1947, the Kaiser Wilhelm Gesellschraft was renamed the Max Planck Society for the Advancement of Science.

Polanyi, Michael: (1891–1976) Hungarian-Jewish physical chemist, philosopher, economist; colleague of Karl-Friedrich Bonhoeffer; 1919, Ph.D., Budapest University; emigrated to Germany and joined Haber Institute; 1923, converted to Christianity; 1926, appointed director at Haber Institute; 1933, to Manchester University in England; eventually changed fields to philosophy of science and social sciences; ended career at Oxford.

Pollnow, Hans: (1902–43) German-Jewish neurologist and psychiatrist; 1929, after completing medical studies, became colleague of Karl Ludwig Bonhoeffer at the Charité Clinic in Berlin; 1933, lost position and went to France; 1943, arrested and sent to Mauthausen concentration camp in Austria where he was executed.

Pompe, Johannes: (1901–45) Dutch pathologist, known for research into glycogen storage disease; during World War II sent messages to British via a transmitter hidden in his laboratory; executed by Nazis.

Powell, Adam Clayton, Sr.: (1865–1953) American minister, leader of Abyssinian Baptist Church in Harlem, NYC; Dietrich Bonhoeffer attended in 1930–31, actively participating in the life of the church, which transformed his ideas about race, prejudice, and social justice.

Q

Quadfasel, Friedrich: (1902–81) German neurologist and psychiatrist; 1914, Ph.D. under Karl Ludwig Bonhoeffer at Berlin University; served as his assistant; 1933, opposed to Nazism and was arrested; came to USA with Bonhoeffer's help, worked at the Hospital of the Veterans Administration in Boston.

R

Rabi, Isidor Isaac: (1898–1988) American physicist, born in Austria-Hungry, raised in USA; Ph.D., Columbia University; 1929, went to Munich University to study under Sommerfeld; returned to Columbia where he spent most of his career; 1944 Nobel Prize in Physics for discovery of nuclear magnetic resonance; during war, part of Manhattan Project; 1950s, chairman of the General Advisory Committee, Atomic Energy Commission.

Rabinowitch, Eugene: (1901–73) Russian-Jewish biophysicist; student of James Franck at Göttingen University, about whom Franck wrote to Karl-Friedrich Bonhoeffer in 1932 imploring him to find him a position; came to USA; during war, worked on Manhattan Project at the Metallurgical Laboratory, University of Chicago; in June 1945, co-authored with Franck the Franck Report which stated that the atomic bomb should not be dropped on Japan but instead it could be used as a publicized demonstration on a desert island, the demonstration alone being enough to force Japan to surrender.

Rathenau, Walther: (1867–1922) German-Jewish statesman, businessman, foreign minister, Weimar Republic; signed treaty opening trade with USSR; 1922, assassinated in Berlin.

Riehl, Nikolaus: (1901–90) German industrial physicist: peripheral member of the Uranium Club; employed by Auer Company; specialist in luminescence; obtained uranium isotopes; at end of war, went to USSR where he worked on atomic bomb; 1957–69, professor Munich Polytech.

Ritter, Gerhard: (1888–1967) German historian; 1925–56, professor Freiburg University; monarchist who opposed Weimar Republic and supported authoritarian rule but who also participated in resistance; 1925, wrote biography of Martin Luther; 1936, wrote biography of Frederich the Great; became an advisor to Carl Goerdeler; arrested November 1944; after war worked to restore German nationalism; 1954, wrote biography of Goerdeler.

Rosbaud, Paul: (1896–1963) Austrian metallurgist; Ph.D. Berlin-Charlottenburg Technische Hochschule; 1930s, scientific editor for *Naturwissenschaften*, a position that gave him extensive knowledge of scientific developments; spy known by the codename the Griffin, passed information to British about atomic research, rocketry, and existence of concentration camps; after war, started Pergamon Press with Robert Maxwell, but a disagreement led to his resignation.

Rust, Bernhard: (1883–1945) German secondary school teacher and school administrator; 1922, joined NSDAP; 1934, Nazi Minister of Science, Education and National Culture; led the purge of Jews from universities; supported Deutsch physik and demanded that all science produced by Jews be rejected; psychologically unstable, he made numerous deleterious changes to the German educational system; May 1945, committed suicide.

Rutherford, Ernest: (1871–1937) New Zealand-born British physicist; known as the father of nuclear physics; famous for his experimental acumen; 1908 Nobel Prize in Chemistry for his research on radioactive half-life; 1919, director of Cavendish Laboratory at Cambridge where he nurtured the careers of many young scientists including Paul Harteck.

S

Saethre, Haakon: (1891–1945) a Norwegian neurologist and psychiatrist who hid Jews and was murdered by the Nazis at the end of the war.

Schacht, Hjalmar: (1877–1970) German economist and banker; 1918, co-founder of German Democratic Party; wrote *The End of Reparations*; during 1930s, president of Reichsbank and minister of economics; initially supported Hitler but was not a Nazi; 1939, dismissed as Reichsbank president; after attempt on Hitler's life in July 1944, arrested but not executed.

Schleicher, Rüdiger: (1895–1945) German lawyer and legal scholar on aviation law; 1923, married Ursula Bonhoeffer; during the 1930s, head of legal department of Reich Transportation Ministry; 1939, head of Institute of Aviation Law, Berlin University; part of resistance; executed April 24, 1945.

Schrödinger, Erwin: (1887–1961) Austrian physicist who made fundamental discoveries in quantum mechanics; 1927–34, professor Berlin University, replacing Planck who had retired; 1933, shared Nobel Prize in Physics with Paul Dirac; 1934, left Germany because he hated Nazism; worked many places including Austria where in 1938 he wrote a letter in support of the Nazi takeover; 1940 took position at the School of Theoretical Physics in Dublin, Ireland, where he remained until 1957.

Seeberg, Erich: (1888–1945) German theologian; son of German theologian Reinhold Seeberg; 1927, professor Berlin University; Dietrich Bonhoeffer wrote to him on behalf of Paul Tillich.

Seidemann, Herta: (1900–84) German-Jewish neurologist and psychiatrist; 1933, worked on habilitation under Karl Ludwig Bonhoeffer at Berlin University but could not complete it after enforcement of Aryan clause; 1938, to USA where, with Karl's recommendation, got position at Montefiore Hospital in NYC.

Simon, Franz (Francis): (1893–1956) German-Jewish physical chemist and physicist; awarded Iron Cross in First World War; Ph.D. Berlin University under Walther Nernst; 1931, professor Breslau University; 1933, forced to leave Germany, going to Oxford, at which point he changed his name from Franz to Francis; worked on low-temperature physics; 1940, member of MAUD; participant in Manhattan Project at Los Alamos.

Soddy, Frederick: (1877–1956) British chemist; 1900, worked with Rutherford at McGill University; 1903, University College London; 1904, Glasgow University; 1919, Oxford; 1921 Nobel Prize in Chemistry for radioactive decay and theory of isotopes.

Sommerfeld, Arnold: (1868–1951) German physicist; 1906–1938, professor Munich University; director theoretical physics; mentored numerous students who became famous, including Heisenberg; on retirement, a fractious battle took place over appointing his successor, with Heisenberg losing out to Wilhelm Müller.

Spielmeyer, Walther: (1879–1935) German neuropathologist; 1928, director at Kaiser Wilhelm Institute in Berlin; after enforcement of Aryan clause, helped Jewish colleagues find jobs; 1935, died of tuberculosis.

Stark, Johannes: (1874–1957) German physicist; 1919 Nobel Prize in Physics; early supporter of Nazism; 1933–39, president of Physikalisch-Technische Reichsanstalt; with Philipp Lenard, strong proponent of Deutsche physik; called Heisenberg a "white Jew" for his defense of relativity; after 1945, stood before de-Nazification tribunal and received four-year sentence, subsequently reduced to a fine.

Stauffenberg, Claus von: (1907–44) Germany military officer; 1930s, supported Hitler but not a Nazi; 1939, his regiment participated in invasion of Poland; 1943, promoted and sent to North Africa where he was severely wounded; while recuperating, joined resistance; 1944, one of the leaders of the plot to kill Hitler; executed.

Stern, Fritz: (1926–2016) German-born American historian of German history and Jewish history; godson of the Nobel Prize winner Fritz Haber; because the family was Jewish, the Sterns left Germany in 1938, coming to USA.

Stern, Otto: (1888–1969) German-Jewish physicist and physical chemist; 1923, director of the Institute of Physical Chemistry, Hamburg University; 1933, lost position; to USA, Carnegie Institute of Technology; his chair at Hamburg was assumed by Paul Harteck; 1943 Nobel Prize in Physics; professor University of California, Berkeley.

Stetter, Georg: (1895–1988) Austrian-German physicist; Ph.D. Vienna University; member of the Uranium Club, measuring atomic constants, cross sections, and transuranic elements; member of the Nazi party; 1953, Vienna University.

Stoecker, Adolf: (1835–1909) German pastor, theologian and court chaplain to Kaiser Wilhelm; pushed for the imposition of Jewish quotas; Harnack and Delbrück opposed the action.

Strassmann, Fritz: (1902–80) German chemist; worked with Lise Meitner and Otto Hahn at the Kaiser Wilhelm Institute for Chemistry in Berlin; 1939, fissioned the atom with Hahn; anti-Nazi, he remained at Kaiser Wilhelm Institute through the war; 1946, Mainz University; 1948, director of the Max Planck Institute for Chemistry.

Straus, Erwin: (1891–1975) German-Jewish neurologist; colleague of Karl Ludwig Bonhoeffer; lost position at the Charité Clinic; came to USA where he taught at Black Mountain College, then practiced in Lexington, Kentucky.

Suess, Hans: (1909–93) Austrian-born German physical chemist; member of the Uranium Club, worked on deuterium; colleague of Paul Harteck in Hamburg; 1950, emigrated to USA.

Sutz, Erwin: (1906–87) Swiss pastor; 1930–1931, attended Union Theological Seminary where he became friends with Dietrich Bonhoeffer.

Svedberg, Theodor: (1884–1971) Swedish physical chemist; Uppsala University, Sweden; member of the Nobel Committee; 1926 Nobel Prize in Chemistry; 1930, nominated Karl-Friedrich for Nobel Prize in Chemistry.

Szilard, Leo: (1898–1964) Hungarian-Jewish physicist; served in First World War; enrolled in Institute of Technology but changed to Berlin University; 1933, to England; 1938, to USA; aware of the consequences of atomic fission, he drafted letter, signed by Einstein, to President Roosevelt; participant through Chicago University in Manhattan Project; after war, researched biochemistry; final years spent as fellow at Salk Institute of Biological Research.

T

Tacke Ida: (1896–1978) German chemist who, with Walter Noddack, discovered Element 75 rhenium and the controversial Element 43; 1926, married Noddack; during war, worked at the Nazified University of Strausbourg.

Taylor, Hugh Stott: (1890–1974) British chemist with specialty in kinetics; 1914–58, spent career at Princeton University where he built chemistry department and oversaw construction of Frick Chemical Laboratory; 1929, Karl-Friedrich Bonhoeffer attended Frick dedication at Taylor's invitation.

Teller, Edward: (1908–2003) Hungarian-Jewish theoretical physicist; studied with Max Born; received Ph.D. under Heisenberg at Leipzig University; 1930, to Göttingen; 1935, to USA. George Washington University; participated in Manhattan Project; after war, involved with hydrogen bomb, advocate for nuclear power and weapons; 1958, director of the Lawrence Livermore National Laboratory and professor University of California.

Thiessen, Peter Adolf: (1899–1990) German physical chemist; 1933, department head Kaiser Wilhelm Institute for Physical Chemistry in Berlin; 1935, professor Munster University; 1935–45, professor Berlin University and director of Kaiser Wilhelm Institute; 1945, willingly went to USSR; received Stalin Prize; 1956, head of the Physical Chemistry Institute of the East German Academy of Sciences.

Tillich, Paul: (1886–1965) German theologian; 1933, lost professorship at Frankfurt University for political reasons; 1935, came to USA with assistance of Reinhold Niebuhr; taught at Union Theological Seminary, and then at Harvard.

Tinel, Jules: (1879–1952) French neurologist who was a participant in French resistance, helping to hide Allied airmen; imprisoned and released; son Jacques, also a resistance fighter, died in a concentration camp.

Tresckow, Henning von: (1901–44) German military officer; initially supported Hitler, but turned against him; 1939, actively involved in resistance within military; 1943, planted bomb to kill Hitler that failed to detonate; following failed 1944 attempt, committed suicide.

Troeltsch, Ernst: (1865–1923) German theologian and historian; politically liberal with a focus on connection between religion and society; taught at Göttingen, Bonn, Heidelberg, and Berlin Universities; 1915, in Berlin, became one of the circle of friends around the Harnacks, Delbrücks, and Bonhoeffers.

Tronstad, Leif: (1903–45) Norwegian chemist and metallurgist, military officer in Norwegian Army; 1920s, studied at Haber Institute under Herbert Freundlich; 1931, Ph.D. Stockholm, spent year at Cambridge; 1936, professor at Norwegian Institute of Technology; worked as consultant for Norsk Hydro; after German invasion, part of Norwegian resistance; 1940, escaped to Sweden and then to England where he helped plan raid on Vemork; 1944, returned to Norway; 1945, killed in action.

U

Urey, Harold: (1893–1981) physical chemist; 1923, Ph.D. University of California; worked with Niels Bohr in Copenhagen; professor at Columbia University; 1931, began work on deuterium; 1934 Nobel Prize in

Chemistry for discovery of deuterium; during war, worked on Manhattan Project; after war, attempted to help Karl-Friedrich Bonhoeffer who had suffered a heart attack; professor University of Chicago.

V

Visser 't Hooft, Willem Adolf: (1900–85) Dutch Reformed theologian: general secretary of the World Council of Churches; Geneva contact for Dietrich Bonhoeffer's resistance efforts.

Vogt, Cécile: (1875–1962) French neurologist, surname also spelled Vogt-Mugnier; married to Oskar Vogt; 1914, the couple conducted research on brain anatomy and function; 1919, department head at Kaiser Wilhelm Institute for Brain Research in Berlin, where Oskar was director; 1936, both lost positions because they were anti-Nazi; opened a small private clinic; hid Jews during World War II.

Vogt, Oskar: (1870–1959) German neurologist, married to Cécile Vogt; 1914, the couple conducted research on brain anatomy and function at the Kaiser Wilhelm Institute for Brain Research in Berlin where he was director; 1924, investigated Lenin's brain after his death; 1936, couple lost their positions because they were anti-Nazi; opened a small private clinic; hid Jews during World War II.

Volmer, Max: (1885–1965) German physical physicist; 1910, Ph.D. Leipzig University; 1916, Berlin University; 1920, Hamburg University; 1922, professor and director of Physical Chemistry and Electrochemistry Institute, Technische Hochschule Berlin; 1945, to USSR, worked on deuterium; 1955, returned to East Germany, professor Humboldt University, president of East German Academy of Sciences.

W

Wach, Joachim: (1898–1955) German-Jewish Lutheran theologian and historian of religion; served in First World War; 1935, lost position at Leipzig University; came to USA, Brown University; 1945–55, University of Chicago Divinity School.

Waerden, Bartel L. van der: (1903–96) Dutch mathematician; studied at Amsterdam and Göttingen Universities; 1931, professor Leipzig University; colleague of Karl-Friedrich Bonhoeffer; known for work on abstract algebra.

Wedemeyer, Maria von: (1924–77) German, daughter of Hans and Ruth von Wedemeyer; 1943, became engaged to Dietrich Bonhoeffer; 1945, searched for Dietrich in the closing days of the war; after the war, attended Göttingen University and then came to USA, attending Bryn Mawr with the assistance of Herbert Jehle; employed by Rand Corp.

Weigert, Fritz: (1876–1947) German-Jewish physical chemist; colleague of Karl-Friedrich Bonhoeffer at Leipzig University where he was associate professor of physiology for photochemistry and scientific photography; 1935, driven out of Leipzig; Bonhoeffer, Heisenberg, Hund, and van der Waerden defended him without success; emigrated to England.

Weizsäcker, Carl Friedrich von: (1912–2007) German physicist; son of Ernst; 1933, Ph.D. under Heisenberg at Leipzig; 1934, Heisenberg's assistant; 1930s, studied nuclear processes in stars; 1936, Kaiser Wilhelm Institute for Physics; 1939, member of the Uranium Club; researched plutonium; his book, *The World View of Physics,* was sent to Dietrich by Karl-Friedrich Bonhoeffer; 1942–44, professor at Strausbourg University; after war, turned to philosophy and peace causes but remained active in physics; 1946–57, professor Göttingen and department head Max Planck Institute of Physics; 1957–69, professor of philosophy, Hamburg University; 1970–80, head of Max Planck Institute for Research into Living Conditions in the Modern World.

Weizsäcker, Ernst von: (1882–1951) German diplomat; father of Carl Friedrich and Richard; 1920, joined German Foreign Service; Nazi Party member; 1938–43, State Secretary of the Foreign Office; 1943–45, ambassador to Vatican; after war, claimed he was part of resistance; found guilty at Nuremberg tribunal.

Wentzel, Gregor: (1898–1978) German physicist; studied under Arnold Sommerfeld in Munich; 1926, Leipzig University; 1928, Zurich University; friend of Karl-Friedrich Bonhoeffer; 1948, to University of Chicago; known for work on quantum mechanics and quantum electrodynamics.

Wernicke, Carl: (1848–1905) German neurologist and psychiatrist; predecessor of Karl Ludwig Bonhoeffer as director of the Breslau Psychiatric Clinic. Karl became his assistant in 1893.

Wigner, Eugene: (1902–95) Hungarian-Jewish theoretical physicist; Ph.D. Berlin University under Polanyi; worked for a while at the Kaiser

Wilhelm Institute and at Göttingen as assistant to David Hilbert; 1930, to Princeton with Von Neumann; remained in USA; during war, participant in Manhattan Project; 1963 Nobel Prize in Physics for contributions to the theory of atomic nucleus and elementary particles, shared with Maria Goeppert-Mayer and J. Hans D. Jensen.

Willstätter, Richard: (1872–1942) German-Jewish organic chemist; 1912, head of Kaiser Wilhelm Institute for Chemistry; 1915 Nobel Prize in Chemistry for research on plant pigments and chlorophyll; when asked by Haber during war to take part in poison gas research, he refused but offered to develop a gas mask; in 1924, resigned from position at Munich as a statement against anti-Semitism, refusing to enter his lab, conducting research by telephone with his assistants; 1939, fled to Switzerland.

Wirtz, Karl: (1910–94) German nuclear physicist and chemist; late 1930s, assistant to Karl-Friedrich at Leipzig University; 1937, on staff at Kaiser Wilhelm Institute for Physics; member of the Uranium Club working on reactor; 1945, interned at Farm Hall in England; 1946, Göttingen; in later life, active in issues related to nuclear power including nonproliferation.

Z

Zinn, Elizabeth: (1908–95) German theologian; theology student with Dietrich Bonhoeffer at Berlin University; 1932, Ph.D. Berlin; for a while, she (a third cousin) and Dietrich were weakly attracted to each other but nothing came of it; 1933, vicar and member of the Confessing Church; 1938, married Günther Bornkamm, a New Testament scholar.

Zutt, Jürg: (1893–1980) German neurologist and psychiatrist; colleague of Karl Ludwig Bonhoeffer; helped conduct psychological assessment of Marinus van der Lubbe, Dutch communist who was executed for setting the Reichstag fire.

Selected Bibliography

Badish, Lawrence, Elizabeth Hodes, and Adolph Tiddens. "Fission: Reaction to the Discovery in 1939." *Proceedings of the American Philosophical Society* 130, no. 2 (June, 1986): 196–231. http://www.jstor.org/stable/987181.

Barnett, Victoria. *For the Soul of the People: Protestant Protest Against Hitler*. Oxford: Oxford University Press, 1998.

Bartosch, Ulrich, ed. *Carl Friedrich von Weizsäcker: Pioneer of Physics, Philosophy, Religion, Politics and Peace Research*. Netherlands: Springer, 2015.

Begun, G.M., ed. *Isotope Separation and Isotope Exchange: A Bibliography with Abstracts*. Oakridge National Laboratory: U.S. Government Report (October 14, 1959). ornl.gov/info/reports/1959/3445603510699.pdf.

Bethge, Eberhard. *Dietrich Bonhoeffer: A Biography*. Minneapolis: Fortress Press, 2000.

Bethge, Eberhard, and Renate Bethge, eds. *Last Letters of Resistance: Farewells from the Bonhoeffer Family*. Translated by Dennis Slabaugh. Philadelphia: Fortress Press, 1984.

Bethge, Eberhard, Renate Bethge, and Christian Gremmels. *Dietrich Bonhoeffer: A Life in Pictures*. Philadelphia: Fortress Press, 1986.

Beyerchen, Alan D. *Scientists Under Hitler: Politics and the Physics Community in the Third Reich*. New Haven: Yale University Press, 1977.

Bonhoeffer, Dietrich. *Barcelona, Berlin, New York: 1928–1931*. Dietrich Bonhoeffer Works, Volume 10. Translated and edited by Reinhart Staats et al. English edition edited by Clifford J. Green. Minneapolis: Fortress Press, 2008.

K. L. Housley, *The Scientific World of Karl-Friedrich Bonhoeffer*, Palgrave Studies in the History of Science and Technology, https://doi.org/10.1007/978-3-319-95801-9

———. *Berlin: 1932–1933*. Dietrich Bonhoeffer Works, Volume 11. Translated by Carsten Nicolaisen and Ernst-Albert Scharffennorth. English edition edited by Larry L. Rasmussen. Minneapolis: Fortress Press, 2009.

———. *Conspiracy and Imprisonment, 1940–1945*. Dietrich Bonhoeffer Works, Volume 16. Translated by Lisa E. Dahill. English edition edited by JØrgen GlenthØj† et al. Minneapolis: Fortress Press, 2006.

———. *Discipleship*. Dietrich Bonhoeffer Works, Volume 4. Translated and edited by Martin Kuske and Ilse Tödt. Minneapolis: Fortress Press, 2001.

———. *Ecumenical, Academic and Pastoral Work, 1931–1932*. Dietrich Bonhoeffer Works, Volume 11. Translated and edited by Eberhard Amelung and Christoph Strohm. English edition edited by Victoria J. Barnett et al. Translated by Anne-Schmidt et al. Minneapolis: Fortress Press, 2012.

———. *Ethics*. Dietrich Bonhoeffer Works, Volume 6. Translated and edited by Ilse Tödt et al. English edition edited by Clifford J. Green. Minneapolis: Fortress Press, 2009.

———. *Fiction from Tegel Prison*. Dietrich Bonhoeffer Works, Volume 7. Translated and edited by Renate Bethge and Ilse Tödt. Minneapolis: Fortress Press, 2000.

———. *Indexes and Supplementary Materials*. Dietrich Bonhoeffer Works, Volume 17. Edited by Victoria J. Barnett and Barbara Wojhoski. Minneapolis: Fortress Press, 2014.

———. *Letters and Papers from Prison*. Dietrich Bonhoeffer Works, Volume 8. Translated and edited by Christian Gremmels et al. English edition edited by John W. De Gruchy. Minneapolis: Fortress Press, 2010.

———. *Letters and Papers from Prison*. Edited by Eberhard Bethge. Translated by Reginald Fuller, Frank Clark, and John Bowden. New York: Touchstone, Simon & Schuster, 1997.

———. *London: 1933–1935*. Dietrich Bonhoeffer Works, Volume 13. Translated and edited by Hans Goedeking, Martin Heimbucher, and Hans-Walter Schleicher. English edition edited by Keith Clements. Minneapolis: Fortress Press, 2007.

———. *Theological Education at Finkenwalde: 1935–1937*. Dietrich Bonhoeffer Works, Volume 14. Translated and edited by Otto Dudzus and Jürgen Henkys et al. English edition edited by H. Gaylon Barker and Marc S. Brocker. Translated by Douglas W. Stott. Minneapolis: Fortress Press, 2013.

———. *Theological Education Underground: 1937–1940*. Dietrich Bonhoeffer Works, Volume 15. Translated and edited by Dirk Schulz. English edition edited by Victoria J. Barnett. Translated by Victoria J. Barnett et al. Minneapolis: Fortress Press, 2012.

———. *The Young Bonhoeffer: 1918–1927*. Dietrich Bonhoeffer Works, Volume 9. Translated and edited by Hans Pfeifer et al. English edition edited by Paul Duane Matheny et al. Minneapolis: Fortress Press, 2003.

Bonhoeffer, Karl-Friedrich. Papers. Max Planck Gesellschraft, Berlin.

Born, Max. *My Life & My Views*. New York: Charles Scribner's Sons, 1968.

———. *The Born-Einstein Letters*. Translated by Irene Born. New York: Walker and Company, 1971.

Bretislav, Friedrich. "How Did the Tree of Knowledge Get Its Blossom? The Rise of Physical and Theoretical Chemistry, with an Eye on Berlin and Leipzig." *Angewandte Chemie*, 55, no. 18 (April 25, 2016). https://www.ncbi.nim.nih.gov/pmc/articles/PMC5071687/.

Brysac, Shareen Blair. *Resisting Hitler: Mildred Harnack and the Red Orchestra*. New York: Oxford University Press, 2000.

Cassidy, David C. "Heisenberg, German Science and the Third Reich." *Social Research* 59, no. 3 (Fall 1992).

———. *Beyond Uncertainty: Heisenberg, Quantum Physics, and the Bomb*. New York: Bellevue Literary Press, 2009.

Chayut, Michael. "From Berlin to Jerusalem: Ladislaus Farkas and the Founding of Physical Chemistry in Israel." *Historical Studies in the Physical and Biological Sciences* 24, no. 2 (1994): 237–263. University of California Press. https://doi.org/10.2307/2775/57724.

Chemist Who Works in Library Wins Coveted Award. *The Science News-Letter* 23, no. 613 (January 7, 1933): 5–6. Published by Society for Science & the Public. https://doi.org/10.2307/3909685. Stable URL: http://www.jstor.org/stable/3909685.

Clark, Ronald. W. *Einstein: The Life and Times*. New York: World Publishing Company, 1971.

Cocks, Geoffrey. "Repressing, Remembering, Working Through: German Psychiatry, Psychotherapy, Psychoanalysis, and the 'Missed Resistance' in the Third Reich." *The Journal of Modern History* 64. Supplement: Resistance Against the Third Reich (December 1992): S204–S216. http://www.jstor.org/stable/2124978.

Cohen, Robert S., and Marx W. Wartofsky, eds. *Proceedings of the Boston Colloquium for the Philosophy of Science 1966–1968*. Dordrecht, Holland: D. Reidel Publishing, 1969.

Colquhoun, David. "Obituary, Bernard Katz, 1911–2003." *Psychology News* 52 (Autumn 2003): 1–4. www-ucl-sib.ucl.ac.uk/pharmacology.

Cornwell, John. *Hitler's Scientists: Science, War and the Devil's Pact*. New York: Viking, 2003.

Dahl, Per F. *Heavy Water and the Wartime Race for Nuclear Energy*. Bristol and Philadelphia: Institute of Physics Publishing, 1999.

Davis, Belinda, J. *Home Fires Burning: Food, Politics, and Everyday Life in World War I Berlin*. Chapel Hill: The University of North Carolina Press, 2000.

De Gruchy, John W., ed. *Bonhoeffer for a New Day: Theology in a Time of Transition*. Grand Rapids: Eerdmans, 1997.

De Kruif, Paul. *Microbe Hunters*. San Diego, New York and London: Harcourt Brace, Jovanovich, 1926, 1954.

Delbrück, Max. Interview by Carolyn Harding, *Archives*, California Institute of Technology (July 14–September 11, 1978). www.oralhistories.library.caltech.edu./16/1/OH-Delbruck_M.pdf.

Drucker, Peter F. *Adventures of a Bystander*. New Brunswick: Transaction Publishers, 1997.

Eigen, Manfred. "Bonhoeffer, Heisenberg's Nobel Prize. Studying Reactions." *Web of Stories*. www.webofstories.com/play/manfred.eigen/25.

Einstein, Albert. Einstein Archives Online. http://alberteinstein.info/vufind1.

———. *Ideas and Opinions*. Translated by Sonja Bergman. New York: Three Rivers Press, 1954.

Engstrom, Eric J. *Clinical Psychiatry in Imperial Germany: A History of Psychiatric Practice*. Ithaca: Cornell University Press, 2003.

Farkas, Adalbert. "Paul Harteck: the Triumphant Decade 1925–1934." *Ambix* 36, part 2 (July 1989).

Fermi, Enrico. Papers. University of Chicago Library, Special Collections and Research Center.

Friedman, Matthew J., Terence M. Keene, and Patricia A. Resick, eds. *Handbook of PTSD*. New York and London: Guilford Press, 2010.

Friedrich, Thomas. *Berlin Between the Wars*. New York: Vendome Press, 1991.

Frisch, Otto R. *What Little I Remember*. Cambridge: Cambridge University Press, 1979.

Frisch-Peierls Memorandum. http://en.wikipedia.org/wiki/Frisch-Peierls_memorandum.

Fritzsche, Peter. *An Iron Wind: Europe Under Hitler*. New York: Basic Books, 2016.

Gearhart, C.A. "'Astonishing Successes' and 'Bitter Disappointment': The Specific Heat of Hydrogen in Quantum Theory." *Archives for History of Exact Sciences* 641 no. 26 (March 2010): 113–202. http://www.jstor.org/stable/4134.

Gerrens, Uwe. *Medizinisches Ethos und theologische Ethik: Karl und Dietrich Bonhoeffer in der Auseinandersetzung um Zwangssterilisation und "Euthanasie" im Nationalsozialismus*. Munich: R. Oldenbourg, 1996.

Glimm, James G., John Impaglizzo, and Isadore Singer, eds. *The Legacy of John von Neumann*. Providence: American Mathematics Society, 2006.

Goudsmit, Samuel A. *Alsos*. Woodbury, NY: AIP Press, 1996.

———. Papers. Series IV Alsos Mission. Niels Bohr Library & Archives with the Center for History and Physics, American Institute of Physics. www.aip.org/history/nbl/collections/goudsmit.

Greenwood, Norman N. "Harry Julius Emeleus, C.B.E." *Biological Memoirs of Fellows of the Royal Society* 42 (November 1996): 124–150. http://www.jstor.org.stable/770201.

Hargittai, István. *The Martians of Science: Five Physicists Who Changed the Twentieth Century*. Oxford: Oxford University Press, 2006.

Harteck, Paul. *In Memoriam: Karl Friedrich Bonhoeffer*. www.ac.eis.com/0095852258900035/1-s2,00958-52258900085-main.pdf.

Harteck, Paul. Papers. MC 17. Institute Archives and Special Collections, Folsom Library. Rensselaer Polytechnic Institute.

Hayes, William. "Max Ludwig Henning Delbruck." 4 September 1906–10 March 1981. *Biographical Memoirs of Fellows of the Royal Society* 28 (November, 1982): 58–90. http://www.jstor.org/stable/769892.

Heilbron, J.L. *The Dilemmas of an Upright Man: Max Planck as Spokesman for German Science*. Berkeley: University of California Press, 1986.

Heller, Adam, Dieter Kolb, and Krishman Rajeshwar. "The Life and Work of Heinz Gerischer." *ECS Classics*. Electrochemical Society (2003): 1.

Helm, Susanne, Carola Sachise, Mark Volker. *The Kaiser Wilhelm Society Under National Socialism*. Cambridge: Cambridge University Press, 2009.

Hentschel. Klaus. *Physics and National Socialism: An Anthology of Primary Sources*. Translated by Ann M. Hentschel. Berlin: Springer, 1996.

———. *The Mental Aftermath*. Oxford: Oxford University Press, 2007.

Herrn, Rainer. "Herta Seidemann," "Edith Jacobson." Institute of the History of Medicine of the Charite. https://gedenkort.charite.de/en/people.

Hippel, von Frank. "James Franck: Science and Conscience." *Physics Today*. American Institute of Physics (June 2010): 41–46.

Hirsch-Heisenberg, Maria, ed. *My Dear Li: Werner Heisenberg, Elisabeth Heisenberg Correspondence 1937–1946*. Translated by Irene Heisenberg. New Haven: Yale University Press, 2016.

Hoffmann, Dieter, Mark Walker, eds. *The German Physical Society in the Third Reich: Physicists Between Autonomy and Accommodation*. Translated by Ann M. Hentschel. Cambridge: Cambridge University Press, 2012.

Holder, Rodney D. "Science and Religion in the Theology of Dietrich Bonhoeffer." *Zygon* 44, no. 1 (March 2009): 115–132.

Horbal, Bogdan. "World's Fair: Enter the World of Tomorrow, A Monument for a King and a Nation." http://exhibition.nypl.org/biblion/worldsfair/moment-time/brink-war/essay-horbal-polish-pavillion.

Housley, Kathleen L. "Daring to Do the Good: The Knight and the Theologian." *Image* 89 (Summer 2016): 72–84.

Ignotus, Paul. "The Hungary of Michael Polanyi." *The Logic of Personal Knowledge: Essays Presented to Michael Polanyi on His Seventieth Birthday*. London: Routledge and Kegan Paul, 1961.

James, Jeremiah, Thomas Steinhauser, Dieter Hoffmann, Bretislav Friedrich. *One Hundred Years at the Intersection of Chemistry and Physics: The Fritz Haber Institute of The Max Planck Society 1911–2011*. Berlin: Walter de Gruyter GmbH, 2011.

Jortner, Joshua. "Reflections on Physical Chemistry: Science and Scientists." *Annual Review Physical Chemistry* 57 (2006): 1–35. http://www.annualreviews.org.

Jost, Wilhelm Karl. "The First 45 Years of Physical Chemistry in Germany." *Annual Review Physical Chemistry* 17 (1966): 1–15. www.annualreviews.org.

———. "Friedrich Bonhoeffer." *Die Naturwissenschaften*, 44. Jahrgang, Heft 24 (Zweites Dezemberheft), 1957.

Kaufmann, Doris. "Psychiatry in the First World War and Weimar Germany." *Journal of Contemporary History* 34. no. 1 (January, 1999): 125–144. London: Sage.

Kinberg, Olof. "Obligatory Psychiatric Examination of Certain Classes of Accused Persons." *Journal of the American Institute of Criminal Law and Criminology* 2, no. 6 (March 1912): 858. Chicago: Northwest University School of Law. http://www.jstor.org/stable/1132807.

Klassen, A. J., ed. *A Bonhoeffer Legacy.* Grand Rapids: Eerdmans, 1981.

Kramish, Arnold. *The Griffin: The Greatest Untold Espionage Story of World War II.* Boston: Houghton Mifflin, 1986.

Laurence, W.L. "New Key Is Found to Atomic Energy," *New York Times*, May 5, 1939, 25.

Leese, Peter. *Shell Shock: Traumatic Neurosis and the British Soldiers of the First World War.* Hampshire, UK: Palgrave Macmillan, 2002.

Lemmel, Birgitta. "The Nobel Medals and the Medal for the Prize in Economic Sciences." www.nobelprize.org/nobel_prizes/about/medals/.

Lemmerich, Jost. *Science and Conscience: The Life of James Franck.* Translated by Ann Hentschel. Stanford: Stanford University Press, 2011.

Lerner, Paul. "Psychiatry and Casualties of War in Germany, 1914–18." *Journal of Contemporary History* 35, no. 1 (2000): 13–28.

———. *Hysterical Men: War, Psychiatry, and the Politics of Trauma in Germany, 1890–1930.* Ithaca: Cornell University Press, 2003.

Leibholz-Bonhoeffer, Sabine. *The Bonhoeffers: Portrait of a Family.* London: Sidgwick and Jackson, 1971.

Logan, Jonothan. "The Critical Mass." *American Scientist* 84, no. 3 (May–June, 1996): 263–277. Sigma Xi, The Scientific Research Society. http://www.jstor.org/stable/29775672.

Lyre, Holger. "C.F. Weizsäcker's Reconstruction of Physics: Yesterday, Today, Tomorrow." In *Time, Quantum and Information: Essays in honor of C.F. von Weizsäcker*, edited by L. Castell and O. Ischebeck. Berlin: Springer, 2003.

Marsh, Charles. *Strange Glory: A Life of Dietrich Bonhoeffer.* New York: Alfred A. Knopf, 2014.

McRae, Kenneth D. *Nuclear Dawn: F. E. Simon and the Race for Atomic Weapons in World War II.* Oxford: Oxford University Press, 2014.

Medawar, Jean, and David Pyke. *Hitler's Gift: The True Story of the Scientists Expelled by the Nazi Regime.* New York: Arcade Publishing, 2000.

Meding, Dorothee von. *Courageous Hearts: Women and the Anti-Hitler Plot of 1944.* Translated by Michael Balfour. Providence, RI and Oxford, UK: Berghahn Books, 1997.

Menninger, Karl A. "Psychoses Associated with Influenza." *The Journal of the American Medical Association* 72, no. 4 (January 25, 1919): 1. Chicago: American Medical Association.

Messow, Ulf. "Karl Friedrich Bonhoeffer." *Saxon Biography.* Institut fur Sächsische Geschichte und Volkskunde (April 2014). http://www.isgv.de/saebi.

Nachmansohn, David. *German-Jewish Pioneers in Science 1900–1933: Highlights in Atomic Physics, Chemistry and Biochemistry.* Berlin: Springer-Verlag, 1979.

Nelson, Anne. *Red Orchestra.* New York: Random House, 2009.

Neumärker, Klaus-Jürgen. "The Kramer-Pollnow Syndrome: A Contribution on the Life and Work of Franz Kramer and Hans Pollnow." *History of Psychiatry* 16, no. 4 (2005): 435–451. http://www.escap.eu/bestanden/Research/kramer_pollnow_background.pdf.

Nye, Mary Jo. *Michael Polanyi and His Generation: Origins of the Social Construction of Science.* Chicago: University of Chicago Press, 2011.

"Obituary: Professor Karl Bonhoeffer." *Journal of Nervous and Mental Diseases* 109, no. 2 (February 1949): 194–197.

Offer, Avner. *The First World War: An Agrarian Interpretation.* Oxford: Clarendon Press, 1989.

Polanyi, Michael. Papers. University of Chicago Library, Special Collections and Research Center.

———. *Personal Knowledge.* Chicago: University of Chicago Press, 1958.

———. "Problem Solving." *The British Journal of Philosophy of Science* VIII, no. 30 (August 1957): 89–100.

———. *Science, Faith and Society.* Chicago: University of Chicago Press, 1964. First Published 1946 by Oxford University Press.

———. *Scientific Thought and Social Reality.* Edited by Fred Schwartz. New York: International Universities Press, 1974.

Pollack, Robert E. "The Religious Obligation to Ask Questions of Nature and the State: Bonhoeffer on the Protection and Dignity of Human Life, in the Context of DNA-Based Genetic Medicine Today." Expanded Version of "Witness to Power: How the Life and Death of Dietrich Bonhoeffer Speak to Us Today." *CSSR News*, Columbia University Earth Institute (Winter 2004): 79–111.

Powers, Thomas. *Heisenberg's War: The Secret History of the German Bomb.* Boston: DaCapo Press, 1993.

Propping, P. "Genetic Disorders Presenting as 'Schizophrenia': Karl Bonhoeffer's Early View of the Psychoses in the Light of Medical Genetics." *Human Genetics* 65, no. 1 (November 1983): 1–10. Berlin: Springer-Verlag.

Rasmussen, Larry. "Dietrich and Karl-Friedrich Bonhoeffer: The Brothers Bonhoeffer on Science, Morality, and Theology." *Zygon* 44, no. 1 (March 2009).

Remarque, Erich Maria. *All Quiet on the Western Front*. Translated by A.W. Wheen. Boston: Little, Brown, 1929.

Renneberg, Monika, and Mark Walker, eds. *Science, Technology and National Socialism*. Cambridge: Cambridge University Press, 1994.

Riehl, Nikolaus, Frederick Seitz. *Stalin's Captive: Nikolaus Riehl and the Soviet Race for the Bomb*. Washington, DC: American Chemical Society, 1996.

Rose, Paul Lawrence. *Heisenberg and the Nazi Atomic Bomb Project*. Berkeley: University of California Press, 1998.

Ruthenberg, Klaus. "Robert Havemann," *HYLE-International Journal for Philosophy of Chemistry* 4, no. 2 (1998). http://www.hyle.org/journal/issues/4/ruthb.htm.

Schlingensiepen, Ferdinand. *Dietrich Bonhoeffer 1906–1945*. Translated by Isabel Best. New York: T & T Clark, 2010.

Schwarz, Alan. "Research Links PTSD to Blasts in Combat." *The New York Times*, June 10, 2016, A19.

Schmuhl, Hans-Walter. *The Kaiser Wilhelm Institute for Anthropology, Human Heredity and Eugenics 1927–1945*. Berlin: Springer, 2008.

Schroer, Bert. "Physicists in Times of War." Berlin: Institute for Theoretical Physics, December 2005. arXiv:physics/0603095v2 [physics.soc-ph] 20 March 2006.

Scott, William Taussig, and Martin X. Moleski, S.J. *Michael Polanyi: Scientist and Philosopher*. Oxford: Oxford University Press, 2005.

Sifton, Elizabeth, and Fritz Stern. *No Ordinary Men: Dietrich Bonhoeffer and Hans von Dohnanyi, Resisters Against Hitler in Church and State*. New York: New York Review Books, 2013.

Sime, Ruth Lewin. *Lise Meitner: A Life in Physics*. Berkeley: University of California Press, 1996.

Soifer, Alexander. *The Scholar and the State: In Search of Van der Waerden*. Berlin: Birkhäuser, 2014.

Stern, Fritz. *Dreams and Delusions: The Drama of German History*. New Haven: Yale University Press, 1987.

———. *Five Germanys I Have Known*. New York: Farrar, Straus and Giroux, 2006.

Stoltzenberg, Dietrich. *Fritz Haber: Chemist, Nobel Laureate, German Jew*. Philadelphia: Chemical Heritage Press, 2004.

Szöllösi-Janze, Margit, ed. *Science in the Third Reich*. Oxford and New York: Berg, 2001.

Taxacher, Gregor. "A Scientist Faced with 'The Ambivalence of Progress'—Carl Friedrich von Weizsäcker." Goethe Institut (June 2012). www.goethe.de/ges/phi/prt/en9544119.htm.

Taylor, Hugh. "Fifty Years of Chemical Kineticists," Woodrow Wilson National Fellowship Foundation, Princeton, NJ. *Annual Review Physical Chemistry* (1962): 11.

Thatje, Sven. "Dr. Arnold Berliner (1862–1942), Physicist and Founding Editor of Naturwissenschaften," *Naturwissenschaften* (2013). https://doi.org/10.1007/s00114-013-1124-4.

Von Weizsäcker, Carl Friedrich. *The History of Nature*. Chicago: The University of Chicago Press, 1949.

———. *Pioneer of Physics, Philosophy, Religion, Politics and Peace Research* (Springer, ebook), edited by Ulrich Bartosch. 2015.

———. *The Structure of Physics*. The Netherlands: Springer, 1985.

———. "Thoughts of a Non-Theologian on Dietrich Bonhoeffer's Theological Development." *The Ecumenical Review* 28 no. 2 (April 1976): 156–173.

———. *The World View of Physics*. Chicago: The University of Chicago Press, 1949.

Von Weizsäcker, Carl Friedrich. Interview by Karl Hufbauer, *Oral History Transcript*, American Institute of Physics (April 18, 1978). www.aip.org/history/ohilist/4948.html.

Walker, Mark. *Nazi Science: Myth, Truth, and the German Atomic Bomb*. New York and London: Plenum Press, 1995.

Weller, Albert. "In Memoriam: Theodor Forster." *EPA Newsletter*. Duesseldorf: EPA, (2011).

Wigner, E.P., and R.A. Hodgkin. "Michael Polanyi." *Biological Memoirs of Fellows of the Royal Society* 23 (November 1977): 413–448. http://rsbm.royalsocietytpublishing.org.

"The World Wars." *Encyclopaedia Britannica*, 15th ed., Vol. 29, 987.

Zeidman, Lawrence A. "Neuroscience in Nazi Europe Part II: Resistance Against the Third Reich." *Canadian Journal of Neurological Sciences* 38, no. 6 (November, 2011): 808–809.

Index

A
Abwehr (German military intelligence), 178, 215
Abyssinian Baptist Church, 93, 182
Adorno, Theodor, 114, 271
Alsos (Allied intelligence mission), 175
Alzheimer, Alois, 7
American Chemical Society, 75, 80
Anderson, Carl, 148, 271
Angell, James W., 92, 271
Antropoff, Andreas von, 83, 271
Ardenne, Manfred von, 202, 253, 271
Aryan clause, 102–103, 142, 144
Atomic bomb, 3, 158, 160, 170–171, 195–196, 201, 206–207, 243
Atomic fission, 155–157
Aubert, Axel, 203, 271
Auer Company, 194, 195

B
Bagge, Erich, 194, 244, 271
Baillie, John, 95
Barth, Karl, 118, 134, 272
Barton, Derek, 154
BASF (German company), 63
Beck, Ludwig, 231, 232, 271
Bell, George (Bishop of Chichester), 162, 166, 179, 272
Benjamin, Georg, 217, 272
Bergh, Sverre, 156
Berg, Otto, 58, 272
Berliner, Arnold, 143–144
Bethe, Hans, 174, 272
Bethel institution, 122, 124
Bethge, Eberhard, 3, 5, 93, 187, 216, 218, 272
Birge, Raymond T., 148, 273
Bodelschwingh, Friedrich von, 124
Bodenstein, Max, 55, 69, 75, 80, 82, 273
Bohr, Niels, 65, 87, 91, 158, 168, 228, 273
Bonhoeffer, Christel (Mrs. Hans Dohnanyi), 14, 68, 83, 179, 213, 216, 232, 273

K. L. Housley, *The Scientific World of Karl-Friedrich Bonhoeffer*, Palgrave Studies in the History of Science and Technology,
https://doi.org/10.1007/978-3-319-95801-9

Bonhoeffer, Dietrich
"After Ten Years", 94, 220
arrest, 213
biographical profile, 273
birth and childhood, 14, 18–21, 41, 42, 44, 46
"*Church and the Jewish Question*", 117
death, 233
Discipleship, 227
Ethics, 224–225
Finkenwalde Seminary, 166
"Führer and Younger Generation,", 46, 116
in London, 118
resistance, 178–180, 213–214
in Tegel prison, 218–220, 225–226
at Union Theological Seminary, 20, 91, 166
in United States (1939 summer), 165, 179–184
Bonhoeffer, Emilie (née Delbrück), 24, 116, 216, 232, 241, 273
Bonhoeffer, Friedrich (son of Karl-Friedrich), 258
Bonhoeffer, Fritz, 30
Bonhoeffer, Julie (née Tafel), 40, 115, 273
Bonhoeffer, Karl-Friedrich
biographical profile, 273
birth and childhood, 14, 18–21
death, 2, 262
Delbrück, 25
discovery ortho and para hydrogen, 70
espionage with Rosbaud, 4, 217
Haber Institute, 60, 107
Haber Memorial, 134–141
Harteck, Paul, 69–72
heavy water, 155–156
Heisenberg, Werner, 161–162
marriage, 83
Max Planck Institute, 258
in military WWI, 39–42
Nobel Prize nominations, 83
Polanyi, 71, 75–84, 250
at Univ. of Berlin, 53
at Univ. of Frankfurt am Main, 88, 114–115
at Univ. of Göttingen, 253, 255
at Univ. of Leipzig, 115
Uranium Club, 194
to U.S. (1929), 75–84, 96
to U.S. (1939), 146–187
Bonhoeffer, Karl Ludwig (father)
biographical profile, 274
Breslau Psychiatric Clinic, 7, 14–15
Charité Clinic, 23, 41, 119
death, 7
education, 10, 12
eugenics and sterilization, 121–122
influenza, 42–43
Jewish colleagues, 119–121
marriage, 13
neurology, 16–17
shell shock, 35, 39
van der Lubbe case, 100–102, 124, 293
Bonhoeffer, Karl Walter Paul (son of Karl-Friedrich), 89, 258
Bonhoeffer, Klaus
arrest, 213
biographical profile, 274
birth and childhood, 14, 23
death, 233
education Tübingen, 53
resistance, 178, 213
response anti-Semitism, 97, 116
Bonhoeffer, Lothar (son of Otto), 30, 274
Bonhoeffer, Margarete (née Dohnanyi), 83, 274
Bonhoeffer, Otto (brother of Karl), 45

Bonhoeffer, Paula (née von Hase), 13, 20, 97, 115–116, 274
Bonhoeffer, Sabine (Mrs. Gerhard Leibholz), 14, 32, 60, 68, 97, 257, 274
Bonhoeffer, Susanne (Mrs. Walter Dreß), 14, 214, 221, 274
Bonhoeffer, Ursula (Mrs. Rüdiger Schleicher), 14, 68, 214, 231, 234, 274
Bonhoeffer, Walter, 14, 20, 39–41
Bonhoeffer, Werner (son of Otto), 30
Born, Max, 67, 88, 168, 170, 172, 174, 261, 275
Bosch, Carl, 63, 275
Bothe, Walther, 149, 159, 194, 201, 275
Braun, Wernher von, 196, 275
Bresch, Carsten, 261, 275
Breslau Psychiatric Clinic, 7, 12–13, 16
Breslau (Wroclaw, Poland), 12–13, 23
Briggs, Lyman, 171
Brodersen, Arvid, 156, 224
Broglie, Louis de, 66, 275
Brown, G.W., 154
Brun, Jomar, 154, 203, 275
Buber, Martin, 96, 114, 276
Bumke, Oswald, 123, 276

C

Canaris, Wilhelm, 215, 276
Chadwick, James, 148, 276
Charcot, Jean-Martin, 11
Charité Clinic, 23, 41, 119, 240
Clusius, Klaus, 194, 276
Coates, J.E., 62
Cockcroft, John, 159, 276
Compton, Arthur, 174, 276
Condon, Edward, 67, 277
Confessing Church, 166, 177
Coster, Dirk, 156, 277
Creutzfeldt, Hans, 7, 123, 277
Curie, Irène (Joliot-Curie), 148, 157, 277
Curie, Marie, 150, 200, 277

D

Debye, Peter, 192, 277
Dehn, Max, 114, 278
Delbrück, Emilie (Mrs. Klaus Bonhoeffer), 24, 97, 116, 234, 241, 278
Delbrück, Hans, 24, 25, 98–99, 278
Delbrück, Justus, 239, 278
Delbrück, Max, 24–26, 66, 98, 137, 168, 258, 261, 278
Dennison, David, 70, 81, 278
Deuterium, 147, 148, 150, 152
Diebner, Kurt, 159, 194, 201, 202, 206, 242, 243, 278
Dimitroff, Georgi, 125
Dirac, Paul, 66, 67, 148, 278
Discipleship (Bonhoeffer book), 177
Dohnanyi, Hans
  arrest, 213
  assassination attempt, 231
  death, 232–234
  diphtheria, 231–232
  influence on Dietrich, 178–179
  influence on Karl-Friedrich, 115
  marriage, 68
  resistance, 115, 178–179, 232
Dohnanyi, Margarete (Mrs. Karl-Friedrich Bonhoeffer), 83, 279
Döpel, Clara, 194
Döpel, Georg Robert, 159, 194, 279
Drucker, Peter F., 113

E

Eddington, Arthur, 103, 279

Ehrenfest, Paul, 33, 104, 279
Ehrlich, Paul, 10, 32, 279
Eigen, Manfred, 258, 280
Einstein, Albert
  atomic bomb, 201
  Berliner, Arnold, 143
  biographical profile, 280
  colloquia, 62
  early education, 19
  entanglement and EPR, 1, 108
  Haber letter, 133
  KWI Physics, 56
  pure science, 51
  quantum mechanics, 66
  relativity, 67
  response to Nazism, 105
  Roosevelt letter, 171
  WWI, 34
Emeléus, Harry Julius, 82, 280
Entanglement, 1, 108
Esau, Abraham, 159, 280
Ethics (book by Dietrich Bonhoeffer), 224–225
Ettal (Benedictine monastery), 224
Eucken, Arnold, 69, 71, 83, 280
Eugenics (sterilization), 16, 39, 121–124
Euthanasia (murder of disabled and mentally ill), 122
Ewald, Paul, 104, 280
Eyring, Henry, 82, 280

F
Farben, I.G. (German company), 154, 202, 209–210
Farkas, Adalbert, 65–66, 84, 89, 107, 114, 281
Farkas, Ladislaus, 84, 89, 281
Farm Hall, 244
Federal Council of Churches (U.S.), 94, 167, 179
Fermi, Enrico, 148, 157, 169, 176, 201, 281
Finkenwalde Seminary, 93, 166
Fischer, Erich, 194, 281
Fisher, Albert Frank, 94
Fission, atomic, 155–157
Flammersfeld, Arnold, 217, 281
Flügge, Siegfried, 194, 281
Foley, Francis, Edward (Frank), 156, 281
Förster, Theodor, 255, 282
Fowler, Ralph, 152, 282
Franck, James, 57, 60, 66, 90, 106, 167, 228, 246, 260, 282
Freud, Sigmund, 11, 15, 17–18, 105
Freundlich, Herbert, 55, 60, 62, 107, 154, 282
Frick Chemical Laboratory (Princeton), 76, 81–82
Friedrichsbrunn (Bonhoeffer summer home), 23, 221
Frisch, Otto, 156–158, 200–201, 282
Frumkin, Aleksander N., 66, 67, 282

G
Gas warfare (WWI), 63–64
Gaupp, Robert, 14, 18, 254, 283
Geiger, Hans, 149, 159, 194, 283
Gentner, Wolfgang, 194, 283
Gerischer, Heinz, 141, 248, 254, 255, 283
Gerischer, Renate, 255
Gerlach, Walther, 194, 242, 244, 283
German Army Ordnance (HWA), 158, 205
German Bunsen Society, 259, 270
Gildemeister, Martin, 144, 222, 283
Goebbels, Joseph, 105, 284
Goerdeler, Carl Friedrich, 178, 284
Goering, Hermann, 125, 284
Goldschmidt, Victor, 90, 284

Göttingen, University of, 56, 61, 67, 75, 88, 90, 253, 255, 257
Goudsmit, Samuel A., 4, 175, 186, 284
Griffin. *See* Rosbaud, Paul (the Griffin)
Groth, Wilhelm, 158, 284
Groves, Leslie Richard, 201, 284

H

Haber, Clara, 64
Haber, Fritz
  Aryan clause, 105–106
  biographical profile, 285
  colloquium, 61–66
  death, 134
  Einstein, Albert, 56, 133
  gas warfare, 64
  Haber-Bosch process, 63
  Haber Institute, 55, 60, 87, 253
  Haber memorial, 137–139
  Nobel Prize, 63
  research 1920s, 64
  resignation, 107
Haber, Hermann, 134, 137
Haber Institute. *See* Kaiser Wilhelm Institute for Physical and Electrochemistry (Haber Institute)
Hahn, Otto
  Aryan clause, 106–107
  atomic fission, 155–157
  Farm Hall, 244
  Haber memorial, 137–139
  Meitner escape, 156
  Meitner letter, 245
  Rosbaud friendship, 161, 217
  Uranium Club, 194, 242
Harnack, Adolf von, 24, 98, 285
Harnack, Arvid von, 97, 239, 285
Harnack Haus (Berlin), 62, 137, 139
Harnack, Mildred, 97, 98, 239
Harteck, Paul
  atomic bomb, letter, 158–159
  biographical profile, 285
  Farm Hall, 244
  Haber Institute, 70
  Hamburg University, 125–126
  research ability, 151
  research with Karl-Friedrich, 68–71
  Rutherford, Ernest, 87, 150–151
  tritium discovery, 150
  Uranium Club, 3, 192–194, 202–203, 205, 242, 243
Harvard University, 76, 79
Hase, Benedikt von (Uncle Bubi), 30, 285
Hase, Clara von, 21
Hase, Paula von, 231, 239, 286. *See also* Bonhoeffer, Paula
Hassel, Odd, 153–154, 286
Heavy water, 3, 147–162, 169, 199–204, 206
Hedgehogs (fraternity), 53
Heidegger, Martin, 100, 286
Heisenberg, Werner
  Aryan clause response, 127
  atomic bomb, 3, 4, 192–194, 198–201, 252–253
  Bohr visit, 228
  Farm Hall, 244
  Göttingen after war, 255
  at Leipzig, 115, 142
  quantum mechanics, 66
  spin isomers, 70
  Uranium Club, 159, 192, 242–244
  U.S. trip (1939), 165, 171–176
  Weizsäcker friendship, 227–228
Held, Robert, 68, 286
Herzog, Fritz, 105, 287
Hevesy, George de, 228, 287
Hildebrandt, Franz, 103, 287
Hill, Archibald, 145
Himmler, Heinrich, 173

Hindenburg, Paul von, 100, 287
Hinshelwood, Cyril, 82, 287
Hitler, Adolf
  anti-Semitism, 96, 97
  assassination and coup attempts, 178, 216, 232
  Beer Hall Putsch, 59
  biographic profile, 287
  death, 234
  Einstein denouncement, 89
  master race, 16
  Planck meeting, 104
  Reich chancellor, 99–103
  Reichstag fire, 100–102
  WWI, 41, 99
Hochberg, J., 161
Hoffmann, Gerhard, 194, 288
Horkheimer, Max, 114, 288
Horney, Karen, 7, 15, 120, 288
Howard University, 93, 94
Hund, Friedrich, 70, 142, 288
Hutton, Robert, 159, 288

I
Influenza epidemic, 16, 42

J
Jacobson, Edith, 15
Jaenicke, Johannes, 139, 288
Jaenicke, Walther, 141, 187, 288
Jander, Gerhart, 107, 125, 289
Jászi, Oszkár, 78–79
Jehle, Herbert, 103, 257–258, 289
John, Hans, 214, 289
John, Otto, 214, 289
Johns Hopkins University, 79, 91
Joliot, Frédéric (Joliot-Curie), 148, 289
Joos, Georg, 194, 289
Jossmann, Paul, 119, 120, 240, 290
Jost, Friedrich Wilhelm, 56, 84, 222, 246, 259, 290

K
Kaiser Wilhelm Institute for Biology, 62
Kaiser Wilhelm Institute for Brain Research, 123
Kaiser Wilhelm Institute for Chemistry, 138, 155, 242
Kaiser Wilhelm Institute for Physical Chemistry and Electrochemistry (Haber Institute), 55, 61, 63, 87
Kaiser Wilhelm Institute for Physics, 56, 107, 205
Kaiser Wilhelm Institute of Anthropology, Human Heredity and Eugenics, 122
Kaiser Wilhelm Institute of Physics, 33
Kaiser Wilhelm Society, 61, 87, 98, 104, 137, 139
Katz, Bernard, 144, 290
Kistiakowsky, George, 76, 290
Kleist-Retzow, Hans Jürgen, 178
Kleist-Retzow, Ruth von, 178, 224, 290
Koch, Robert, 10
Koeth, Joseph, 137, 290
Kopfermann, Hans, 87, 194, 290
Korsching, Horst, 194, 244, 291
Kraepelin, Emil, 11, 13, 15, 35, 291
Kramer, Franz, 119, 291
Krieck, Ernst, 115, 291
Kröger, Martin, 245, 291
Kruif, Paul de, 226–227

L
Ladenburg, Rudolf, 61, 107, 291
Landsberger, Benno, 143, 291
Lange, Ruth, 217

Langmuir, Irving, 75, 82, 291
Lasserre, Jean, 92, 292
Laue, Max von, 105–106, 136, 161, 194, 217, 228, 242–244
Lawrence, Ernest, 149
Le Blanc, Max, 115, 222, 292
Lehmann, Paul, 116, 183–184, 292
Leibholz, Gerhard, 24, 43, 60, 96–97, 117, 257, 292
Leiper, Henry Smith, 167, 183, 292
Lenard, Philipp, 105, 136, 292
Levi, Friedrich, 143, 293
Lewis, Gilbert, 152, 293
Liebknecht, Karl, 44, 293
Life Together (book by Dietrich Bonhoeffer), 177
Lind, Samuel C., 80, 293
Livingston, Milton Stanley, 149, 293
Lobetal Institute, 122
Locke, Alain, 94
London, Fritz, 70, 78, 293
Lubbe, Marinus van der, 100–102, 124
Ludloff, Hanfried, 54
Lufthansa, 97, 213–214
Luxemburg, Rosa, 44, 294
Lynd, Helen Merrell, 78
Lynd, Robert Staughton, 78

M

Manhattan Project, 169–171, 196, 201
Manifesto to the Civilized World, 32
Mattauch, Josef, 159, 161, 194, 217
MAUD (British committee on atomic research), 159, 252
Max Planck Institute of Physical Chemistry (Karl Friedrich Bonhoeffer Institute), 258
Max Planck Society (formerly Kaiser Wilhelm Society, 253
Meitner, Lise, 126, 137, 139, 156–157, 186, 197–205, 245
Menninger, Karl, 43
Menzel, Donald H., 148, 294
Milch, Erhard, 206
Minkowski, Hermann, 67, 294
Mitscherlich, Alexander, 123
Müller, Josef, 213, 295
Murphy, George, 149, 295

N

Nachmansohn, David, 61, 295
Naturwissenschaften, Die, 71, 143, 156
Nazi Party (National Socialist German Workers Party, NSDAP), 97, 99, 100, 102
Neisser, Albert and Toni, 21
Nernst, Walther, 47, 55–56, 61–62, 75, 295
Neumann, John von, 59, 78, 82, 258, 295
Neurology, 11, 35
New York Academy of Sciences, 261
New York World's Fair (1939), 179–180
Niebuhr, Reinhold, 91, 95, 167, 295
Niemöller, Martin, 177, 295
Noddack, Walter, 58, 296
Nonne, Max, 36, 123, 296
Nordheim, Gertrud, 174, 296
Nordheim, Lothar, 174, 296
Norsk Hydro (Vemork), 152–155, 204–205, 208–209, 242

O

Oberlin College, 78
Oliphant, Marcus, 150, 296
Oppenheimer, J. Robert, 67, 171, 297
Oppenheim, Hermann, 15, 36, 296

Ortho hydrogen, 70–71
Osram (German company), 197, 223, 236
Oster, Hans, 178, 214, 297
Ostwald, Carl Wilhelm Wolfgang, 222, 297
Ostwald, Friedrich Wilhelm, 55, 115, 297

P
Para hydrogen, 70–71
Pasteur, Louis, 10
Pauli, Wolfgang Ernst, 148, 297
Peenemünde (German rocket facility), 196
Peierls, Rudolf, 174, 200, 297
Perrin, Francis, 81, 297
Perrin, Jean, 81, 297
Perrin, Michael, 4, 252, 298
Phillip, Lee Cato, 94, 298
Photochemistry, 56
Physical chemistry, 55–56, 58
Planck, Erwin, 231
Planck, Max, 32, 47, 51, 56, 65, 104, 137–139, 253, 298
Polanyi, Karl, 76–78
Polanyi, Michael
  after war, 248–249
  Aryan clause response, 106
  Bonhoeffer discovery response, 71
  books and essays, 87, 248
  at Haber Institute, 60, 62, 87
  in Hungary, 58–59
  scientific autonomy, 5, 251
  scientific language, 150
  U.S. trip (1929), 76–78, 82, 83
  visit Bonhoeffer, 250–251
Pollnow, Hans, 119–120, 298
Pompe, Johannes, 123, 298
Powell, Adam Clayton, Sr., 93, 298
Princeton University, 75, 81
Psychiatry, 11
Pure science, 5, 51–52, 222

Q
Quadfasel, Friedrich, 120–121, 299
Quantum mechanics, 1, 81, 108

R
Rabi, Isidor Isaac, 174, 299
Rabinowitch, Eugene, 90–91, 299
Racism, U.S., 79, 94–97
Rathenau, Walther, 60, 299
Reich Research Council, 159, 206
Reichstag fire, 100–102
Remarque, Erich Maria, 91, 105
Renn, Ludwig, 91
Reply of the Cultural Federation of German Scholars and Artists, 33
Riehl, Nikolaus, 194, 195, 299
Ritter, Gerhard, 226, 299
Roberts, David, 182
Rockefeller Foundation, John D., 61, 87
Roentgen, Wilhelm, 32
Rosbaud, Paul (the Griffin)
  Aryan clause response, 106
  atomic bomb, 206–207
  atomic fission, 155–157
  Berliner and *Naturwissenschaften*, 143
  British Secret Service, 156
  espionage, 4, 217
  Final Solution response, 217
  Nazi denounciation of Bonhoeffer, 142
  Pergamon Press, 251
  post-war Germany, 186, 245
  rescue of Meitner, 156
  Tate Medal, 262

Rust, Bernhard, 119, 138, 300
Rutherford, Ernest, 87, 150, 151

S
Sachs, Alexander, 171–173
Saethre, Haakon, 123, 300
Schacht, Hjalmar, 92, 300
Schleicher, Renate (Mrs. Eberhard Bethge), 216
Schleicher, Rüdiger, 2, 68, 213, 231, 234, 301
Schneider, Paul, 185
Schrödinger, Erwin, 66, 70, 108, 149, 301
Scottsboro case, 95
Secret Intelligence Service (U.K., MI6), 4, 156, 252
Seeberg, Erich, 114, 301
Seidemann, Herta, 119–120, 301
Seitz, Frederick, 195–196, 207
Shell Shock, 16, 34–39
Sherriff, Robert C., 91
Simon, Franz (Francis), 4, 126, 246, 252, 259–261, 301
Soddy, Frederic, 150, 151, 301
Sommerfeld, Arnold, 173, 301
Spartacist uprising, 43–44
Speer, Albert, 193
Spielmeyer, Walther, 123, 302
Springer, Ferdinand, 144
Springer, Julius, 143
Stark, Johannes, 136, 142, 302
Stauffenberg, Claus von, 231, 302
Sterilization, 121–122, 124, 215
Stern, Fritz, 11, 22, 127, 302
Stern, Otto, 126, 302
Stern, Richard, 21, 22
Stern, Rudolf, 54, 134, 246
Stetter, Georg, 194, 302
Stewart, Henry, 76
Stifter, Adalbert, 225
Stoecker, Adolf, 98
Strassmann, Fritz, 139, 156–157, 161, 302
Straus, Erwin, 119–120, 303
Suess, Hans, 209, 303
Sutz, Erwin, 103, 303
Svedberg, Theodor, 83, 303
Szilard, Leo, 59, 78, 105, 157, 169, 200, 201, 303

T
Tacke, Ida, 58, 303
Taylor, Hugh Stott, 75, 80, 303
Tegel prison, 214, 223, 232
Teller, Edward, 67, 168, 171, 303
Thiessen, Peter Adolf, 253, 304
Thomson, George, 6
Tillich, Paul, 114, 304
Tinel, Jules, 123, 304
Tresckow, Henning von, 216, 304
Troeltsch, Ernst, 99
Tronstad, Leif, 154–155, 204, 207–208, 304

U
Union of German Chemists, 137
Union Theological Seminary, 20, 91, 115, 116, 166
University of Berlin (Friedrich-Wilhelms, Humboldt), 23, 52–54, 55, 56, 138, 214, 248, 255
University of Frankfurt am Main (Goethe), 88–89, 96, 113–115
University of Göttingen, 56, 67, 88, 90, 253, 255
University of Leipzig, 56, 115, 142–144
University of Minnesota, 77, 80
University of Tübingen, 12, 53, 60, 96

Uranium, 199–200
Uranium Club, 3, 159, 191–210, 242
Urey, Harold, 83, 148–149, 249–250

V
Vemork. *See* Norsk Hydro (Vemork)
Versailles Treaty, 38, 42, 99
Visser 't Hooft, Willem Adolf, 186, 305
Vogt, Cécile, 123, 305
Vogt, Oskar, 123, 305
Volmer, Max, 253, 305

W
Wach, Joachim, 143, 305
Waerden, Bartel L. van der, 142, 247, 305
Walcha, Helmut, 216
Wedemeyer, Maria von, 178, 218, 239–240, 257, 306
Weigert, Fritz, 143, 306
Weizsäcker, Carl Friedrich von, 100, 142, 159–160, 194, 199, 226–231, 242, 244, 306
Weizsäcker, Ernst Heinrich von, 227–228, 306
Wentzel, Gregor, 89, 306
Wernicke, Carl, 7, 12–14
Wiedemann, Fritz, 162
Wigner, Eugene, 77, 82, 201, 306
Willstätter, Richard, 106, 138–139, 307
Wirtz, Karl, 193, 203, 242, 244, 307

Z
Zinn, Elizabeth, 92, 307
Zutt, Jürg, 124, 307

The manufacturer's authorised representative in the EU is Springer Nature Customer Service Centre GmbH, Europaplatz 3, 69115 Heidelberg, Germany. If you have any concerns regarding our products, please contact ProductSafety@springernature.com

Printed and bound by CPI Group (UK) Ltd, Croydon, CR0 4YY

30/09/2026

02225653-0013